Radar and Electronic Warfare Principles for the Non-Specialist

4th edition

Paul Hannen
Science Applications International Corporation (SAIC)

SciTech
PUBLISHING
an imprint of the IET

Edison, NJ
scitechpub.com

SCITECH
PUBLISHING
an imprint of the IET

Published by SciTech Publishing, an imprint of the IET.
www.scitechpub.com
www.theiet.org

Third edition 2004
Fourth edition 2013

Editor: Dudley R. Kay
Cover Design: Brent Beckley

10 9 8 7 6 5 4 3 2 1

ISBN 978-1-61353-011-5 (paperback)
ISBN 978-1-61353-196-9 (PDF)

Typeset in India by MPS Limited
Printed in the USA by Sheridan Books, Inc
Printed in the UK by Hobbs the Printers Ltd

To Kim, Amanda, and Adam

Contents

8   Electronic Warfare Overview   231

9   Electronic Warfare Receivers   243

10   Self-Protection Jamming Electronic Attack   259

In Memoriam

Major General John C. Toomay
U.S. Air Force
1922–2008

Preface

What This Book Is

This book is about radar and electronic warfare (EW). It will teach you the essentials of radar and EW, the underlying principles, in a clear and consistent way. It is not like an engineering handbook that provides detailed design equations without explaining either derivation or rationale. It is not like a graduate school textbook that may be abstruse and esoteric to the point of incomprehensibility. Moreover, it is not like an anthology of popular magazine articles that may be gaudy but superficial. It is an attempt to distill the very complex, rich technology of radar and EW into its fundamentals, tying them to the laws of nature on one end and to the most modern and complex systems on the other.

Radar Principles for the Non-Specialist by John C. Toomay provided the foundation of this book. For over 20 years, Toomay (Major General, U.S. Air Force, retired) was associated with scientists, engineers, mathematicians, business administration graduates, and other college-educated people, all of whom aspired to jobs with broader responsibilities. One of the principal technical areas was radar. Surprisingly, even the electrical engineers with recent bachelor's degrees had no knowledge of radar, which is really a specialty topic to be taken up in graduate school. In response, in the 1970s Toomay put together a rudimentary version of this book. However, his subsequent experience revealed an even greater need for radar knowledge, leading to the first two editions of the book in 1982 and 1989 (reissued in 1998 by SciTech Publishing), respectively. I wrote the third edition in 2004, building on the unique style and solid foundation of the previous ones. Every section of each chapter was revised and enhanced in some way. These enhancements were aimed at making the book easier to learn from and teach out of, which I do—both undergraduate and graduate level college classes and professional short courses.

What Is New in the Fourth Edition

Several years of using the third edition, combined with numerous comments and suggestions from interested parties, revealed what was needed to make it even better. The most obvious addition is the EW-related material, which addresses electronic support (ES), electronic attack (EA), and electronic protection (EP) and is consistent in both scope and depth with the radar-related material. All the radar-related material was reviewed, revised, and enhanced as necessary, with the following significant updates: target signal-to-noise ratio, target detection theory, array antennas, radar measurements and tracking, and target signatures. The advanced radar concepts chapter was revised, including the addition of a section on modern multifunction, multimode, multimission radar systems. The fourth edition remains true to the traditional strength of the book, providing radar principles for the non-specialist while now also offering EW principles.

Who It Is For

If you are intellectually curious about the way radar systems and EW systems work and their inherent interrelationships, if you are not satisfied with a casual explanation, yet you are without the time to take a course leading to a master's degree, this book will provide you the level of comprehension you seek. If you have taken a radar or EW short course but find you need a better foundation or want the concepts to be tied together better, then this book will do that.

If your work requires you to supervise or meet as coequals with radar systems or EW engineers or designers, this book will allow you to understand them, to question them intelligently, and perhaps to provide them with a perspective (a dispassionate yet competent view) that they lack. If you are trained in another discipline but have been made the manager of a radar or EW project or of a system program with one or more radar systems as subsystems, this book will provide you with the tools you need, not only to give your team members confidence but also to make a substantive technical contribution yourself.

How It Approaches Radar and Electronic Warfare

This book is focused on imparting whole pieces of knowledge, developed with an evolutionary approach and tied together with a consistent thread of logic through the radar and EW material. It starts with electromagnetic propagation, describes a radar system of the utmost simplicity, and derives the associated radar equation. Once the radar equation is available, the book enhances the meaning of each term in it and concept behind it, moving through target detection, antennas, measurements and tracking, radar cross section, and system applications in an orderly progression. Similarly, the book continues with an introduction to EW, moving through ES, EA, and EP. At the finish, the reader should be able to perform a respectable, first-order radar and EW system design or analysis, including trading radar parameters or concepts. However, more importantly, the reader should know enough to critique the designs or analysis of others and to understand which technical issues are fundamental and which are secondary. While clever design ideas, and acronyms for them, are rampant, radar and EW functions do not change. Thus, while there is no way to keep up with each new design wrinkle, knowledge of the principles governing them will give instant critical understanding.

What Is Unusual

This book does three unusual, although perhaps not unique, things. It presents a comprehensive set of radar and EW principles—including many of the latest applications—in a relatively modest volume. It outlines these principles with their underlying derivations, using the simplest mathematics possible, explaining the steps, and using only popularly tabulated functions, integrals, and other expressions. In addition, it uses the same method of derivation, the same mathematics, and the same conceptual approach to discussing both radar and EW.

What Is Useful

This book presents information in logical chunks, which are meant to be self-contained. Most chapters stand alone so the reader may be selective and still benefit from their content. The chapters are scaled to their information content rather than to the time required absorbing them. Some readers will require much more effort than others to master a particular chapter. There are two levels of comprehension provided. The reader may simply memorize key relationships, which are always identified in the text, or may master the principle and its derivation. Useful (mostly generally available) references are provided throughout. Exercises at the end of each chapter are calculated not to stump the reader but to reinforce the concepts presented and illustrate their applications. The answers to the exercises at the end of each chapter have been double-checked, and a solution set (Mathcad and PDF) is available for download from the publisher's website.

How It Is Organized

The book goes from the fundamental toward the more complex and from philosophy to quantification. It starts with an introduction to radar and EW and then moves to an incremental discussion first of radar systems and then of EW as the following outline shows:

	Chapter	Summary
1	INTRODUCTION TO RADAR AND ELECTRONIC WARFARE	Introduction to radar and electronic warfare
2	RADAR SYSTEMS	History, technical fundamentals, deriving the radar equation
3	TARGET DETECTION	Building on the radar equation and explaining how a radar systems detects the presence of targets
4	RADAR ANTENNAS	Addressing and expanding on an important term in the radar equation
5	RADAR MEASUREMENTS AND TARGET TRACKING	Explaining how a radar system measures characteristics of the target—range, range rate, and angle—and tracks the target
6	TARGET SIGNATURE	Addressing and expanding on the only target-related term in the radar equation
7	ADVANCED RADAR CONCEPTS	Explaining advanced radar concepts, both common and uncommon
8	ELECTRONIC WARFARE OVERVIEW	Introduction to the components of electronic warfare: electronic support, electronic attack, and electronic protection
9	ELECTRONIC WARFARE RECEIVERS	Defining and discussing electronic warfare receivers, the most common element of electronic support
10	SELF-PROTECTION JAMMING	Defining and discussing self-protection jamming electronic attack
11	SUPPORT JAMMING	Defining and discussing support jamming electronic attack
12	ELECTRONIC PROTECTION CONCEPTS	Defining and discussing electronic protection concepts
13	LOOSE ENDS OF RADAR AND/OR ELECTRONIC WARFARE LORE	Cleaning up loose ends

Acknowledgements

A special thanks to my colleagues and friends, Mr. Bruce Esken, Mr. James Helton, and Mr. Mike Sutton, for their countless suggestions, detailed discussions, and thorough technical review of the manuscript. Your scrutiny and big-picture focus are invaluable. If this book looks like it was written by someone who actually has a command of the English language, as opposed to just radar and electronic warfare, you can thank Miss Jackie Sansavera; I know I do. Also, thanks to my family, friends, colleagues, students, and anyone who at one time or another said, "You should just write your own book."

Paul J. Hannen

Publisher's Acknowledgments

SciTech Publishing – IET gratefully acknowledges the manuscript reviewing efforts from the following members of the international radar and electronic warfare community. Refinements to content and expression for the benefit of all readers represent the blessing of 'a community effort'.

Dr A.M. (Tony) Ponsford – Raytheon Canada Ltd.
John Erickson – Wright Patterson Air Force Base, USA
Neal Brune – Esterline Defense Technologies, USA
Dr Thodoris G. 'Ted' Kostis – University of the Aegean, Greece
Margaret M. 'Peggy' Swassing – Edwards Air Force Base, USA
Dr Pinaki S. Ray – University of Adelaide, Australia
Stephen Harman – QinetiQ, UK
Dr Firooz Sadjadi – Lockheed Martin Corporation, USA
Dr Edward R. Beadle – Harris Corporation, USA

Symbols

A = Area of the flat plate, square meters (m^2)

A = Effective intercept area of the target, square meters (m^2)

A = Phasor amplitude, volts

A = Physical area of the antenna, square meters (m^2)

a = Dimension of the square, meters

a = Refractive index gradient

$A(y)$ = Current distribution in the y–z plane

A_c = Area of the clutter intercepted by the radar resolution cell, square meters (m^2)

A_e = Effective aperture area, square meters (m^2)

A_e = Effective area of the receive antenna, square meters (m^2)

A_e = Effective area of the antenna, square meters (m^2)

A_e = Electromagnetic area of the object as seen by the radar, square meters (m^2)

A_e = Radar receive antenna effective area in the direction of the target/jammer, square meters (m^2)

A_e = Radar receive antenna effective area in the direction of the jammer, square meters (m^2)

A_e = Radar warning receiver antenna effective area in the direction of the radar, square meters (m^2)

a_{nt} = Radius of the nose tip, meters

B = Number of different phase shifts, no units

B_J = Jammer noise bandwidth, hertz

B_{pc} = Pulse compression modulation bandwidth, hertz

B_{pc} = Radar pulse compression modulation bandwidth, hertz

B_R = Radar receiver filter bandwidth, hertz

B_R = Radar receiver matched filter bandwidth, hertz

B_R = Radar receiver processing bandwidth, hertz

B_R = Receiver filter bandwidth – Matched Filter bandwidth, hertz

B_{RWR} = Radar warning receiver bandwidth, hertz

c = Speed of light, 3×10^8 meters/second

C/N = Single pulse clutter signal-to-noise ratio, no units

CR = Cancellation ratio, no units

d = Diameter of the cylinder, meters

d = Element spacing, meters

d = Physical distance between array elements, meters

$D =$ Antenna dimension, meters

$D =$ Dimension of the array, meters

$D =$ Dimension of the object, meters

$D_0 =$ Signal-to-noise ratio required for detection (detection threshold) for a constant target signal, no units

$D_{0dB} =$ Signal-to-noise ratio required for detection (detection threshold) for a constant target signal using Albersheim's formula, decibels

$D_{0n} =$ Equivalent single-pulse signal-to-noise ratio required for detection after noncoherent integration of multiple pulses for a Swerling Case 0 target, no units

$D_{0ndB} =$ Equivalent single-pulse signal-to-noise ratio required for detection after noncoherent integration of multiple pulses for a Swerling Case 0 target using Albersheim's formula, dB

$D_1 =$ Signal-to-noise ratio required for detection (detection threshold) for a Swerling Case 1 target, no units

$d_a =$ Cross-range (azimuth) resolution, meters

$d_{a\ min} =$ Finest cross-range (azimuth) resolution for a focused synthetic aperture radar, meters

$d_{a\ min} =$ Finest cross-range (azimuth) resolution for an unfocused synthetic aperture radar, meters

$dB =$ decibel, no units

$dBi =$ Decibels relative to an isotropic antenna, no units

$dBm =$ Decibels relative to one milliwatt, carries the units of milliwatts

$DBS_a =$ Doppler beam sharpening cross-range (azimuth) resolution, meters

$dBsm =$ Decibels relative to 1 square meter, carries the units of square meters

$dBW =$ Decibels relative to one watt, carries the units of watts

$D_c =$ Ideal signal-to-noise ratio required for coherent detection, no units

$d_i =$ Distance of the i-th array element from the start of the array, meters

$d_{max} =$ Maximum element physical separation to steer the beam to θ_0 without any grating lobes, meters

$d_r =$ Range resolution, meters

$d_t =$ Transmit duty cycle, no units

$E =$ Energy in one pulse, watt-seconds, or joules

$E(\theta) =$ E-field produced by the current distribution, perpendicular to A(y), voltage antenna gain pattern, no units

$\mathtt{erfc}(T) =$ Complementary error function

$\mathtt{erfc}^{-1}(T) =$ Inverse complementary error function

$ERP_J =$ Jammer effective radiated power, watts

$ERP_R =$ Radar effective radiated power, watts

$f =$ Frequency, hertz

$f =$ Frequency of the laser light, hertz

$f =$ Radar frequency, megahertz

$f =$ Transmitted frequency, hertz

$f(t) =$ Time domain representation of the voltage pulse envelope, volts

$f_0 =$ Frequency corresponding to the beam pointing at broadside, hertz

$FAR =$ Average false alarm rate; average number of false alarms per second, no units

$f_c =$ Radar carrier frequency, 100's megahertz (MHz) ~10's gigahertz (GHz), hertz (cycles per second)

$f_c =$ Radar carrier frequency, hertz

$f_c =$ Radar carrier frequency, radio frequency, hertz

$f_c =$ Radar transmitted carrier frequency, hertz

$f_c + f_{IF} =$ Local oscillator frequency, hertz

$f_{cr} =$ Received carrier frequency, hertz

$f_d =$ Doppler shift, hertz

$f_d =$ Target Doppler shift, hertz

$f_{dmax} =$ Maximum Doppler spread across the mainbeam of the radar antenna, hertz

$f_{du} =$ Unambiguous Doppler shift, hertz

$f_{IF} =$ Radar receiver intermediate frequency, hertz

$f_{IF} =$ Receiver intermediate frequency, hertz

$f_J =$ Jammer transmitted frequency, hertz

$F_R =$ Radar receiver noise figure (≥ 1), no units

$F_{RWR} =$ Radar warning receiver noise figure, no units

$G =$ Antenna gain, no units

$G =$ Antenna mainbeam gain, no units

$G =$ Boresight mainbeam antenna gain for an array antenna, no units

$G =$ Gain back in the direction of the lidar, no units

$G =$ Mainbeam array antenna gain with the beam pointing at boresight, no units

$G =$ Radar antenna gain, no units

$G =$ Reradiation gain in the direction of the radar, no units

$G(\theta) =$ Normalized antenna gain pattern (power) for a uniform current distribution, no units

$G(\theta) =$ Normalized array antenna gain pattern as a function of angle off boresight for a uniform current distribution, no units

$G(\theta) =$ Normalized array antenna gain pattern with the beam steered to θ_0 as a function of angle off boresight for a uniform current distribution, no units

$G(\theta_0) =$ Mainbeam array antenna gain as a function of beam steering angle θ_0, no units

$g(\omega) =$ Frequency spectrum of the pulse, volts

$G(\omega) =$ Frequency spectrum of the pulse, watts

$G_a(\theta) =$ Normalized array factor as a function of angle off boresight (θ) for a uniform current distribution, no units

$G_a(\theta) =$ Normalized array factor with the beam steered to θ_0 as a function of angle off boresight (θ) for a uniform current distribution, no units

$G_{AJ} =$ Auxiliary antenna gain in the direction of the jammer, no units

$g_e =$ Antenna gain of each element, no units

$G_e(\theta) =$ Normalized element factor (element antenna gain) as a function of angle off boresight, no units

$G_I =$ Integration gain, no units

$G_I =$ Radar integration gain, no units

$G_{I0} =$ Noncoherent integration gain for a Swerling Case 0 target, no units

$G_{If} =$ Noncoherent integration gain for a Swerling Case 1, 2, 3, or 4 target, no units

$G_{JR} =$ Jammer transmit antenna gain in the direction of the radar, no units

$G_{JRr} =$ Jammer receive antenna gain in the direction of the radar, no units

$G_{JRt} =$ Jammer transmit antenna gain in the direction of the radar, no units

$G_p(f) =$ Single-delay canceler power frequency response, no units

$G_p(f) =$ Multiple delay canceler power frequency response, no units

$G_{RE} =$ Radar antenna gain in the direction of the passive expendable, no units

$G_{RJ} =$ Radar antenna gain in the direction of the jammer (where ever the jammer is in the radar antenna pattern: mainbeam, sidelobes, back-lobe), no units

$G_{RJ} =$ Radar antenna gain in the direction of the jammer, no units

$G_{RT} =$ Radar antenna gain in the direction of the target, no units

$G_{RT} =$ Radar antenna gain in the direction of the target/jammer, no units

$G_{RT} =$ Radar receive antenna gain in the direction of the target/jammer, no units

$G_{RT} =$ Radar transmit antenna gain in the direction of the target, no units

$G_{RT} =$ Radar transmit antenna gain in the direction of the target/radar warning receiver (anywhere in the radar antenna pattern, mainbeam or sidelobes), no units

$G_{RrT} =$ Receive antenna gain in the direction of the target, no units

$G_{RtT} =$ Transmit antenna gain in the direction of the target, no units

$G_{RWR} =$ Radar warning receiver antenna gain in the direction of the radar, no units

$G_s =$ Jammer system gain, no units

$G_{sp} =$ Radar signal processing gain, no units

$G_{sp} =$ Signal processing gain, no units

$g_x =$ Antenna gain of a half-wave dipole on a ground plane, no units

$h =$ Height of the target or height of the radar, meters

$h =$ Plank's constant, 6.6×10^{-34} joule-seconds

$h_R =$ Radar height, meters

$h_T =$ Target height, meters

$I =$ In-phase phasor of the signal, volts

$I =$ Interference signal power, watts

$I_{fMTI} =$ Moving target indicator improvement factor, no units

$I_0(x) =$ Modified Bessel function of the first kind and zero order

$J =$ Received self-protection jammer peak power, watts

$J =$ Received support jammer peak power, watts

$J/N =$ Single pulse false target jamming-to-noise ratio, no units

$(J/N)_n =$ False target jamming-to-noise ratio after integration of multiple pulses, no units

$J_{AE} =$ Received active expendable signal peak power, watts

$J_{AE}/N =$ Single pulse active expendable jamming-to-noise ratio, no units

$(J_{AE}/N)_n =$ Active expendable jamming-to-noise ratio after integration of multiple pulses, no units

$(J_{AE}/S)_n =$ Active expendable jamming-to-target signal ratio after integration of multiple pulses, no units

$J_{cg} =$ Received constant gain jammer power, watts

$J_N =$ Jammer noise power out of the radar receiver, watts

$(J_N/S)_n =$ Noise jamming-to-target signal ratio after integration of multiple pulses, no units

$J_{PE} =$ Received passive expendable signal peak power, watts

$J_{PE}/N =$ Single pulse passive expendable signal-to-noise ratio, no units

$(J_{PE}/N)_n =$ Passive expendable signal-to-noise ratio after integration of multiple pulses, no units

$(J_{PE}/S)_n =$ Passive expendable jamming-to-target signal ratio after integration of multiple pulses, no units

$k =$ Boltzmann's constant, 1.38×10^{-23} joule/Kelvin

$k =$ Boltzmann's constant, 1.38×10^{-23} watt-seconds/Kelvin

$k_r =$ Refraction factor, no units

$l =$ Distance over which measurements are made, meters

$L =$ Distance from the last measurement to the prediction point, meters

$L =$ Electrical distance between array elements, meters

$L =$ Length of the cylinder, meters

$L =$ Synthetic aperture (array) length, meters

$L_e =$ Maximum effective synthetic aperture (array) length, meters

$L_J =$ Total jammer-related losses, no units

$L_{Jpol} =$ Jammer-radar polarization mismatch loss, no units

$L_{JRa} =$ Jammer-to-radar atmospheric attenuation loss, no units

$L_{Jt} =$ Jammer transmit loss, no units

$L_{max} =$ Maximum synthetic aperture (array) length, meters

L_R = Bistatic radar system losses, no units

L_R = Radar-related losses, no units

L_R = Total radar-related losses, no units

L_{Rpol} = Radar-radar warning receiver polarization mismatch loss, no units

L_{Rr} = Radar receive loss, no units

L_{Rsp} = Radar signal processing loss, no units

L_{Rt} = Radar transmit loss, no units

L_{RTa} = Radar-to-target/radar warning receiver atmospheric attenuation loss, no units

L_{RTa} = Target/jammer-to-radar atmospheric attenuation loss, no units

L_{RTRa} = Radar-to-target and target-to-radar atmospheric attenuation loss, no units

L_{RWR} = Total radar warning receiver-related losses, no units

L_{RWRr} = Radar warning receiver receive loss, no units

L_{RWRsp} = Radar warning receiver signal processing loss, no units

M = Number of times the detection threshold is crossed

n = Integer, n = 0, 1, 2, ...

n = Integer number of the pulse repetition frequency, no units

$\bar{n}$ = Variance of fluctuations in the Poisson distribution, no units

$\bar{n}_s$ = Average number of signal photons arriving at the output of an optical receiver, no units

N = Mean noise power at the linear envelope detector output, watts

N = Number of array elements, no units

N = Number of chaff dipoles in the cloud, no units

N = Number of delay lines, no units

N = Number of detection attempts

N = Number of elements, no units

N = Number of individual detections attempts

N = One sample of receiver thermal noise power, watts

N = Radar main and auxiliary receiver thermal noise power, assumed to be the same, watts

N = Radar receiver thermal noise power, watts

N_a = Thermal noise power generated in an actual receiver at the standard reference temperature, watts

n_b = Number of antenna beam positions in which detection decisions are made (≥ 1), no units

N_d = Number of detection decisions performed over a time period of interest, such as radar search time or tracking interval, no units

n_{df} = Number of Doppler filters in which detection decisions are made (≥ 1), no units

N_e = Electron density, electrons/cm^3

N_f = Number of Doppler filters, no units

N_{fa} = Average number of false alarm in one time period of interest, no units

N_i = Thermal noise power generated in an ideal receiver at the standard reference temperature, watts

N_n = Receiver thermal noise power after integration of n_p noise samples, watts

n_p = Number of pulses integrated, no units

n_{rg} = Number of range gates in which detection decisions are made (≥ 1), no units

N_{RWR} = Radar warning receiver thermal noise power, watts

N_s = Surface refractivity

$N\phi$ = Number of phase coded segments or chips, no units

$\cap_n$ = Value of the n-th observation

= New observation

p = Probability of crossing the detection threshold for each attempt

P = Peak transmit power, watts

P = Power, watts

$P(M,N,p)$ = Probability of M detection threshold crossing out of N attempts

$P(mW)$ = Power, milliwatts

$P(v \geq V_T)$ = Probability of noise exceeding a threshold, no units

$P(W)$ = Power, watts

$P(x \geq T)$ = Probability of x exceeding the value T

$p(x)$ = Probability density function

P_{ave} = Average transmit power, watts

PCR = Pulse compression ratio, no units

P_d = Probability of detection, no units

P_d = Probability of detection (the same for all detection attempts), no units

$p_d(x)$ = Probability density function for a single dice

P_{dc} = Cumulative probability of detection, no units

P_{di} = Probability of detection for the i-th individual detection attempt, no units

P_{fa} = Probability of false alarm, no units

P_{fa} = Probability of false alarm (the same for all detection attempts), no units

$P_{fa}(1)$ = Probability of noise alone exceeding the detection threshold once (P_{fa})

$P_{fa}(2)$ = Probability of noise alone exceeding the detection threshold twice

$P_{fa}(N)$ = Probability of noise alone exceeding the detection threshold N times

P_{fac} = Cumulative probability of false alarm, no units

P_{fai} = Probability of false alarm for the i-th individual detection attempt, no units

$P_i =$ Power in, watts

$P_J =$ Jammer peak transmit power, watts

$P_{Js} =$ Jammer saturation peak power, watts

$p_N(v) =$ Probability of v (noise) occurring for the Rayleigh probability density function

$P_o =$ Power out, watts

$P_r =$ Received energy, watts-seconds or joules

$P_R =$ Radar peak transmit power, watts

$P_R =$ Radar transmitter peak power, watts

$PRF =$ Pulse repetition frequency, 100's hertz ~100's kilohertz (kHz), hertz (pulses per second)

$PRF =$ Pulse repetition frequency, hertz

$PRF =$ Pulse repetition frequency to avoid both range and Doppler ambiguities, hertz

$PRF =$ Radar pulse repetition frequency, hertz

$PRF_{max} =$ Maximum pulse repetition frequency which will avoid range ambiguities, hertz

$PRF_{min} =$ Minimum pulse repetition frequency which will avoid Doppler ambiguities, hertz

$PRI =$ Pulse repetition interval, 0.01's seconds ~0.01's milliseconds (msec), seconds

$PRI =$ Pulse repetition interval, seconds

$p_{S+N}(v) =$ Probability of v (target signal plus noise) occurring for the Rician probability density function

$P_Z =$ Residual noise, jamming and thermal, power after cancellation, watts

$Q =$ Quadrature phasor of the signal, volts

$Q(E) =$ Q probability integral

$Q^{-1}(E) =$ Inverse Q probability integral

$r =$ Radius of the flat back, meters

$r =$ Radius of the sphere, meters

$R =$ Distance to the far field point, meters

$R =$ Lidar-to-target slant range, meters

$R =$ Radar-to-clutter slant range, meters

$R =$ Radar-to-target slant range, meters

$R =$ Radar-to-target/clutter slant range, meters

$R =$ Range from the radar to the object, meters

$R =$ Range to the outer edge of the image, meters

$R_1 =$ Measured radar-to-target range at time t_1, meters

$R_1 =$ Range mcasurement at time t_1, meters

$R_1 =$ Transmitter-to-target slant range, meters

$R_1 R_2 =$ Bistatic detection range product, square meters (m^2)

R_2 = Measured radar-to-target range at time t_2, meters

R_2 = Range measurement at time t_2, meters

R_2 = Receiver-to-target slant range, meters

R_3 = Predicted radar-to-target range at time Δt in the future, meters

R_{bt} = Radar burnthrough range, meters

R_{dot} = Object-to-radar range rate; positive for an opening target, radar-to-target range increasing, and negative for a closing target, radar-to-target range decreasing, meters/second

R_{dot} = Range rate, meters/second

R_{dotC} = Clutter range rate, meters/second

R_{dotu} = Unambiguous range rate, meters/second

R_{dt} = Radar detection range, meters

R_{dt1} = Old radar detection range, meters

R_{dt1} = Radar detection range for radar cross section σ_1, meters

R_{dt2} = New radar detection range, meters

R_{dt2} = Radar detection range for radar cross section σ_2, meters

R_{dtRWR} = Radar warning receiver detection range, meters

R_E = Radius of the earth, 6371 kilometers

R_{ff} = Far field distance, meters

R_h = Range to the horizon, meters

R_{hC} = Clutter horizon, meters

R_{hT} = Target horizon, meters

R_{LOS} = Radar line of sight, meters

R_p = Range to the patch on the ground, meters

R_{RE} = Radar-to-active expendable slant range, meters

R_{RE} = Radar-to-passive expendable slant range, meters

R_{RJ} = Radar-to-jammer slant range, meters

R_{RT} = Radar-to-target slant range, meters

R_{RT} = Radar-to-target/jammer slant range, meters

R_{RT} = Radar-to-target/radar warning receiver slant range, meters

R_{sw} = Range extent of the image or swath width, meters

R_u = Unambiguous range, meters

S = Constant amplitude target signal power, watts

S = Received single-pulse target signal peak power, watts

S = Received target signal power from one pulse, watts

S = Target signal power, watts

S/C = Competing target signal-to-clutter ratio, no units

$(S/C)_{in}$ = Target signal-to-clutter ratio into the moving target indicator, no units

$(S/C)_{out}$ = Target signal-to-clutter ratio out of the moving target indicator, no units

S/I = Single pulse target signal-to-interference ratio, no units

$S/I =$ Target signal-to-interference ratio, no units

$(S/I)_n =$ Target signal-to-interference ratio after integration of multiple pulses, no units

$(S/I)_n =$ Post-cancellation target signal-to-interference ratio after integration of multiple pulses, no units

$S/N =$ Power signal-to-noise ratio required for detection (detection threshold), no units

$S/N =$ Single-pulse target signal-to-noise ratio, no units

$S/N =$ Signal-to-noise ratio, no units

$(S/N) =$ Single-pulse target signal-to-noise ratio, no units

$(S/N)_n =$ Target signal-to-noise ratio after integration of n_p coherent pulses, no units

$(S/N)_n =$ Target signal-to-noise ratio (power) after the integration of multiple pulses, no units

$(S/N)_p =$ Processed target signal-to-noise ratio, no units

$(S/N)_{RWR} =$ Single-pulse radar signal-to-noise ratio of the radar warning receiver, no units

$(S/J_N)_n =$ Target signal-to-jamming ratio after integration of multiple pulses, no units

$S_{min\ dis} =$ Minimum discernible signal sensitivity, watts

$S_{min\ dt} =$ Minimum detectable signal sensitivity, watts

$S_n =$ Received target signal power after integration of n_p coherent pulses, watts

$SNR_{dt} =$ Radar detection threshold, no units

$SNR_{dt} =$ Single pulse signal-to-noise ratio required for detection, no units

$SNR_{dt} =$ Signal-to-noise ratio required for detection (detection threshold), no units

$SNR_{dt} =$ Target signal-to-noise ratio required for detection (detection threshold), no units

$SNR_{dt2} =$ New detection threshold, no units

$SNR_{dt1} =$ Old detection threshold, no units

$SNR_{dtRWR} =$ Radar warning receiver detection threshold, no units

$S_{RWR} =$ Received single-pulse radar signal peak power, watts

$S_{RWR} =$ Received radar signal power, watts

$t =$ Duration of the measurement, seconds

$T =$ Time ahead the predication is made, seconds

$T =$ Time it takes for the object to rotate one time, seconds

$T_0 =$ IEEE standard reference temperature for radio frequency receivers, 290 Kelvin (62 °F)

$T_0 =$ Receiver standard reference temperature, 290 Kelvin

$T_d =$ Time to make an individual detection decision, seconds

$T_e =$ Receiver effective noise temperature, Kelvin

$T_{fa} =$ Average time between false alarms, seconds

T_I = Coherent integration time (time to form the synthetic aperture, time a patch on the ground is illuminated, and time it takes the leading edge to the trailing edge of the radar antenna mainbeam to transit across the patch), seconds

T_I = Integration time, seconds

T_I = Radar integration time or pulse burst duration or coherent processing interval, seconds

T_{ill} = Target illumination time, seconds

T_S = Radar scan time or tracking interval, seconds

T_s = Receiver system noise temperature, Kelvin

T_s = Time between measurements, seconds

T_{sp} = Signal processing interval, seconds

v = Collision frequency of the electrons with particles

v = Linear envelope detector output voltage, volts

V = Pulse amplitude, volts

V = Voltage, volts

V = Volume of the sphere, cubic meters (m^3)

V_{ac} = Aircraft speed, magnitude of the velocity vector, meters/second

V_i = Voltage in, volts

V_o = Voltage out, volts

V_R = Radar speed, meters/second

V_T = Threshold voltage, volts

$V_T{}^2/N$ = Threshold-to-noise power ratio, no units

v_n = n-th blind speed, meters/second

$\hat{v}_n$ = Smoothed target velocity, meters/second

W = Weighting function

X_n = Measured target position, meters

X_{pn} = Predicted target position, meters

$\hat{X}_n$ = Smoothed target position, meters

$\bar{X}_n$ = Average after n observations

$\bar{X}_{n-1}$ = Average after n − 1 observations

$\bar{X}'$ = Estimate

$\bar{X}$ = Current mean

B = Entire phase progression across the array, radians

ΔCR = Cross-range resolution, meters

Δf = Frequency excursion, hertz

Δf_d = Difference in Doppler shift between two patches on the ground, hertz

Δf_d = Doppler filter bandwidth, hertz

Δf_d = Doppler resolution (Doppler filter bandwidth), hertz

Δf_d = Doppler resolution; single simple pulse, hertz

ΔM = Measurement resolution

ΔN = Noise power added by an actual receiver above what is generated in an ideal receiver, watts

$\Delta R =$ Difference in range, meters

$\Delta R =$ Range resolution, meters

$\Delta R =$ Root mean square range error, meters

$\Delta R_{dot} =$ Range rate resolution, meters/second

$\Delta R_{dot} =$ Range rate resolution; single simple pulse, meters/second

$\Delta R_f =$ Error in the final range measurement, meters

$\Delta R_i =$ Error in the initial range measurement, meters

$\Delta R_{pc} =$ Range resolution with pulse compression, meters

$\Delta SNR_{f1} =$ Additional signal-to-noise ratio, relative to a Swerling Case 0 target, required for detection for a Swerling Case 1 target, no units

$\Delta SNR_{f1} =$ Additional signal-to-noise ratio, relative to a Swerling Case 0 target, required for detection for a Swerling Case 1 target, no units

$\Delta SNR_{f2} =$ Additional signal-to-noise ratio, relative to a Swerling Case 0 target, required for detection for a Swerling Case 2 target, no units

$\Delta SNR_{f3} =$ Additional signal-to-noise ratio, relative to a Swerling Case 0 target, required for detection for a Swerling Case 3 target, no units

$\Delta SNR_{f4} =$ Additional signal-to-noise ratio, relative to a Swerling Case 0 target, required for detection for a Swerling Case 4 target, no units

$\Delta t =$ Elapsed time or time delay from when the pulse was transmitted and when the reflected pulse is received, seconds

$\Delta t =$ Time delay from when a pulse is transmitted and when the reflected pulse is received, seconds

$\Delta t =$ Time it takes for the lobe in the RCS pattern to cross the radar line of sight, seconds

$\Delta v =$ Error in the velocity estimate, meters/second

$\Delta \phi =$ Difference in phase, radians or degrees

$\Delta \phi =$ Incremental phase shift between elements necessary to steer the beam to an angle θ_0, radians

$\Delta \theta =$ Angle resolution, degrees or radians

$\Delta \theta =$ Angular separation between two patches on the ground, radians

$\Delta \theta_0 =$ Maximum beam steering error, radians or degrees

$\Delta \theta_{dot} =$ Rate of increase in root mean square angular error after the final measurement, radians/second

$\Omega =$ Faraday rotation, degrees

$\Omega =$ Total solid angle scan coverage, radians squared or degrees squared

$\alpha =$ Attenuation, dB/Nmi

$\alpha =$ Grazing angle of the radar wave relative to the ground, radians or degrees

$\alpha =$ Position smoothing parameter, no units

$\beta =$ Velocity smoothing parameter, no units

$\beta_i =$ Phase shift at the i-th element, radians

$\delta_M =$ Root mean squared measurement accuracy, same units as Δ_M

$\varepsilon =$ Arbitrary phase difference from center to edge of the antenna, meters

$\varepsilon_R =$ Range error at the prediction point, meters

$\varepsilon_\theta =$ Angular error at the prediction point, radians

$\phi =$ Elevation, vertical plane, angle from the aircraft velocity vector to the clutter, radians or degrees

$\phi =$ Phasor angle from the in-phase axis, radians or degrees

$\phi =$ Random polarization angle, radians

$\phi_{3dB} =$ Antenna half-power (–3 dB) beamwidth in the elevation (vertical) plane, radians or degrees

$\phi_p =$ Elevation (vertical plane) angle relative to the velocity vector to the patch on the ground, degrees

$\phi_{RT} =$ Radar-to-target elevation angle, degrees

$\phi_S =$ Angular scan coverage in the elevation (vertical) plane, radians or degrees

$\gamma_f =$ Swerling case fluctuation exponent, no units

$\eta =$ Index of refraction

$\eta =$ Quantum efficiency (the percentage of incident photons converted into electrons), no units

$\lambda =$ Radar transmitted wavelength (c/f_c), meters

$\lambda =$ Transmitted wavelength, meters

$\lambda =$ Wavelength, meters

$\lambda_0 =$ Wavelength corresponding to the beam pointing at broadside, meters

$\theta =$ Azimuth, horizontal plane, angle from the aircraft velocity vector to the clutter, radians or degrees

$\theta =$ Peak to first null in the radar cross section pattern, radians

$\theta =$ Polarization angle, radians

$\theta^2 =$ Effective aperture beam area, square radians

$\theta_0 =$ Desired mainbeam pointing angle relative to the array boresight, radians

$\theta_0 =$ Beam steering angle, radians or degrees

$\theta_1 =$ Beam steering limits, radians or degrees

$\theta_{3dB} =$ Antenna half-power (–3 dB) beamwidth, degrees or radians

$\theta_{3dB} =$ Antenna half-power (–3 dB) beamwidth for a uniform current distribution, radians or degrees

$\theta_{3dB} =$ Antenna half-power (–3 dB) beamwidth in the azimuth (horizontal) plane, radians or degrees

$\theta_{3dB} =$ Array antenna half-power beamwidth with the beam pointing at boresight, radians or degrees

$\theta_{3dB} =$ Radar antenna azimuth half-power (–3 dB) beamwidth, radians

$\theta_{3dB}(\theta_0) =$ Array antenna half-power (–3 dB) beamwidth as a function of beam steering angle θ_0, radians or degrees

$\theta_{4dB} =$ Antenna two-way –4 dB beamwidth, radians

$\theta_{dot} =$ Antenna scan rate, degrees/second or radians/second

$\theta_g =$ Angle at which a grating lobe will appear, radians or degrees

$\theta_{nn} =$ Antenna null-to-null beamwidth for a Uniform current distribution, radians or degrees

$\theta_{nn} =$ Array null-to-null beamwidth for a Uniform current distribution, degrees

$\theta_{nn} =$ Radar antenna null-to-null beamwidth, radians

$\theta_p =$ Azimuth (horizontal plane) angle relative to the velocity vector to the patch on the ground (squint angle), degrees

$\theta_S =$ Angular scan coverage in the azimuth (horizontal) plane, radians or degrees

$\theta_0 =$ Beam steering angle, radians or degrees

$\rho =$ Antenna efficiency, no units

$\rho =$ Antenna efficiency, 30% ~ 70% for most practical radar antennas, no units

$\rho =$ Reflectivity of the target to the lidar wavelength, no units

$\sigma =$ Radar cross section for nonisotropic object, square meters (m^2)

$\sigma =$ Radar cross section at a polarization angle, square meters (m^2)

$\sigma =$ Target radar cross section, square meters (m^2)

$\sigma\,(m^2) =$ Radar cross section, square meters (m^2)

$\sigma_0 =$ Clutter reflectivity, square meters/square meters (m^2/m^2)

$\sigma_0 =$ Radar cross section of a non-depolarizing target, square meters (m^2)

$\sigma_0^{\,2} =$ Variance of the most current measurement

$\sigma_1 =$ Target radar cross section for radar detection range R$_1$, square meters (m^2)

$\sigma_2 =$ Target radar cross section for radar detection range R$_2$, square meters (m^2)

$\sigma_a =$ Standard deviation of the antenna modulation power spectral density, hertz

$\sigma_b =$ Target bistatic radar cross section, square meters (m^2)

$\sigma_c =$ Clutter radar cross section, square meters (m^2)

$\sigma_{cly} =$ Radar cross section of a cylinder, square meters (m^2)

$\sigma_{Cpsd} =$ Standard deviation of the clutter power spectral density, hertz

$\sigma_d =$ Radar cross section of a dihedral corner reflector, square meters (m^2)

$\sigma_{flatback} =$ Radar cross section of the flat back, square meters (m^2)

$\sigma_{fp} =$ Radar cross section for a flat plate, square meters (m^2)

$\sigma_i =$ Standard deviation of the radar instabilities power spectral density, hertz

$\sigma_{nosetip} =$ Radar cross section of nose tip, square meters (m^2)

$\sigma_p =$ Peak chaff radar cross section, square meters (m^2)

σ_{pe} = Passive expendable radar cross section, square meters (m^2)

$\sigma_{Rayleigh}$ = Radar cross section of a sphere in the Rayleigh region, square meters (m^2)

σ_{Rdots} = Standard deviation of the clutter source range rate, meters/second

σ_s = Radar cross section of sphere, $2\pi a/\lambda > 10$, square meters (m^2)

σ_s = Standard deviation of the clutter source power spectral density, hertz

σ_t = Radar cross section of a trihedral corner reflector, square meters (m^2)

σ_w = Radar cross section of a long wire, square meters (m^2)

σ^2/n = Variance of all measurements that have gone before

$\bar{\sigma}$ = Average radar cross section across all polarization angles, square meters (m^2)

$\bar{\sigma}$ = Average chaff radar cross section, square meters (m^2)

$\bar{\sigma}$ = Average radar cross section across all polarization angles, square meters (m^2)

$\bar{\sigma}_{HH}$ = Average radar cross section (horizontal transmit–horizontal receive), square meters (m^2)

$\bar{\sigma}_{HV}$ = Average radar cross section (horizontal transmit–vertical receive), square meters (m^2)

$\sum \sigma_{0i}$ = Sum of the clutter reflection coefficients from all contributing scatterers, square meters/square meters (m^2/m^2)

τ = Pulse duration, seconds

τ = Pulse duration, or pulse width, or pulse length, ≈ 1 microsecond (μsec), seconds

τ = Pulse width, seconds

τ = Radar transmitted pulse width, seconds

τ = Transmitted pulse duration, or pulse width, or pulse length, seconds

τ_c = Compressed pulse width, seconds

ω = Radar frequency, radians/second

ω_o = Angular frequency of the pulse, radians/second, $2\pi f_c$, where f_c = carrier frequency, hertz

ψ = Reflection coefficient, square meters/square meters (m^2/m^2)

Introduction to Radar and Electronic Warfare

HIGHLIGHTS

- Introduction to radar systems and the target signature
- Introduction to electronic warfare

▮▮ 1.1 ▮ RADAR SYSTEMS

Radar is an acronym for "RAdio Detection And Ranging." A radar system detects the presence of objects in the environment and measures characteristics of the object. The word "radio" comes from the radio frequency (RF) or microwave portion of the electromagnetic spectrum. While this covers a very wide frequency range, most radar systems operate at 100's of megahertz (MHz or a million or 10^6 Hz) to 10's of gigahertz (GHz or a billion or 10^9 Hz). A measurement is a characteristic of a detected object as determined by the radar system: range, range rate, angle. The signature of the object affects detection based on its radar cross section (RCS) and range rate measurements based on its spectrum.

If the range is not too long or too short, less than several thousand kilometers (km) in the first case and a kilometer or so in the second, radar is the preeminent sensor in any application requiring night and bad weather operations. This is a very broad mission including vehicle traffic on the highways and at state or national borders, aircraft on runways and in the air, ships at sea and in harbors, and satellites and other objects in outer space. Radar systems like altimeters and collision avoidance are also on-board the vehicles themselves.

Radar is preeminent in many missions for two reasons: it is active; and it is usually all-weather. Being active allows night operations since the radar provides its own illumination. A radar system's measuring stick is the speed of light. Nothing in the universe is faster or more efficient. Even if measurements of position could be made by other means (and they can in many cases), the radar measurement is simpler and easier. The use of the RF portion of the electromagnetic spectrum provides the radar system's all-weather capability. Although radar is thought of as all-weather, this is not the case for all radar systems. The lower the frequency, the less problem there is with weather.

As ranges get longer or the need for very fine measurement resolution emerges, radar begins to falter. The first condition is deep-space surveillance; an example of the second is perimeter security of a plant, base, or other facility. For example, a radar system covering the outside perimeter of a security fence would need to differentiate among vehicles, animals, and human beings. The measurement resolution required for recognition of various classes of target is a matter of intense debate, but it is clear that the finer the better. At these comparatively short ranges, weather and night conditions do not affect passive systems appreciably, making infrared sensors and low-light-level TV very competitive, even when there is considerable dust, fog, and smoke.

Radar systems are traditionally divided into three main functions or modes: search, track, and image. Many radar systems are optimized for one function, while others perform multiple functions. This is particularly true for airborne radar systems.

Search

Search, also commonly called surveillance, radar systems are designed to detect the presence and provide measurements of a large number of targets such as aircraft, ships, and missiles over a large volume—range, azimuth (horizontal plane) angle, and elevation (vertical plane) angle—within a desired time. Examples of search radar systems are early warning (EW), surveillance, target acquisition radar (TAR), air traffic control (ATC), and weather.

Target detection must occur before these targets reach a specified range, often called the "detection range." While for some search radar systems this range is modest, for many search radar systems it is a few to several hundreds of kilometers (km). Detection range is a function of the physical size of the antenna and transmit power of the radar system. The search radar designer will select the frequency of the radar at which the combination of the physical size of the radar antenna and the generation of radar power are least expensive. Big antenna and high-powered transmitters are characteristic of surveillance radar systems. Because both these characteristics favor the use of lower microwave frequencies, the vast majority of surveillance radar systems are less than 3 GHz. Radar antennas may range in diameter from one to tens of meters for both reflectors and arrays. Peak transmit power of a few kilowatts (kW) to a few megawatts (MW) are common for search radar systems.

Track

Tracking radar systems are designed to provide continuous, or near continuous, measurements of one, a few, or several targets so they can be "tracked" or followed over time. Examples of tracking radar systems are single target, multiple targets, military fire control, target tracking radar (TTR), target engagement radar (TER), weapon guidance, and test and training range instrumentation radar (RIR).

Target measurements (range, range rate, and angle), capability, and performance are important for target tracking, both for individual measurements and target states (position, velocity, and acceleration). The driving parameter in good target tracking is measurement resolution, which for range rate and angle measurements requires a high frequency. Not surprisingly, tracking radar systems (including the ground-controlled approach radar systems at airports and missile trackers of the military) are at the high microwave frequencies, generally from 3 to 20 GHz.

Image

Imaging radar systems produce an image of a scene or an individual target using a combination of fine measurement resolution and signal processing algorithms. Examples of imaging radar systems are real beam, synthetic aperture radar (SAR), and Doppler beam sharpening (DBS). Radar systems provide an all-weather imagery capability. Again, the military applications dominate, although remote sensing (e.g., land management, natural disasters) and navigation have substantial relevance. The military commander needs to know the status of an enemy's deployed forces and cannot wait for daylight or good weather; attacking aircraft must be able to identify enemy targets through snow, rain, fog, smoke, and dust.

Additional Radar Functions

While there are many other radar functions beyond search, track, and image, a few require a quick discussion, navigation, and science. Radar systems can support platform navigation by providing Doppler information to update an inertial navigation system and radar-generated maps (images), which can provide references by which a vehicle can locate itself. Both Doppler updates and radar mapping are essential in military operations. In such an emergency, it is necessary to be self-contained, independent of time of day, weather, and the status of outside navigation aids like global positioning systems (GPSs). In this situation, radar shines. Because we can guide these vehicles in several other ways, radar's role could be described as important but not unique.

Speculation on the potential uses of radar in science is for the futurists. Current and past uses are substantial, however. Cases in point are the studies of the ionosphere and of our own and neighboring planets, the latter of which—particularly of the auroral region in the Northern Hemisphere—have proceeded vigorously for over thirty years. Much knowledge about its propagation has been exploited in radar system design.

The Haystack radar was used in the 1960s in a test of relativity by making measurements of the planet Mercury as the radar line of sight to Mercury neared the edge of the sun. The Arecibo system has examined all the planets inside Jupiter.

NASA has flown mapping radar systems several times in space. Because electromagnetic waves have some penetrating power, images of subsurface features in both land and water were obtained that predicted new dimensions in Earth resources research. Visualizing planets like Venus (shrouded in an atmosphere impenetrable to many of the higher frequencies) are made to order for radar systems. We have seen high-resolution SAR maps of the surface of Venus. That planet is being explored from earthbound radar systems and from radar systems on-board the space vehicles sent there. In fact, virtually all explorers of the solar system that expect to land will have at least a radar altimeter on board.

Our discussion of radar systems starts with the fundamentals of radio frequency waves in Chapter 2. From there we go into more detail: target detection in Chapter 3; radar antennas in Chapter 4; radar measurements and target tracking in Chapter 5; target signature in Chapter 6; and advanced radar concepts in Chapter 7.

In summary, the qualities making radar systems essential to modern life also limit its utility. The fact that radar furnishes its own illumination makes it valuable for night operations against passive or noncooperative targets but limits its ultimate detection range. The fact that the frequencies at which radar systems operate can penetrate bad weather, smoke, and dust and can see over the earth's horizon makes them essential to emergency operations, particularly military operations. However, it also prevents them from obtaining the high resolution necessary to provide the fine-grained images associated with optical systems. The vigorous application of new technology and human ingenuity will tend to expand radar's applications and mitigate its shortcomings. However, the natural laws cannot be overcome, inevitably consigning radar to its niche, an interesting and important technology with multifaceted uses.

1.2 | ELECTRONIC WARFARE

As we just discussed, radar systems detect the presence of targets in the environment and measure characteristics of the targets. For military applications this information is used to assess the situation, determine number and types of targets, their heading, etc. and support weapon systems to engage select targets. Electronic warfare (EW) is the reaction to an adversary's use of radar systems. Its three components—electronic support, electronic attack, and electronic protection—are shown in Figure 1-1. The purpose of ES is to intercept, identify,

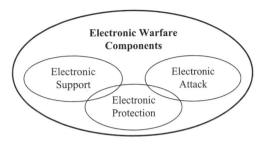

FIGURE 1-1 ■
Electronic Warfare
Components

and locate radar systems, and it is often provided by dedicated EW receivers. The purpose of EA is to "attack" radar systems, and it is often provided by self-protection and support jammers. The purpose of EP is to protect radar systems from ES or EA, and it is often provided by enhanced radar system characteristics, operation, and signal processing. Traditionally, ES has been known as electronic support measures (ESM), EA as electronic countermeasures (ECM), and EP as electronic counter-countermeasures (ECCM).

There is a very wide range of EW concepts in reaction to the very wide range of radar systems and their applications. Our discussion of EW systems starts with the fundamentals in Chapter 8. From there we go into more detail: EW receivers in Chapter 9; self-protection jamming EA in Chapter 10; support jamming EA in Chapter 11; and EP concepts in Chapter 12. All along we will rely on the radar system information in the earlier chapters.

Radar Systems

Radar is an acronym for "RAdio Detection And Ranging." A radar system detects the presence of objects in the environment and measures characteristics of the object: range, range rate, and angle. A radar system works by sending out (transmitting) radio frequency (RF) waves and collecting (receiving) the signals that reflect from objects in the environment. Reflection is the reradiating (scattering) of the incident RF wave from the object. Objects the radar is designed to detect are called "targets," while all other objects are called "clutter." Before I develop the principles of radar, I will review the characteristics of RF waves the basis of radar systems.

2.1 | RADIO FREQUENCY WAVES

RF waves occupy a portion of the electromagnetic spectrum from frequencies of a few kilohertz (a few thousand cycles per second) to a few million megahertz (over 10^{12} cycles per second). The total electromagnetic spectrum embraces all the frequencies to cosmic rays, beyond 10^{16} megahertz (MHz). RF waves represent less than one billionth of the total spectrum, see Appendix 2 for a definition of the RF bands. The overwhelming majority of radar systems use RF waves ranging from about 500 MHz to slightly over 10 GHz.

Although electromagnetic energy can be described either as waves or as quanta, the lower frequencies are much better suited to explanations by wave theory. RF waves are certainly thought of in those terms.

The definitive experiments in electromagnetism were performed by Michael Faraday in a period of ten days in 1831 [Encyclopedia Britannica, 1984, Volume 7, pg. 174; Williams, 1971, pp. 535–537]. Using Faraday's work as his foundation, James Clerk Maxwell succeeded, by the early 1860s, in synthesizing the properties of electricity and magnetism into a set of equations achieving a unified theory for electromagnetics [Encyclopedia Britannica, 1984, Volume 11, pg. 718; Everitt, 1974, pp. 204–217].

In Maxwell's time, it was only dimly appreciated that light was electromagnetic energy or that all electromagnetic energy propagated with the same velocity in free space. Yet Maxwell's equations, solved for the speed of light, give the correct result and serve as the foundation for the theory and design of modern communication (e.g., radio, television, cell phone) and radar systems. Faraday and Maxwell noted that time-varying electric currents produced time-varying electric and magnetic fields in free space. These fields induce time-varying electric currents in materials they encountered. These currents, in turn, generate electric and magnetic fields of their own. These fields "propagate" in free space at the speed of light.

In 1886, Heinrich Hertz conducted a number of experiments showing that RF waves reflected, refracted, were polarized, interfered with each other, and traveled at high velocity. Hertz is credited with verifying Maxwell's theories [Encyclopedia Britannica, 1984, Volume 6, pp. 647–648; McCormach, 1972, pp. 341–350]. These characteristics of reradiation and propagation at a known velocity portended the invention of radar.

The first use of RF waves was for communication, and the means of generating RF waves was with spark gaps, which generated short, intense pulses of current to achieve the needed electromagnetic radiation. The creation of sinusoidal waves, arising first from the use of alternators and later from oscillators designed using the vacuum tubes invented by Lee DeForest in 1906, revolutionized communications [Susskind, 1971, pp. 6–7]. A radio system works by transmitting a modulated continuous wave (CW) waveform. One or more receivers collect the propagated waveform and demodulate the signal to extract the information, such as music or video. When radar was invented, the use of CW oscillators was adopted from communications, but the radar transmitters sent periodic bursts of these sinusoidal waves, carefully counting time between them. More complex modulation schemes for radar came later.

The increasing use of radio in the early 1900s led to observations of objects passing between the transmitters and receivers producing interference patterns (exactly as aircraft do today to broadcast television reception). Bistatic (noncollocated transmitter and receiver) CW "radars" that could detect targets in this manner were explored by many countries at the beginning of the 1930s.

Monostatic (collocated transmitter and receiver) radar systems were developed shortly after. The first successful pulsed radar experiments were conducted by the U.S. Naval Research Laboratory (NRL) in 1934. By 1937, NRL had demonstrated a radar system at sea, but deployment was delayed until 1940. In the meantime Great Britain, which earlier had trailed in radar development, succeeded in deploying the first operational system (the Chain Home radar systems) by 1938. Concurrently, France, Germany, and the Soviet Union had substantial radar programs under way [Skolnik, 2001, pp. 14–19].

2.2 | A BASIC RADAR SYSTEM

The principles of a primitive radar are now clear, transmission, propagation, reflection, and reception. A functional diagram of a radar system is shown in Figure 2-1. A pulse of electromagnetic energy, oscillating at a predetermined carrier frequency, duration, and repetition interval, is generated by the wave-form generator. The waveform parameters are supplied to the receiver and signal processor, allowing them to efficiently and accurately operate. The waveform generator can vary all the waveform parameters, allowing the radar system to perform multiple functions. Changes in waveform parameters can be made by the radar operator or automatically by the radar system.

The transmitter amplifies the waveform to a very high power: about 1 kW to slightly over 1 MW. The overwhelming majority of radar systems use vacuum tube technology such as magnetron, cross-field amplifiers, klystron, traveling wave tube (TWT), and microwave power modules (MPM). Solid-state amplifier technology like bipolar transistors and field-effect transistors (FET) is used by some radar systems. Most radar systems use a single high-power transmitter, though some use several modest power transmitters operating in parallel. The high-power RF pulse is routed through a transmit/receive switch to an antenna. The transmit/receive switch, or duplexer, allows the same antenna to be used for transmit and receive and protects the sensitive receiver from the high-power transmitted RF pulse. The overwhelming majority of

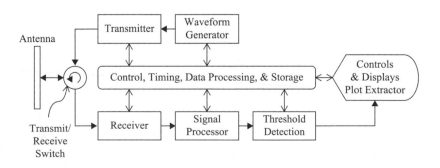

FIGURE 2-1 ■
Radar Block
Diagram

radar systems use the same antenna for transmitting and receiving due to size, weight, and cost constraints. As shown in Figure 2-1, the duplexer acts as an RF electromagnetic switch. It connects the transmitter to the antenna when an RF pulse is being transmitted and connects the receiver to the antenna when reflected RF signals are being received. The antenna is sometimes referred to by the generic term "aperture."

The RF pulse is radiated into free space through an antenna. The antenna focuses the transmitted RF pulse in a specific angular direction, similar to the reflector of a flashlight. The RF pulse propagates outward at the speed of light, scattering (i.e., reradiating) from objects it encounters along the way. Part of the scattered electromagnetic wave returns to the radar system. The scattered wave is collected by the antenna and routed through the transmit/receive switch to the receiver. The received signal efficiently passes through the receiver because it is nearly identical to the frequency and the duration of the transmitted pulse. The received signal is enhanced and interfering signals are reduced in the signal processor. The presence of the received signal can be determined by thresholding, passing along only signals that exceed the detection threshold. The resulting detections of received signals are presented to radar operators on displays and/or automatic plot extractors.

Detecting the presence of an object is good, but the real value of radar is being able to measure characteristics of the object that caused the reflection. A key characteristic of the object is the range between the radar and the object. The desire to simply measure range led to the development of pulsed radar systems. A pulsed radar system takes advantage of a basic physics principle; distance equals speed times time (Equation 2-1). The distance traveled by the pulse is twice the range: once to the object and once back to the radar. The radar system measures the elapsed time, or time delay, between when the pulse was transmitted and when it is received. The time delay can be measured using timing circuits or range gates (see Chapter 5). The range to a detected object can be calculated based on the measured time delay (Equation 2-1).

$$R + R = c\,\Delta t \quad \Rightarrow \quad R = \frac{c\,\Delta t}{2} \tag{2-1}$$

where:

   R = Range from the radar to the object, meters
   c = Speed of light, 3×10^8 meters/second
   Δt = Elapsed time or time delay from when the pulse was transmitted and
      when the reflected pulse is received, seconds

The transmitted pulsed radar waveform is shown in Figure 2-2. Fundamental parameters of the transmitted pulsed radar waveform are the carrier frequency (f_c) and duration (τ). The wavelength of the propagated wave is

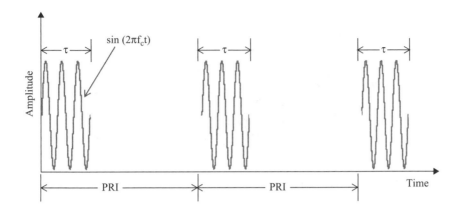

FIGURE 2-2 ■
Transmitted Pulsed
Radar Waveform

given in Equation (2-2) and shown in Figure 2-3. The angular frequency, ω (radians/sec), is $2\pi f_c$. The time between radar pulses is the interpulse period, or the pulse repetition interval (PRI). The number of pulses sent per unit time is the pulse repetition frequency (PRF). PRI and PRF are reciprocal, PRF $= 1/\text{PRI}$. This follows linear systems theory; the frequency (1/seconds) is one over the interval (seconds).

$$\lambda = \frac{c}{f_c} \tag{2-2}$$

where:

$\lambda =$ Wavelength, meters

$c =$ Speed of light, 3×10^8 meters/second

$f_c =$ Radar carrier frequency, 100's megahertz (MHz) to 10's gigahertz (GHz), hertz (cycles per second)

$\tau =$ Pulse duration, or pulse width, or pulse length, ≈ 1 microsecond (μsec), seconds

$\text{PRI} =$ Pulse repetition interval, 0.01's seconds to 0.01's milliseconds (msec), seconds

$\text{PRF} =$ Pulse repetition frequency, 100's hertz to 100's kilohertz (kHz), hertz (pulses per second)

The phase (the starting point in the wave) relationship between successive pulses is known as coherence. A coherent waveform either has the same starting point of the carrier frequency of each pulse with the others; in other words, they are "in phase," as in Figure 2-2. Or the phase relationships between pulses are known in the receiver, and then successive pulses can be aligned in phase. A coherent waveform provides the best signal processing performance: integration, Doppler processing, and pulse compression, which are discussed in future chapters. A noncoherent waveform has randomly varying starting points of the carrier frequency for each pulse or the phase relationship between pulses in

FIGURE 2-3 ■
Polarization of an
Electromagnetic
Wave

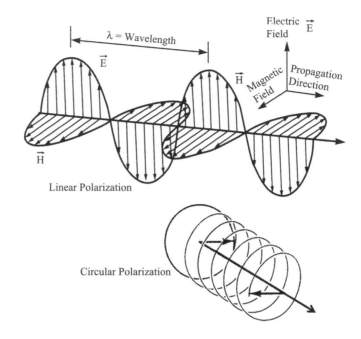

unknown in the receiver. A noncoherent radar system is much less expensive
and complex than a coherent radar system.

Polarized RF waves are equivalent to polarized light waves [Sears, 1948,
pp. 167–185], and the singly polarized radar wave is analogous to polarized
optical glasses. Polarization is the orientation of the electromagnetic wave
relative to the direction of its propagation. An electromagnetic wave consists of
two perpendicular components, the electric field (E-field) and the magnetic
field (H-field), as shown in Figure 2-3. Polarization is defined by the alignment
of the E-field. Linear polarization describes a linear alignment of the E-field,
usually either horizontal or vertical (as shown in Figure 2-3). Circular polarization describes either a left-hand or right-hand rotating vector (one full
revolution each radio frequency cycle).

All the previously mentioned transmitted waveform parameters are
important in radar design and performance. The remainder of this chapter and
Chapters 3 and 6 discuss the parameters' impact on radar detection; their effect
on radar measurements is covered in Chapter 5.

2.3 | THE RADAR EQUATION

The fundamental determinant of radar performance (detection and measurements), in any of the missions prescribed for it, is the radar equation, which
relates radar, target, and geometry/environment characteristics. The first step in

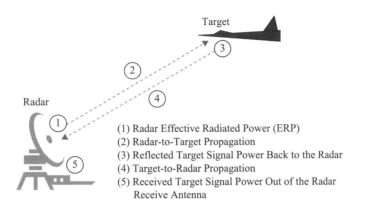

FIGURE 2-4 ■
Incremental
Build-Up of the
Received Target
Signal Power

(1) Radar Effective Radiated Power (ERP)
(2) Radar-to-Target Propagation
(3) Reflected Target Signal Power Back to the Radar
(4) Target-to-Radar Propagation
(5) Received Target Signal Power Out of the Radar
     Receive Antenna

developing the radar equation is determining the received single-pulse target signal power. We will derive the equation for the received target signal power from fundamental principles. As shown in Figure 2-4, the received target signal power is computed using incremental steps:

Step 1: transmitted radar effective radiated power (ERP)
Step 2: radar-to-target propagation
Step 3: reflected target signal power back to the radar
Step 4: target-to-radar propagation
Step 5: received target signal power out of the radar receive antenna

Imagine a point source of an electromagnetic pulse radiating isotropically, equally everywhere, into free space. The finite transmit power of the electromagnetic pulse radiates uniformly away from the source along an ever expanding imaginary sphere, as shown in Figure 2-5. The power density (watts per square meter, W/m^2) at a distance away from the point source can be determined by dividing the transmit power by the surface area of a sphere, $4\pi R^2$. The vast majority of radar systems use a highly directional antenna to concentrate or focus the transmitted power from isotropic in a specific angular direction, similar to a flashlight focusing the light in one direction, as shown in

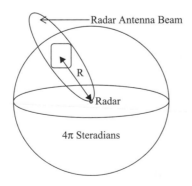

FIGURE 2-5 ■
Radar Spherical
Geometry

Figure 2-5. The ratio of this focusing over isotropic radiation is the gain of the antenna. Radar antennas are discussed in detail in Chapter 4.

Based on the transmitter power and antenna gain a radar system has an effective radiated power in the direction of the target (Equation 2-3) that accounts for the use of a directional antenna to focus the electromagnetic wave in a specific angular direction. For example, a 1 kW transmitter coupled with an isotropic, gain of one everywhere, antenna provides an ERP of 1 kW $\times$ 1 = 1 kW. The same 1 kW transmitter coupled with an antenna gain of 30 provides an ERP of 1 kW $\times$ 30 = 30 kW. Thus, it's "effectively" the same as having a 30 kW transmitter coupled with an isotropic antenna. There is a loss in power from the transmitter to the antenna, which is represented by a transmit loss term in the denominator of the ERP equation.

$$ERP_R = \frac{P_R\, G_{RT}}{L_{Rt}} \tag{2-3}$$

where:

ERP_R = Radar effective radiated power, watts
P_R = Radar peak transmit power, watts
G_{RT} = Radar transmit antenna gain in the direction of the target, no units
L_{Rt} = Radar transmit loss, no units

The transmit loss is one of many losses we will include in the radar equation. Because a loss is in the denominator of the equation, numerically its value is greater than or equal to 1. A loss of 1 indicates all the power passes through, whereas a loss greater than 1 indicates only a portion of the power passes through. In subsequent equations other losses are not included for clarity. Total radar-related losses are included at the conclusion of the received target signal power equation development.

The power-aperture (PA) product, the peak transmit power times the antenna (aperture) area (watt-square meters, W-m^2), is similar to the ERP. Very large ground-based radar systems are sometimes described in terms of power-aperture product. The relationship between antenna area, used for the power-aperture product, and antenna gain, used for ERP, is discussed shortly.

The radar ERP propagates from the radar to the target, resulting in a power density at the target (Equation 2-4). The incident radar power density is intercepted by a target, a portion of which is reradiated back to the radar, based on the radar cross section (RCS) of the target. Target RCS is discussed in detail in Chapter 6. The power reradiated from the target back to the radar is given in Equation (2-5).

$$\frac{P_R\, G_{RT}}{L_{Rt}} \left(\frac{1}{(4\pi)\, R_{RT}^2} \right) = \frac{P_R\, G_{RT}}{(4\pi)\, R_{RT}^2\, L_{Rt}} \tag{2-4}$$

$$\frac{P_{RT} \, G_{RT}}{4\pi \, R_{RT}{}^2 \, L_{Rt}} (\sigma) \tag{2-5}$$

where:

R_{RT} = Radar-to-target slant range, meters
σ = Target radar cross section, square meters (m^2)

The reflected power from the target propagates back to the radar system resulting in a power density at the radar antenna. The power density arriving back at the radar antenna from the target is given in Equation (2-6), assuming the transmit and receive antenna are collocated (monostatic). This incident target power density is collected by the effective area of the radar receive antenna in the direction of the target, resulting in the received target signal power at the output of the radar antenna (Equation 2-7).

$$\frac{P_R \, G_{RT} \, \sigma}{(4\pi) \, R_{RT}{}^2 \, L_{Rt}} \left(\frac{\perp}{(4\pi) \, R_{RT}{}^2} \right) = \frac{P_R \, G_{RT} \, \sigma}{(4\pi)^2 \, R_{RT}{}^4 \, L_{Rt}} \tag{2-6}$$

$$\frac{P_R \, G_{RT} \, \sigma}{(4\pi)^2 \, R_{RT}{}^4 \, L_{Rt}} (A_e) \tag{2-7}$$

where:

A_e = Effective area of the receive antenna, square meters (m^2)

Because of size, weight, and cost considerations, the vast majority of radar systems uses the same antenna to transmit and receive. Thus, the received target signal power equation contains two terms for the same antenna: gain and effective area. The tradition is to use only one term, the antenna gain, in the received target signal power equation. The gain of an antenna is directly related to its effective area (Equation 2-8). We will discuss radar antennas in more detail in Chapter 4.

$$G = \frac{4\pi \, A_e}{\lambda^2} \quad \Rightarrow \quad A_e = \frac{G \, \lambda^2}{4\pi} \tag{2-8}$$

$$A_e = \rho \, A$$

where:

G = Radar antenna gain, no units
ρ = Antenna efficiency, no units
A = Physical area of the antenna, square meters (m^2)

Therefore, when the radar's transmit and receive antenna are the same, the resulting single-pulse received target signal power is given in Equation (2-9). We will now add the other losses to the received target signal power equation.

Losses are factors not explicitly taken into consideration in determining the received target signal power. Examples of individual losses include power lost from the radar transmitter to the antenna (discussed earlier), lost power as the transmitted and reflected waveforms propagate through the atmosphere (see Chapter 13), power lost from the antenna to the receiver, power lost in the radar receiver, and power lost in the radar signal processor. These and numerous others are discussed in detail in Richards et al. [2010, Section 2.7] and Skolnik [2001, Section 2.12]. Different radar–target–environment combinations have different losses, so we collect them up as we need them into a single term (Equation 2-10). When all radar-related losses are considered, the resultant received single-pulse target signal power is given in Equation (2-11).

$$\frac{P_R \, G_{RT} \, \sigma}{(4\pi)^2 \, R_{RT}^4 \, L_{Rt}} \left(\frac{G_{RT} \, \lambda^2}{(4\pi)} \right) = \frac{P_R \, G_{RT}^2 \, \lambda^2 \, \sigma}{(4\pi)^3 \, R_{RT}^4 \, L_{Rt}} \tag{2-9}$$

$$L_R = L_{Rt} \, L_{RTRa} \, L_{Rr} \, L_{Rsp} \tag{2-10}$$

$$S = \frac{P_R \, G_{RT}^2 \, \lambda^2 \, \sigma}{(4\pi)^3 \, R_{RT}^4 \, L_R} \tag{2-11}$$

where:

L_R = Total radar-related losses, no units

L_{Rt} = Radar transmit loss, no units

L_{RTRa} = Radar-to-target and target-to-radar atmospheric attenuation loss, no units

L_{Rr} = Radar receive loss, no units

L_{Rsp} = Radar signal processing loss, no units

S = Received single-pulse target signal peak power, watts

We often represent some of the terms in the received target signal power equation in decibels (dB): antenna gain, noise figure, losses. Also, the received target signal power is often in decibels relative to 1 watt (dBW) or relative to 1 milliwatt (dBm). A decibel is a logarithmic representation of a number. Decibels are discussed in detail in Appendix 1. Of course, we never mix decibels and absolute (or linear) numbers in the same equation, except for very rare exceptions. The received target signal power equation is not one of those exceptions.

2.3.1 Receiver Thermal Noise

Unfortunately, there is always noise (random amplitude and phase) power contaminating the target signal power arriving at the receiver. Some of the noise is generated in the transmitter; some of it is added by the cosmos (galactic noise), a minor source at frequencies above 1 GHz but potentially a major source below 1 GHz; some of it is contributed by the earth's atmosphere (spherics) and some by the earth itself; some of it comes from manufactured sources (automobiles, power facilities, or other radar systems); and some is intentionally transmitted by jammers (see Chapter 8). However, for most radar systems the vast majority of the noise is generated in the front end of the radar receiver, particularly by the first amplifier and mixer stages (see Section 2.4).

The source of the noise generated in the front end of the radar receiver is the thermal heating of its electronic components. Thermal noise is a stochastic process resulting in a random amplitude and phase signal with a uniform power spectral density. We will use the mean amplitude of the receiver thermal noise. Basic chemistry tells us that when atoms are heated their electrons flow. Flowing electrons produce current flow, and we have noise energy (watt-seconds or joules), also called power spectral density (watts/hertz) (Equation 2-12). Since we cannot build ideal devices, we account for the additional noise generated by an actual device using a receiver noise figure (see Section 2.3.2) (Equation 2-13). The thermal noise power generated by an actual receiver is given in Equation (2-14). The receiver thermal noise can be measured at the receiver output in the absence of an input signal.

$$k \, T_0 \tag{2-12}$$

$$F_R \, k \, T_0 \tag{2-13}$$

$$N = F_R \, k \, T_0 \, B_R \tag{2-14}$$

where:

k = Boltzmann's constant, 1.38×10^{-23} watt-seconds/Kelvin
T_0 = IEEE standard reference temperature for radio frequency receivers, 290 Kelvin ($\approx 62°F$)
F_R = Radar receiver noise figure (≥ 1), no units
N = Radar receiver thermal noise power, watts
B_R = Radar receiver processing bandwidth, hertz

Many radar systems define the receiver processing bandwidth using matched filter theory (see Section 5.1.1). A matched filter maximizes the received target signal power relative to the receiver thermal noise power. For radar systems with a pulsed constant carrier frequency waveform, the matched filter bandwidth is approximately inversely proportional to the transmitted pulse width (Equation 2-15). Some radar systems frequency or phase modulate the carrier frequency to provide fine range resolution using pulse compression signal processing (see Section 5.2.2). For radar systems with a modulated carrier frequency waveform, the matched filter bandwidth is the pulse compression modulation bandwidth (Equation 2-16).

$$B_R \approx \frac{1}{\tau} \quad \text{Constant radar carrier frequency} \qquad (2\text{-}15)$$

$$B_R = B_{pc} \quad \text{Modulated radar carrier frequency} \qquad (2\text{-}16)$$

where:

B_R = Radar receiver processing bandwidth, hertz
τ = Radar transmitted pulse width, seconds
B_{pc} = Radar pulse compression modulation bandwidth, hertz

We often represent the receiver thermal noise power in decibels relative to 1 watt (dBW) or relative to 1 milliwatt (dBm). The single-pulse received target signal power and receiver thermal noise as a function of radar-to-target range are shown in Figure 2-6. Notice how as radar-to-target range increases the received target signal power approaches and then is less than the receiver thermal noise. As seen in this figure, the received target signal power varies as a function of $1/R^4$. Thus, the received target signal power decreases by 12 dB (1/16) when the radar-to-target range doubles and increases by 12 dB (16 times) when the radar-to-target range halves. Note that the y-axis grid for this figure is in 10 dB steps, one order of magnitude (power of ten), so the reader can easily interpret the change in power with range. Other figures in this and later chapters will use the same approach.

Exercises at the end of this chapter will show how little power is received compared with what was transmitted, often 10's to 20's of orders of magnitude (power or 10) less. The vast majority of the power is lost by the propagation out

FIGURE 2-6 ■
Single-Pulse
Received Target
Signal Power (S) and
Receiver Thermal
Noise (N) vs. Radar-
to-Target Range

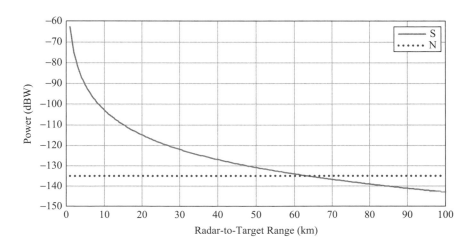

to and back from the target—the $1/R^4$ term that drives the received target signal power (Equation 2-11).

2.3.2 Receiver Noise Figure

The noise figure is a measure of the thermal noise power generated in an actual receiver compared with the thermal noise power generated in an ideal receiver at a standard reference temperature (Equation 2-17) [Blake, 1986, pp. 146–147; Edde, 1993, pp. 113–114]. The noise power added by an actual receiver above what is generated by an ideal receiver can be expressed in terms of the effective noise temperature (Equation 2-18). The effective noise temperature is not a physically measurable temperature; it is just a way to account for the additional noise added by an actual receiver. The relationships between noise figure and effective noise temperature are given in Equations (2-19) and (2-20). As seen in these equations, an ideal receiver has a noise figure of 1 and an effective noise temperature of 0 Kelvin. Using the relationships between effective noise temperature and noise figure, the thermal noise power generated in an actual receiver is given in Equation (2-21). This equation simplifies first in terms of effective noise temperature and then in terms of noise figure (Equation 2-22). After further simplification, the thermal noise power generated in an actual receiver is given in Equation (2-23).

$$F_R = \frac{N_a}{N_i} = \frac{k\,T_0\,B_R + \Delta N}{k\,T_0\,B_R} \tag{2-17}$$

$$\Delta N = k\,T_e\,B_R \tag{2-18}$$

$$F_R = \frac{k\,T_0\,B_R + k\,T_e\,B_R}{k\,T_0\,B_R} = \frac{T_0 + T_e}{T_0} = 1 + \frac{T_e}{T_0} \tag{2-19}$$

$$T_e = T_0\,(F_R - 1) \tag{2-20}$$

$$N_a = k\,T_0\,B_R + k\,T_e\,B_R \tag{2-21}$$

$$N_a = k\,B_R\,(T_0 + T_e) = k\,B_R\,[T_0 + T_0\,(F_R - 1)] \tag{2-22}$$

$$N_a = k\,B_R\,T_0\,(1 + F_R - 1) = k\,B_R\,T_0\,F_R \tag{2-23}$$

where:

N_a = Thermal noise power generated in an actual receiver at the standard reference temperature, watts

N_i = Thermal noise power generated in an ideal receiver at the standard reference temperature, watts

ΔN = Noise power added by an actual receiver above what is generated in an ideal receiver, watts

T_e = Receiver effective noise temperature, Kelvin

2.3.3 Target Signal-to-Noise Ratio

The radar equation is the ratio of the received target signal power to the receiver thermal noise (Equation 2-24). The radar equation tells us how strong, or weak, the received target signal power is relative to the receiver noise. The radar equation often includes the effect of signal processing techniques used by radar systems (Equation 2-25). Examples of radar signal processing techniques are pulse compression (see Chapter 5), Doppler filters (see Chapter 5), and moving target indicator (MTI) (see Chapter 7). The specifics of these signal processing techniques and their associated signal processing gain are discussed in future chapters.

$$\frac{S}{N} = \frac{P_R \, G_{RT}{}^2 \, \lambda^2 \, \sigma}{(4\pi)^3 \, R_{RT}{}^4 \, F_R \, k \, T_0 \, B_R \, L_R} \tag{2-24}$$

$$\frac{S}{N} = \frac{P_R \, G_{RT}{}^2 \, \lambda^2 \, \sigma \, G_{sp}}{(4\pi)^3 \, R_{RT}{}^4 \, F_R \, k \, T_0 \, B_R \, L_R} \tag{2-25}$$

where:

S/N = Single-pulse target signal-to-noise ratio, no units

G_{sp} = Radar signal processing gain, no units

The radar equation is dominated by the R^4 factor in the denominator, as shown in Figure 2-7. As shown in this figure the signal-to-noise ratio (S/N) is often represented in decibels. A positive S/N (in dB) indicates that the target signal is larger than the receiver noise. A negative S/N (in dB) indicates the opposite situation. For example, a 10 dB S/N indicates the target signal is 10 times the receiver noise, a 0 dB S/N indicates the target signal equals the receiver noise, and a –10 dB S/N indicates the target signal is 1/10 of the receiver thermal noise.

There is no magic way to achieve a high-performance radar system. If low RCS targets are to be engaged, a combination of high power, high antenna gain, and low noise seems to be dictated. Fortunately, if needed, and it most often is, we can "integrate" multiple received target pulses to improve the S/N. Considerable S/N benefits can result from integration of multiple pulses [Barton, 1988, Section 2.3]. We will discuss integration in detail in Chapter 3.

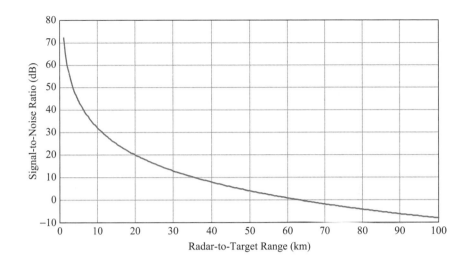

2.3.4 Detection Range

The single-pulse radar equation can be solved for the range at which the radar will detect the presence of a target with a given S/N. The radar detection range is given in Equation (2-26). We will discuss how to determine the S/N required for detection in detail in Chapter 3. Very few radar systems attempt to detect the presence of a target based on the single-pulse S/N. Thus, Equation (2-26), while valid, is limited in its application. We will develop a comprehensive detection range equation incorporating the integration of multiple pulses in Chapter 3.

$$R_{dt} = \sqrt[4]{\frac{P_R \, G_{RT}^2 \, \lambda^2 \, \sigma \, G_{sp}}{(4\pi)^3 \, SNR_{dt} \, F_R \, k \, T_0 \, B_R \, L_R}} \qquad (2-26)$$

where:

R_{dt} = Radar detection range, meters
SNR_{dt} = Single-pulse signal-to-noise ratio required for detection, no units

2.3.5 Other Forms of the Radar Equation

Equation (2-25) is just one of many forms of the radar equation. There are, however, several ways of manipulating the radar equation to illustrate various radar uses: search, track, image, and/or to provide computational ease: average power, energy, accumulation of constants, and unit conversions factors. Unfortunately, each one of these other forms of the radar equation remove insights or contain one or more assumptions that are all too easy for us to lose sight of, often leading to erroneous applications and results. Therefore, we will consistently use the form of the radar equation in Equation (2-25) throughout this book.

2.4 | A BASIC RADAR RECEIVER

When the return target pulse clears the antenna, it is routed to the receiver chain, a block diagram of which is shown in Figure 2-8. Richards et al. [2010, Chapter 11] and Stimson [1998, p. 28] discuss radar receivers in detail. Most radar systems use a modified form of a superheterodyne receiver, similar to an AM/FM radio. The RF amplifier is a moderate-gain low-noise amplifier (LNA) inserted as close as possible to the receive antenna to minimize the noise in the receiver. The mixer down-coverts, or shifts, the frequency of the received signals from RF to an intermediate frequency (IF) while preserving other signal characteristics, as shown in Figure 2-9. Having the receive signals at IF allows for an easier implementation of rest of the receiver components.

The math associated with down-conversion, the multiplication of the RF and local oscillator signals, is given in Equations (2-27) through (2-31), including the necessary trigonometry identities. Most superheterodyne receivers use more than one mixer stage. For simplicity we show only one. The IF amplifier passes and amplifies those signals within its bandwidth, as shown in

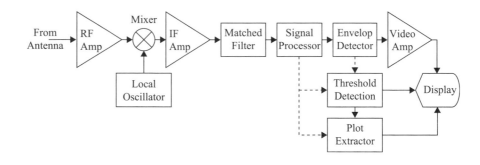

FIGURE 2-8 ■ A Basic Radar Receiver

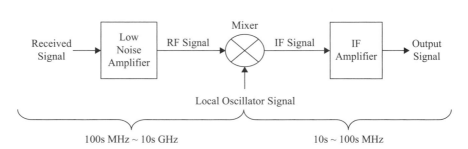

FIGURE 2-9 ■ Superheterodyne Mixer

Figure 2-10. Notice how the high-frequency component of the IF signal is "discarded" because it is outside the bandwidth of the IF amplifier.

$$\sin(2\pi\, f_c\, t) \times \sin\left(2\pi\,(f_c + f_{IF})\, t\right) \qquad (2\text{-}27)$$

$$\sin(\alpha) \times \sin(\beta) = \frac{1}{2}\left[\cos(\alpha - \beta) - \cos(\alpha + \beta)\right] \qquad (2\text{-}28)$$

$$\frac{1}{2}\left[\cos\left(2\pi\,(f_c - f_c - f_{IF})\, t\right) - \cos\left(2\pi\,(f_c + f_c + f_{IF})\, t\right)\right]$$
$$\frac{1}{2}\left[\cos\left(2\pi\,(-f_{IF})\, t\right) - \cos\left(2\pi\,(2f_c + f_{IF})\, t\right)\right] \qquad (2\text{-}29)$$

$$\cos(-\theta) = \cos(\theta) \qquad (2\text{-}30)$$

$$\frac{1}{2}\left[\cos(2\pi\, f_{IF}\, t) - \cos\left(2\pi\,(2f_c + f_{IF})\, t\right)\right] \qquad (2\text{-}31)$$

where:

f_c = Radar carrier frequency, radio frequency, hertz
f_{IF} = Radar receiver intermediate frequency, hertz
$f_c + f_{IF}$ = Local oscillator frequency, hertz

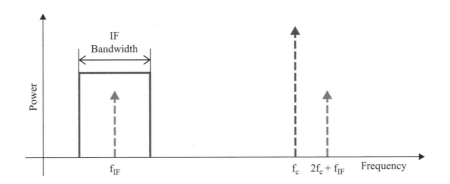

FIGURE 2-10 ■ Intermediate Frequency Amplifier and Bandpass Filter

Matched Filter

After the IF amplifier, a matched filter is used to optimally pass the most received signal power relative to the receiver noise power (see Section 5.1.1). All operations on the signal requiring preservation of the phase inside a pulse or from pulse to pulse must be conducted here. Examples of operations requiring the preservation of phase are coherent integration (see Section 3.3), Doppler processing (see Section 5.3), pulse compression (see Section 5.2.2), moving MTI clutter rejection (see Section 7.1.1), and coherent sidelobe cancellation (SLC) electronic protection (EP) (see Chapter 8).

Signal Processor

Signal processing is the sequence of operations done on the received signal to extract and display or store essential information. Although powerful digital computers have become the major force in signal processing in modern sophisticated radar systems, a substantial portion remains analog in older radar systems—for both practical and historical reasons. Due to the complexity of the subject and the proliferation of clever ideas, separating concept from technique is often difficult. Many of the multitudes of signal processing capabilities available in modern radar systems are discussed conceptually elsewhere in this book. Examples of such signal processing capabilities include multiple-pulse integration, range gates (see Section 5.2), pulse compression, Doppler filter banks, sophisticated target tracking (see Section 5.6), clutter rejection, and EP techniques.

Performing signal processing using digital hardware, special purpose programmable microprocessors, microcomputers, and field-programmable gate arrays (FPGAs) has become customary in most radar systems, first at video frequencies and then at IF. A digital signal processing flow diagram is shown in Figure 2-11. The quadrature detector, sometimes called a phase detector or synchronous detector, detects the magnitude of the IF signal in two channels [Richards et al., 2010, Chapter 14; Stimson, 1998, pp. 29–30]. These channels are in phase quadrature, 90° apart in phase, and called in-phase and quadrature (I/Q) channels as given in Equation (2-32). In this way, digital sampling of the signals will never be zero in both channels, the direction the phase is advancing (toward the radar or away from it) will always be known, and the angle and magnitude of the resultant are trivial to calculate (Equation 2-33). The analog-to-digital (A/D) converter samples the analog signals and tags each sample with a binary word telling time and voltage level. By Nyquist's theorem, there must be at least two samples per hertz of the highest frequency present in the signal not to lose information [Richards et al., 2010, Section 14.2.1].

FIGURE 2-11 ■
Digital Signal
Processing Flow

$$I = A\cos\phi \qquad\qquad Q = A\sin\phi \qquad\qquad (2\text{-}32)$$

$$A = \sqrt{I^2 + Q^2} \qquad\qquad \phi = \tan^{-1}\left(\frac{Q}{I}\right) \qquad\qquad (2\text{-}33)$$

where:

I = In-phase phasor of the signal, volts
A = Phasor amplitude, volts
ϕ = Phasor angle from the in-phase axis, radians or degrees
Q = Quadrature phasor of the signal, volts

Digital processing has many advantages. Once a signal has been converted into a series of computer words, it can be stored in computer memory, thereby allowing the preservation of amplitude and phase indefinitely. The problems of instability and unreliability in analog circuits are nonexistent with digital processing. Digital processing also has unmatched flexibility. To revise digital processor designs often requires only software modifications.

Envelope Detector

At the envelope detector, the signal is further filtered and the signal phase information is lost, leaving only the modulation envelope (the video signal). The modulation envelope is amplified by the video amplifier to further process it or to drive the radar display. Often, one or more of the feedback loops necessary to permit the receiver to operate over very large dynamic ranges is applied here. With the $1/R^4$ received signal strength dependency (see Section 2.3) and large potential fluctuations in target RCS (see Chapter 6), signals in the receiver may vary in amplitude by 80 dB or more.

Different aspects of the problem of large signal dynamic range are attacked by two popular techniques: automatic gain control (AGC) and sensitivity time control (STC) processing [Richards et al., 2010, Section 11.6]. AGC is a lot like an automatic volume control for a car radio. Part of the signal voltage in an amplifier stage is tapped off and fed back to preceding stages as a negative bias, making the chain self-damping at a tolerable signal level. Because AGC involves a feedback loop operating continuously and adaptively, real signal levels are lost, with obvious effects on a radar system designed for precision measurements. STC is synchronized with the radar system's master clock, applying exponential attenuation (or gain) to the receiver chain based on time elapsed since a pulse was transmitted. In this way, received signals from nearby targets are heavily attenuated (or have minimal gain) while those from far away have no attenuation (or maximum gain). Fourth- and third-power exponential attenuation (or gain) functions are common, similar to the $1/R^4$ received signal dependency.

Threshold Detection

Originally, and still common in many ground-based radar systems, an operator declares target detection by assessing what he sees on a radar display. Extensive testing revealed that operators' target detection performance is best when there are a small number of closely spaced targets. To accurately and consistently handle a large number of spatially diverse targets required an automatic detection process. The automatic detection process used by the vast majority of radar systems is thresholding. The radar system declares a detection when the receiver output exceeds the detection threshold. The output of the receiver is the sum of the received target signal and receiver thermal noise (S+N). A "target detection" is when the S+N exceeds the detection threshold. A "false alarm" is when the noise alone exceeds the detection threshold. The detection threshold is set to maximize target detections and minimize false alarms. Target detection is discussed in detail in Chapter 3.

Radar Displays

It is easy to see how the output of the radar receiver might be hooked into an oscilloscope to display the range of a target to a radar operator. The vertical sweep is connected to the output of the receiver and the horizontal sweep to elapsed time. Each horizontal sweep begins at the instant the radar sends out a pulse (each PRI), as shown in Figure 2-12. This kind of display is called an A-scope and was the original radar display; it is still often used in radar systems. The horizontal axis is often calibrated as range, instead of time, so the operator can directly read range off the scope.

FIGURE 2-12 ■ A Radar A-Scope Display

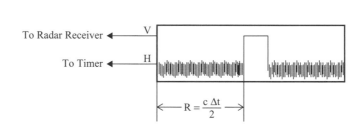

An A-scope shows receiver output, S+N, on the ordinate (y-axis) with the range, time delay within the PRI, on the abscissa (x-axis) of the display. In like manner, the B-scope displays range (y-axis) as a function of azimuth angle (x-axis). A C-scope shows target angular position as a function of elevation angle (y-axis) and azimuth angle (x-axis). An E-scope displays range as a function of elevation angle. The plan position indicator (PPI) is one of the most common radar displays. Used by search radar systems, a PPI gives a plan view (polar plot) of the radar coverage: radar in the center of the scope, several concentric rings marking range, with azimuth read directly from a compass rose around the edge of the display. With a B-scope, C-scope, E-scope, and PPI, the magnitude of the S+N is indicated by intensity. "Official" displays are numerous, but many are not widely used or have fallen into disuse [Skolnik, 2001, Section 11.5; Stimson, 1998, p. 21].

In older radar systems, but still numerous and in very wide use, the receiver output drives the radar displays. In modern radar systems the graphics for modern displays are synthesized from the receiver output by the radar system's computer, threshold detection, or plot extractor. With modern computer graphics, all of the previously described displays could be made available on a single screen, either in time multiplex or with a split-screen presentation. In fact, the PPI could be combined with the E-scope in a three-dimensional (3D) perspective display. More advanced concepts, such as holographic displays providing what seems to be a true 3D presence, are also receiving attention.

Plot Extractors

Many modern radar systems use an automatic target detection (see Chapter 3) and measurement (see Chapter 5) process called a plot extractor [Kingsley and Quegan, 1999, pp. 118–119]. The operator runs the overall radar system and in some systems aids in target detection. The plot extractor transfers the receiver output into target information, detection, and measurements. A plot extractor performs automatic target detection and assigns the associated measurements to each detection and to detections made by the operator. The combination of the detection and associated measurements is called a plot. A plot extractor can assist the operator by identifying detected targets on the radar display. The plots from the extractor are passed along to target trackers providing the information to initiate, maintain, and drop target tracks.

2.5 | SUMMARY

This chapter provided an introduction to radar system, starting with an overview of RF electromagnetic waves and how they provide the "detection" and "ranging" behind RAdio Detection and Ranging (RADAR). Then a discussion was given of the components of a basic radar system: transmitter, antenna, and receiver, and how they work together allowing the radar system to detect the presence of objects in the environment and to measure its characteristics: range, range rate, and angle. The radar equation was then developed, which relates radar, target, and geometry/environment characteristics together to allow us to determine radar system performance (detection and measurements). We will discuss detection in detail in Chapter 3 and measurements in Chapter 5. We concluded this chapter by discussing basic radar receiver components, functions, and associated math.

2.6 | EXERCISES

2-1. What is the range associated with a time delay $\Delta t = 0.67$ msec? What is the time delay associated with the range to the moon (Earth-to-moon range $R = 3.84 \times 10^8$ meters)?

2-2. A radar with transmit peak power $P_R = 1$ MW and radar antenna gain in the direction of the target $G_{RT} = 40$ dBi irradiates a target with a radar cross section $\sigma = 1$ m^2 target at a radar-to-target range $R_{RT} = 500$ km. If the radar transmit loss $L_{Rt} = 0.5$ dB, what power density arrives back at the radar antenna?

2-3. What is the receiver thermal noise power, N (watts, dBW, and dBm), for the following receiver characteristics: noise figure $F_R = 6$ dB and receiver bandwidth $B_R = 500$ kHz? What is the receiver equivalent noise temperature, T_e (Kelvin), and noise power added by an actual receiver above what is generated by an ideal receiver, ΔN (watts, dBW, and dBm)? What do these results tell us about the receiver thermal noise power?

2-4. The radar in Exercises 2 and 3 is transmitting at $f_c = 1$ GHz and has a signal processing gain $G_{sp} = 1$. What total radar-related losses, L_R (no units and dB), is required to give a single-pulse signal-to-noise ratio of unity (S/N = 1)?

2-5. A radar system has the following characteristics: peak transmit power $P_R = 800$ kW, antenna gain in the direction of the target $G_{RT} = 38$ dBi, carrier frequency $f_c = 3$ GHz, signal processing gain $G_{sp} = 1$, receiver noise figure $F_R = 6$ dB, receiver bandwidth $B_R = 750$ kHz, total radar-related losses $L_R = 13$ dB, and radar transmit loss $L_{Rt} = 2$ dB. A target with a radar cross section $\sigma = 5$ m^2 is at a radar-to-target range $R_{RT} = 150$ km. Compute the following: (a) the transmitted effective radiated power, ERP_R (watts and dBW); (b) the radar power density at the target; (c) the power reflected off the target back to the radar; (d) the received power density at the radar receive antenna; (e) the received single-pulse target signal power, S (watts and dBW); (f) compare the received single-pulse target signal power to the radar's transmitted effective radiated power; (g) the radar receiver thermal noise power, N (watts and dBW); and (h) the single-pulse target signal-to-noise ratio, SNR (no units and dB).

2-6. The Anti-Ballistic Missile Defense Treaty of the Strategic Arms Limitations accords with Russia limits radar systems at ABM sites in the two countries to 3×10^6 watts/m^2 of power-aperture (PA) product. The radar designers wants to use a peak transmit power $P_R = 250$ kW operating at a frequency $f_c = 6$ GHz. Assuming such a radar requires a signal-to-noise ratio for detection $SNR_{dt} = 13$ dB and has a receiver bandwidth $B_R = 300$ kHz, receiver noise figure $F_R = 5$ dB, signal processing gain $G_{sp} = 1$, and total radar-related losses $L_R = 8$ dB, what would be its maximum possible detection range against a target with a radar cross section $\sigma = 10$ m^2?

2-7.   The cost of a radar system is the cost of power plus the cost of aperture plus a constant. The cost of power is the cost/kilowatt multiplied by the number of kilowatts: $C_p = C_{kW} N_{kW}$. In addition, the cost of aperture is the cost/square meter multiplied by the number of square meters: $C_A = C_{m2} N_{m2}$. Show for the minimum cost $2C_p = C_A$. (Hint: Assign a design power-aperture product.)

2.7 | REFERENCES

Barton, D. K., 1988, *Modern Radar System Analysis,* Norwood, MA: Artech House.

Blake, Lamont V., 1986, *Radar Range-Performance Analysis*, Norwood, MA: Artech House.
   Covers a wide range of topics, with detailed supporting equations for radar detection and propagation.

Edde, Byron, 1993, *Radar Principles, Technology, Applications*, Upper Saddle River, NJ: Prentice-Hall.
   Covers a wide range of topics, with supporting equations and numerical examples.

Encyclopedia Britannica, 15th Edition, 1984, Chicago: Benton.
   Maxwell and Faraday were Englishmen, and *Britannica* does well by them.

Everitt, C. W. F., 1971, James Clerk Maxwell, in C. C. Gillespie (ed.), *Dictionary of Scientific Biography*, New York: Scribner's.

Kingsley, Simon, and Quegan, Shaun, 1999, *Understanding Radar Systems*, Raleigh, NC, SciTech Publishing.
   An easy-to-read, wide-ranging guide to the world of modern radar systems.

McCormach, R., 1972, Heinrich Hertz, in C. C. Gillespie (ed.), *Dictionary of Scientific Biography*, New York: Scribner's.

Richards, M., Scheer, J., and Holm, W. (editors), 2010, *Principles of Modern Radar*, Volume I: Basic Principles, Raleigh, NC: SciTech Publishing.
   A comprehensive and modern textbook for college students, professional training, self-study, and professional reference book.

Sears, F. W., 1948, *Principles of Physics*, Volume 3, Optics, Cambridge, MA: Addison-Wesley.

Stimson, George W., 1998, *Introduction to Airborne Radar*, 2nd Edition, Raleigh, NC: SciTech Publishing.
   Just as its title says, an introduction to airborne radar systems, concepts, and applications.

Skolnik, M., 1985, Fifty Years of Radar, *Proceedings of the IEEE*, February.

Skolnik, M., 2001, *Introduction to Radar Systems*, 3rd Edition, New York: McGraw-Hill.
   The latest is a long line of books (original publication 1962; 2d edition 1980) on radar systems from one of the radar community's most established authors.

Susskind, C., 1971, Lee DeForest, in C. C. Gillespie (ed.), *Dictionary of Scientific Biography*, New York: Scribner's.

Watson-Watt, R., 1957, *Three Steps to Victory*, London: Odham's Press.

Watson-Watt, R., 1959, *The Pulse of Radar*, New York: Dial Press.
    Watson-Watt's books are both written in popular style but with important technical issues discussed.
Williams, L. P., 1971, Michael Faraday, in C. C. Gillespie (ed.), *Dictionary of Scientific Biography*, New York: Scribner's.
    Provides biographical information, brief and in a narrative style, and also includes technical contributions and some mathematical developments.

Target Detection

Assume radar systems having some of the parameters described in the preceding chapters can be built and operated. The output of the receiver is the complex (amplitude and phase) combination of the received target signal and the receiver noise which varies randomly. A vitally important question is how the operator decides when a target is present on the display or how the automatic detection system decides when to declare a target. Based on an operator's training and experience, they detect the presence of a target based on how strong a "target" appears on the radar display relative to the background interference. Figure 3-1 shows the receiver output (target signal plus noise) for several different target signal-to-noise ratios (S/N). As seen in this figure, when the S/N is high it is easy to accurately detect the presence of a target signal in the receiver output. However, as the S/N decreases it is difficult to reliably detect the target signal.

Well-trained and experienced operators are extremely good at detecting targets, especially in clutter (see Chapter 6) or jamming (see Chapter 8) environments. However, target detection by operators is best for a small ($\approx$ 4–6) number of closely spaced targets. Of course, detection performance by operators varies from person to person. To accurately and consistently handle a large

FIGURE 3-1 ◾
Target Signal-Plus-
Noise Examples

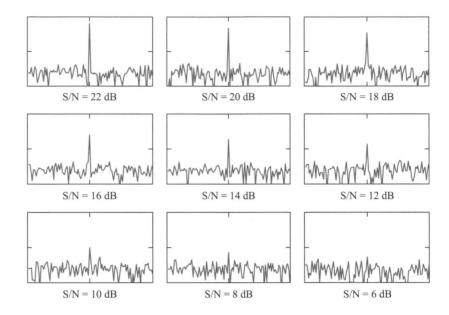

FIGURE 3-1 ◾
Target Signal-Plus-
Noise Examples

number and/or spatially diverse targets requires an automated detection process. We will discuss automatic target detection in detail in this chapter.

3.1 | THE PROBLEM OF TARGET DETECTION

Automatic target detection by radar systems is performed using the "thresholding" concept. As shown in Figure 3-2, the radar system declares a detection when the receiver output (target signal plus noise) exceeds the detection

FIGURE 3-2 ◾
Threshold Detection
of Target Signals

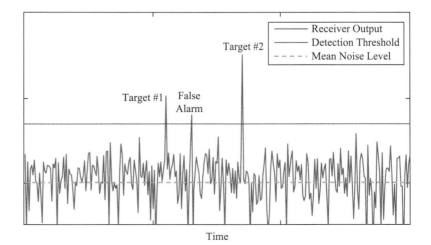

threshold. A target detection is when the threshold is crossed by the target signal plus noise (S+N). When the threshold is crossed by a receiver thermal noise sample it is called a false alarm. The problem of detection is therefore reduced to finding an appropriate threshold, one above which noise samples seldom rise and below which target signals seldom fall. As seen in Figure 3-2, we could raise the detection threshold to avoid the false alarm, but we have to be careful not to raise it too much or we will miss detecting Target #1.

■■■ 3.2 ■ | DETECTION THEORY

The detection threshold is set to maximize detections of received target signals and minimize false alarms. The random nature of the receiver thermal noise, and as we will see possibly the received target signal, complicates determining the detection threshold. Therefore, we will use probability theory to help us determine the detection threshold.

3.2.1 Basic Probability Theory

Basic probability theory tells us a random signal is defined by a probability density function (PDF). The PDF defines the probability any given event will occur. A very simple PDF is the one for a single die, as given in Equation (3-1) and shown in Figure 3-3.

$$p_d(x) = \frac{1}{6} \qquad (3\text{-}1)$$

where:

$p_d(x)$ = Probability density function for a single die

There are a few probability theory mathematical relationships we take advantage of for radar detection. First, if we integrate a PDF over its range of values we get the value one (1). When all values are greater than zero, this mathematical relationship is given in Equation (3-2). Second, the probability of

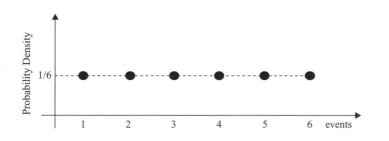

FIGURE 3-3 ■
Probability Density Function for a Single Die

exceeding a specific value is given in Equation (3-3). We will use this mathematical relationship to determine the radar detection threshold.

$$\int_{-\infty}^{\infty} p(x)\, dx = 1 \qquad \int_{0}^{\infty} p(x)\, dx = 1 \qquad (3\text{-}2)$$

$$P(x \geq T) = \int_{T}^{\infty} p(x)\, dx = 1 - \int_{0}^{T} p(x)\, dx \qquad (3\text{-}3)$$

where:

$\quad p(x)$ = Probability density function

$P(x \geq T)$ = Probability of x exceeding the value T

3.2.2 Receiver Noise Probability Density Function and Probability of False Alarm

To get at detection quantitatively, one must define the characteristics of the noise at the output of a radar receiver. At most radar frequencies, the principal contributor has been found to be receiver thermal noise. Receiver thermal noise has a Gaussian (or normal) PDF. After passing through the intermediate frequency (IF) filter and a linear envelope detector, the noise has a Rayleigh PDF, which is given in Equation (3-4), and shown in Figure 3-4. Linear envelope detectors linearly respond to signal voltage and are common in modern radar systems. There are many detailed discussions of linear envelope detectors,

FIGURE 3-4 ■ Rayleigh Probability Density Function and Probability of False Alarm

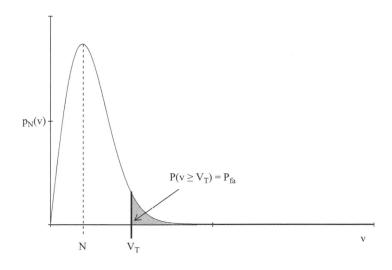

detected noise, and the associated Rayleigh PDF [Barton, 1988, p. 59; Richards et al., 2010, p. 98; Skolnik, 2001, p. 40].

$$p_N(v) = \frac{v}{N} e^{\left(-\frac{v^2}{2N}\right)} \quad v > 0$$
$$= 0 \qquad\qquad \text{elsewhere}$$

(3-4)

where:

$p_N(v)$ = Probability of v (noise) occurring for the Rayleigh probability density function

v = Linear envelope detector output voltage, volts

N = Mean noise power at the linear envelope detector output (see Section 2.3.1), watts

Other references as well as past editions of this book describe the receiver noise using an exponential PDF, which is applicable for a square-law envelope detector where the receiver responds linearly to signal power [Barton, 1988, p. 59]. The use of different PDFs is primarily to simplify mathematically working through the details of detection theory. The overall single-pulse detection theory results are the same for any envelope detector type. However, for integration of multiple noncoherent pulses, coming up shortly in Section 3.3, the envelope detector type has a significant effect [Blake, 1986, p. 38]. Because of its popularity in modern radar systems we will consider linear envelope detectors in our detection theory development.

It is straightforward to calculate the probability of noise exceeding a given threshold using the Rayleigh PDF. We can use Equations (3-3) and (3-4) to calculate the probability of noise alone exceeding a threshold (Equation 3-5). The threshold-to-noise power ratio (V_T^2/N) is related to the power signal-to-noise ratio (S/N) from Chapter 2 (Equation 3-6) [Barton, 1988, p. 61; Skolnik, 2001, p. 44]. The probability of the noise alone exceeding the threshold is called the probability of false alarm (P_{fa}), as given in Equation (3-7) and shown in Figure 3-4.

$$P(v \geq V_T) = \int_{V_T}^{\infty} \frac{v}{N} e^{\left(-\frac{v^2}{2N}\right)} dv = e^{\left(-\frac{V_T^2}{2N}\right)}$$

(3-5)

$$\frac{S}{N} = \frac{V_T^2}{2N}$$

(3-6)

$$P_{fa} = e^{\left(-\frac{S}{N}\right)}$$

(3-7)

where:

$P(v \geq V_T)$ = Probability of noise exceeding a threshold, no units

V_T = Threshold voltage, volts

$V_T{}^2/N$ = Threshold-to-noise power ratio, no units

S/N = Power signal-to-noise ratio (see Chapter 2) required for detection (detection threshold), no units

P_{fa} = Probability of false alarm, no units

From a statistical perspective, the P_{fa} is the *conditional* probability that, given no target signal is present, the noise alone will exceed the detection threshold. For example, if the S/N threshold equals 10, the target signal is 10 times the noise, the P_{fa} is 4.54×10^{-5} (Equation 3-8). Equation (3-7) can be solved for the S/N required for detection, the detection threshold, in terms of the P_{fa} (Equation 3-9). For example, a P_{fa} of 10^{-6} requires an S/N of 13.8 (11.4 dB) (Equation 3-10).

$$P_{fa} = e^{(-10)} \cong 4.54 \times 10^{-5} \qquad (3\text{-}8)$$

$$\frac{S}{N} = -\ln(P_{fa}) \qquad (3\text{-}9)$$

$$\frac{S}{N} = -\ln(10^{-6}) = 13.82 \qquad 10\log(13.82) = 11.4 \text{ dB} \qquad (3\text{-}10)$$

Often the probability of false alarm by itself does not provide a tangible indication of false alarms. The average false alarm rate (FAR), the average number of false alarms per unit time, and subsequent average time between false alarms (T_{fa}) provide this tangible indication [Richards et al., 2010, pp. 99–100, 553–554; Jeffery, 2009, p. 147]. The FAR and T_{fa} are a function P_{fa}, the number of detection decisions performed over a time period of interest, and the time to make an individual detection decision. For a search radar system, the time period of interest is often the time it takes to scan the required angular coverage once. For a tracking radar system, the time period of interest is often a tracking interval relative to overall time a target is being tracked. In the very rare case where detection is declared based on a single pulse, the time to make an individual detection decision can be the radar receiver response time. Most often the radar receiver response time is the time it takes to integrate multiple pulses (which we will discuss shortly in Section 3.3).

The number of detection decisions performed (number of range gates and/or number of Doppler filters and/or antenna beam positions) over a time period of interest is given in Equation (3-11). We will discuss the concept of range gates, Doppler filters, and antenna beam positions in Chapter 5 and how to determine their associated numbers when discussing the radar resolution cell in Section 5.5.3. The average number of false alarms in one time period of interest is given in Equation (3-12). The number of detection decisions per unit time is a function of the time it takes to make an individual detection decision (Equation 3-13). The FAR is a function of P_{fa} and the number of detection decisions per unit time (Equation 3-14). The average time between false alarms is the inverse of the FAR (Equation 3-15). This equation can be solved for P_{fa} (Equation 3-16). The overall radar scan time or tracking interval is the product of the number of antenna beam positions and the time to make a detection decision (Equation 3-17).

$$N_d = n_{rg} \, n_{df} \, n_b \qquad (3\text{-}11)$$

$$N_{fa} = N_d \, P_{fa} \qquad (3\text{-}12)$$

$$\frac{N_d}{T_d} \qquad (3\text{-}13)$$

$$FAR = \frac{N_d}{T_d} P_{fa} \qquad (3\text{-}14)$$

$$T_{fa} = \frac{1}{FAR} = \frac{T_d}{N_d \, P_{fa}} \qquad (3\text{-}15)$$

$$P_{fa} = \frac{T_d}{N_d \, T_{fa}} \qquad (3\text{-}16)$$

$$T_s = n_b \, T_d \qquad (3\text{-}17)$$

where:

N_d = Number of detection decisions performed over a time period of interest (e.g., radar search time or tracking interval), no units

n_{rg} = Number of range gates in which detection decisions are made (≥ 1), no units

n_{df} = Number of Doppler filters in which detection decisions are made (≥ 1), no units

n_b = Number of antenna beam positions in which detection decisions are made (≥ 1), no units

N_{fa} = Average number of false alarm in one time period of interest, no units

T_d = Time to make an individual detection decision, seconds

FAR = Average false alarm rate; average number of false alarms per second, no units

T_{fa} = Average time between false alarms, seconds

T_s = Radar scan time or tracking interval, seconds

A few of the exercises at the end of this chapter demonstrate the use of FAR or T_{fa} to determine P_{fa} based on tangible radar system characteristics and performance. FAR or T_{fa} is actually the primary requirement for the use of the radar system, thus establishing the required P_{fa}.

The number of detection decisions per second varies widely from radar system to radar system. Thus, we will look at the most general case where the receiver output is continuously applied to the detection threshold to show the relationship between FAR, T_{fa}, and P_{fa} [Barton, 1988, pp. 59–60]. When the

FIGURE 3-5 ■
Probability of False
Alarm as a Function
of Average Time
Between False
Alarms and Radar
Receiver
Bandwidth—General
Case

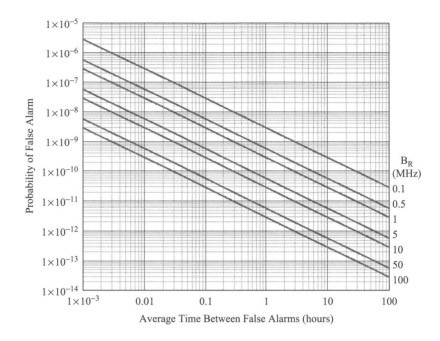

FIGURE 3-5 ■ Probability of False Alarm as a Function of Average Time Between False Alarms and Radar Receiver Bandwidth—General Case

receiver output is continuously applied to the detection threshold, there are no discrete range gates, Doppler filters, and antenna beam positions ($n_{rg} = n_{df} = n_b = 1$). For this general case, independent samples of receiver noise are applied to the detection threshold at a rate equal to the receiver bandwidth (B_R), one over the response time of the radar receiver, or $T_d = 1/B_R$. Thus, the FAR is B_R times P_{fa}, and the corresponding T_{fa} is $1/(B_R P_{fa})$. A plot of P_{fa} as a function of T_{fa} and radar receiver bandwidth for this general case is shown in Figure 3-5.

3.2.3 Signal Plus Noise Probability Density Function and Probability of Detection

The target signal almost invariably has different statistics from the receiver noise. For simplicity, we will start with a target signal represented by a sine wave of constant amplitude. A constant amplitude signal has an impulse PDF, as shown in Figure 3-6. In a noiseless environment, it would appear with value S (target signal power from Chapter 2) at a calculable location, and threshold detection would become elementary. However, the constant amplitude target signal has added to it the noise admitted into and generated by the receiver. The resulting target signal-plus-noise (S+N) PDF is the convolution of the two individual PDFs. Because the target signal PDF is an impulse, the S+N PDF is the noise PDF displaced to the right by the constant target signal amplitude.

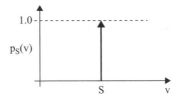

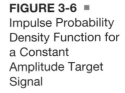

FIGURE 3-6 ▪
Impulse Probability
Density Function for
a Constant
Amplitude Target
Signal

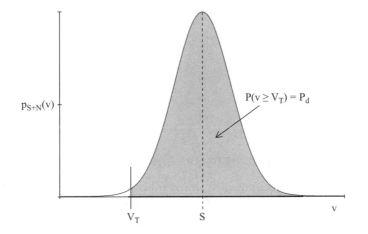

FIGURE 3-7 ▪
Rician Probability
Density Function
and Probability of
Detection

After passing through the IF filter and a linear envelope detector, the constant amplitude target signal plus noise has a Rician PDF, after Stephen Rice, the pioneer of modern signal detection theory. The Rician PDF is given in Equation (3-18) and shown in Figure 3-7. When no target signal is present ($S = 0$ and $I_0(0) = 1$), the Rician PDF becomes a Rayleigh PDF just as one would expect.

$$p_{S+N}(v) = \frac{v}{N} e^{\left(-\frac{v^2 + S}{2N}\right)} I_0\left(\frac{v\sqrt{S}}{N}\right) \qquad (3\text{-}18)$$

where:

$p_{S+N}(v)$ = Probability of v (target signal plus noise) occurring for the Rician probability density function

v = Linear envelope detector output voltage, volts

N = Mean noise power at the linear envelope detector output (see Section 2.3.1), watts

S = Constant amplitude target signal power, watts

$I_0(x)$ = Modified Bessel function of the first kind and zero order

Where previously the idea was to set a threshold for an arbitrary probability, the noise alone would exceed it; now the threshold may be placed to ensure that the S+N exceeds it for an arbitrary probability. We use Equations (3-3) and (3-18) to calculate the probability of the S+N exceeding a threshold (Equation 3-19). The probability that the S+N will exceed the threshold is called the probability of detection (P_d), as shown in Figure 3-7. From a statistical perspective, P_d is the *conditional* probability that, given a target signal is present, the target signal plus noise exceeds the detection threshold. Unfortunately, the integral in Equation (3-19) does not have a simple closed-form solution.

$$P_d = \int_{V_T}^{\infty} p_{S+N}(v)\, dv = \int_{V_T}^{\infty} \frac{v}{N}\, e^{\left(-\frac{v^2+S}{2N}\right)} I_0\left(\frac{v\sqrt{S}}{N}\right) dv \qquad (3\text{-}19)$$

where:

P_d = Probability of detection, no units

3.2.4 Designing Detection Thresholds

A choice of detection threshold affects both the P_{fa} and P_d. The actual threshold chosen must be based on a trade-off of the two with the radar capability, that is, the target S/N. Given an existing radar system, the achievable S/N is determined, and thresholds are then calculated based on the mission. Or if a mission is defined, a P_{fa} and P_d can be assigned (and a threshold thereby implicitly defined), and the radar must then be designed to achieve the S/N required. With the fundamental detection theory previously developed, both operations can be conducted, as shown in Figure 3-8.

FIGURE 3-8 ■
Probability of False
Alarm and
Probability of
Detection

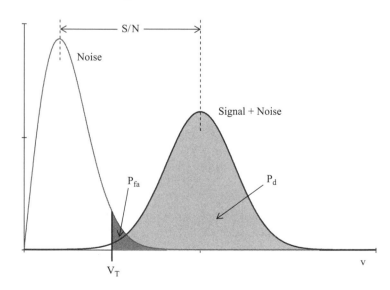

As seen in this figure, there is a relationship between the threshold, S/N, P_d, and P_{fa}. The P_d is found by integrating the Rician probability distribution based on an S/N threshold providing a desired P_{fa}. Unfortunately, the integral of the Rician probability distribution does not result in a simple closed-form solution. An accurate approximation to this integral was presented by North [1963] (Equation 3-20), based on the complementary error function (Equation 3-21) [and discussed in detail in Barton, 1988, pp. 65–67]. North also developed an equation for the signal-to-noise ratio required (detection threshold) to achieve a desired detection criteria (P_d and P_{fa}) (Equation 3-22), based on the Q probability integral or function (Equation 3-23). The relationship between the complementary error and Q functions is given in Equation (3-24). The complementary error and Q functions are included in many of our favorite computer tools, such as Mathcad and MATLAB®.

$$P_d \approx Q\left[\sqrt{-2\ln(P_{fa})} - \sqrt{2\,D_0 + 1}\right]$$

$$\approx 0.5\,\text{erfc}\left(\sqrt{-\ln(P_{fa})} - \sqrt{D_0 + 0.5}\right) \qquad \begin{array}{l}\text{Constant Target Signal}\\ \text{(3-20)}\end{array}$$

$$\text{erfc}\,(T) = 1 - \frac{2}{\sqrt{\pi}}\int_0^T e^{-x^2}\,dx \qquad (3\text{-}21)$$

$$D_0 \approx \left[\sqrt{-\ln(P_{fa})} - \frac{1}{\sqrt{2}}Q^{-1}(P_d)\right]^2 - 0.5$$

$$\approx \left[\sqrt{-\ln(P_{fa})} - \text{erfc}^{-1}(2\,P_d)\right]^2 - 0.5 \qquad \begin{array}{l}\text{Constant Target Signal}\\ \text{(3-22)}\end{array}$$

$$Q(E) = \frac{1}{\sqrt{2\pi}}\int_E^{\infty} e^{-\frac{x^2}{2}}\,dx \qquad (3\text{-}23)$$

$$Q(E) = \frac{1}{2}\text{erfc}\left(\frac{E}{\sqrt{2}}\right) \qquad \text{erfc}(T) = 2\,Q(\sqrt{2}\,T) \qquad (3\text{-}24)$$

where:

P_d = Probability of detection, no units

P_{fa} = Probability of false alarm, no units

D_0 = Signal-to-noise ratio required for detection (detection threshold) for a constant target signal, no units

$\text{erfc}(T)$ = Complementary error function

$Q(E)$ = Q probability integral

$\text{erfc}^{-1}(T)$ = Inverse complementary error function

$Q^{-1}(E)$ = Inverse Q probability integral

Detailed curves of the relationship between P_d, P_{fa}, and S/N (detection threshold) have been worked out by many [Barton, 1988, p. 62; Marcum, 1960; Skolnik, 2001, p. 44; DiFranco and Rubin, 1980]. I have used Equation (3-20) to generate such curves. The result is shown in Figure 3-9 and tabulated in Table 3-1. As seen in Figure 3-8 and Figure 3-9, as the S/N threshold is raised or lowered, both P_d and P_{fa} will decrease (higher threshold) or increase (lower threshold). To provide a higher P_d without increasing the P_{fa} requires increasing the separation between the noise and signal-plus-noise PDFs, which is achieved by having a higher target S/N.

The required P_d and P_{fa} are based on operational aspects of the radar system. As previously discussed in Section 3.2.2, the P_{fa} is often based on a required FAR or average time between false alarms (T_{fa}). Search radar systems, with the emphasis on long detection range, typically have $P_d \approx 50\%$ and $P_{fa} \approx 10^{-6}$, whereas tracking radar systems, with the emphasis on highly reliable detections, typically have $P_d \approx 90\%$ and $P_{fa} \approx 10^{-8}$.

Equations (3-20) and (3-22) provide very accurate P_d and D_0 values. However, to solve them we need a computer tool with the complementary error and/ or Q functions. Albersheim [1981] developed simple empirical formulas for D_0 (Equation (3-25) and P_d (Equation 3-26), we can compute using a scientific calculator. Equation (3-25) is said to be accurate to within 0.2 dB for $10^{-7} \leq P_{fa} \leq 10^{-3}$ and $0.1 \leq P_d \leq 0.9$. Albersheim's formula is probably suitable

FIGURE 3-9 ▪
Probability of
Detection (P_d) as a
Function of Signal-
to-Noise Ratio (S/N)
and Probability of
False Alarm (P_{fa}) —
Constant Target
Signal

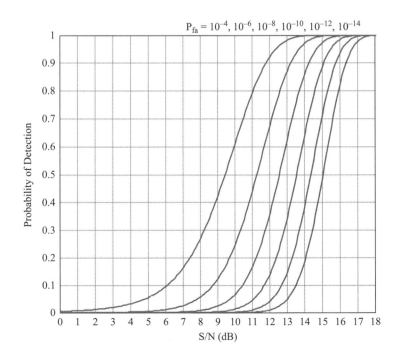

TABLE 3-1 ■ S/N (dB) Required for Detection as a Function of P_d and P_{fa}— Constant Target Signal

P_d	P_{fa}										
	10^{-4}	10^{-5}	10^{-6}	10^{-7}	10^{-8}	10^{-9}	10^{-10}	10^{-11}	10^{-12}	10^{-13}	10^{-13}
0.05	4.8	6.5	7.8	8.8	9.7	10.4	11.0	11.6	12.1	12.6	13.0
0.10	6.1	7.5	8.7	9.6	10.4	11.1	11.7	12.2	12.7	13.1	13.5
0.15	6.8	8.2	9.2	10.1	10.9	11.5	12.0	12.6	13.0	13.4	13.8
0.20	7.4	8.7	9.7	10.5	11.2	11.8	12.3	12.8	13.3	13.7	14.0
0.25	7.8	9.0	10.0	10.8	11.5	12.1	12.6	13.1	13.5	13.9	14.2
0.30	8.2	9.4	10.3	11.1	11.7	12.3	12.8	13.3	13.7	14.1	14.4
0.35	8.5	9.7	10.6	11.3	11.9	12.5	13.0	13.5	13.9	14.2	14.6
0.40	8.8	9.9	10.8	11.5	12.2	12.7	13.2	13.6	14.0	14.4	14.7
0.45	9.1	10.2	11.0	11.7	12.3	12.9	13.4	13.8	14.2	14.5	14.9
0.50	9.4	10.4	11.2	11.9	12.5	13.1	13.5	13.9	14.3	14.7	15.0
0.55	9.7	10.7	11.5	12.1	12.7	13.2	13.7	14.1	14.5	14.8	15.2
0.60	9.9	10.9	11.7	12.3	12.9	13.4	13.9	14.3	14.6	15.0	15.3
0.65	10.2	11.1	11.9	12.5	13.1	13.6	14.0	14.4	14.8	15.1	15.4
0.70	10.5	11.4	12.1	12.7	13.3	13.8	14.2	14.6	14.9	15.3	15.6
0.75	10.7	11.6	12.3	12.9	13.5	13.9	14.4	14.8	15.1	15.4	15.7
0.80	11.0	11.9	12.6	13.2	13.7	14.1	14.6	14.9	15.3	15.6	15.9
0.85	11.4	12.2	12.9	13.4	13.9	14.4	14.8	15.2	15.5	15.8	16.1
0.90	11.8	12.5	13.2	13.8	14.2	14.7	15.1	15.4	15.7	16.0	16.3
0.95	12.3	13.1	13.7	14.2	14.7	15.1	15.4	15.8	16.1	16.4	16.7
0.99	13.3	14.0	14.5	15.0	15.4	15.8	16.1	16.4	16.7	17.0	17.3

for fairly accurate calculations for lower P_{fa} values and/or greater P_d values. This accuracy is much better than we could read from a graph. An exercise at the end of this chapter will demonstrate the impact of detection threshold differences on radar detection range. In case we have forgotten in all the excitement, this is the reason we are discussing detection theory.

$$D_{0dB} \approx 10 \log (A + 0.12\, A\, B + 1.7\, B) \quad \text{Constant Target Signal} \quad (3\text{-}25)$$

$$P_d \approx \frac{1}{1 + e^{-C}} \quad \text{Constant Target Signal} \quad (3\text{-}26)$$

$$A = \ln\left(\frac{0.62}{P_{fa}}\right) \qquad B = \ln\left(\frac{P_d}{1 - P_d}\right) \qquad C = \frac{10^{(D_{0dB}/10)} - A}{1.7 + 0.12\, A}$$

where:

D_{0dB} = Signal-to-noise ratio required for detection (detection threshold) for a constant target signal using Albersheim's formula, decibels

If the received target signal does not have constant amplitude it cannot be represented by an impulse PDF. The target signal may have amplitude modulation added to it by either a fluctuating target radar cross section (RCS) or the intervening propagation medium or both. The PDF associated with a fluctuating target signal has a variance and, when combined with the PDF associated with the receiver noise, results in a PDF for the S+N that is more dispersed as well. From elementary statistics, we know that for independent random variables the variance of the sum of these PDFs is the sum of the variances. Accordingly, higher S/N values are required for high P_d, but the fluctuating target has better P_d at low S/N values.

A great deal of work has been done to understand radar detection theory for fluctuating target signals in receiver noise. The most widely used work is that performed by Dr. Peter Swerling [DiFranco and Rubin, 1980, Chapter 11; Richards et al., 2010, pp. 103–104, 559–560; Skolnik, 2001, pp. 66–70]. Swerling developed four statistical models for representing fluctuating target signals that account for the variation of the target RCS values and the correlation of the target RCS values with time. The four Swerling fluctuating target RCS models or cases are shown in Table 3-2. Often the cases are simply called Swerling 1, 2, 3, and 4 (SW1, SW2, SW3, and SW4, respectively). Sometimes a nonfluctuating target is called Swerling 0 (SW0).

For each of these four cases, Swerling computed the fundamental detection theory relationship between S/N (detection threshold), P_d, and P_{fa}. A fluctuating target signal results in a wider (higher variance) PDF for the S+N; thus, the S + N and noise PDF curves will overlap more than a nonfluctuating target. This results in a lower P_d than for a nonfluctuating target at the same S/N and P_{fa}. Or conversely, it requires a higher S/N than for a nonfluctuating target for the same P_d and P_{fa}. We will see this relationship graphically and numerically shortly.

For a Swerling Case 1 target signal the P_d is found by integrating the fluctuating target signal-plus-noise Rayleigh PDF based on an S/N threshold providing a desired P_{fa}, resulting in Equation (3-27). The S/N required to provide a desired detection criteria (P_d and P_{fa}) for a Swerling Case 1 target signal can be found using Equation (3-28). Integrating the S+N PDFs for the other

TABLE 3-2 ■ Swerling Target Fluctuation Cases

Radar Cross Section Defined By	Radar Cross Section Fluctuation Period	
	Slow or Scan-to-Scan or Dwell-to-Dwell	Fast or Pulse-to-Pulse
Many similar scatterers, none of which are dominant	Case 1	Case 2
Many similar scatterers, with one dominant	Case 3	Case 4

Swerling cases unfortunately does not result in such simple equations. Thus, a common approach for determining the S/N required for detection of Swerling target signals is based on using an additional S/N to account for the fluctuating target signal. With this approach we start with the S/N required to provide a desired detection criteria for a nonfluctuating target signal from Figure 3-9, Table 3-1, Equation (3-22), or Equation (3-25) and add (when all values are in decibels) S/N to account for a specific Swerling fluctuating target signal. The additional S/N for a Swerling Case 1 target signal fluctuation is given in Equation (3-29). The additional S/N for all Swerling cases is given in Equation (3-30). The additional S/N as a function of P_d for $P_{fa} = 10^{-6}$ for all Swerling cases is shown in Figure 3-10 and tabulated in Table 3-3 for Swerling Cases 1 and 2 and Table 3-4 for Swerling Cases 3 and 4.

$$P_d = (P_{fa})^{1/(1+D_1)} \quad \text{Swerling Case 1} \qquad (3\text{-}27)$$

$$D_1 = \frac{\ln(P_{fa})}{\ln(P_d)} - 1 \quad \text{Swerling Case 1} \qquad (3\text{-}28)$$

$$\Delta SNR_{f1} = \frac{D_1}{D_0} \quad \text{Swerling Case 1} \qquad (3\text{-}29)$$

$$\begin{aligned} \Delta SNR_{f2} &= \Delta SNR_{f1} && \text{Swerling Cases 1 \& 2} \\ \Delta SNR_{f3} &= \Delta SNR_{f4} = \Delta SNR_{f1}{}^{1/2} && \text{Swerling Cases 3 \& 4} \end{aligned} \qquad (3\text{-}30)$$

where:

D_1 = Signal-to-noise ratio required for detection (detection threshold) for a Swerling Case 1 target, no units

ΔSNR_{f1} = Additional signal-to-noise ratio, relative to a Swerling Case 0 target, required for detection for a Swerling Case 1 target, no units

ΔSNR_{f2} = Additional signal-to-noise ratio, relative to a Swerling Case 0 target, required for detection for a Swerling Case 2 target, no units

ΔSNR_{f3} = Additional signal-to-noise ratio, relative to a Swerling Case 0 target, required for detection for a Swerling Case 3 target, no units

ΔSNR_{f4} = Additional signal-to-noise ratio, relative to a Swerling Case 0 target, required for detection for a Swerling Case 4 target, no units

As shown in Figure 3-10, to provide the same P_d and P_{fa} fluctuating targets required a larger detection threshold (S/N) than nonfluctuating targets when the P_d is greater than about 30%, which is the case for radar systems in realistic conditions. This is especially true for Swerling Cases 1 and 2 (many similar scatterers) targets and also to a lesser extent with Swerling Cases 3 and 4 (one dominant scatterer) targets. For example, to provide a $P_d = 90\%$ and $P_{fa} = 10^{-8}$ requires an S/N = 14.2 dB for a Swerling Case 0 target (Table 3-1). To provide the same P_d and P_{fa} for Swerling Cases 1 and 2 requires 8.2 dB more S/N (Table 3-3) for a detection threshold of $14.2 + 8.2 = 22.4$ dB. Likewise, to

FIGURE 3-10 ▪
Additional Signal-to-
Noise Ratio
Required for
Detection for
$P_{fa} = 10^{-6}$ Based on
Swerling Case

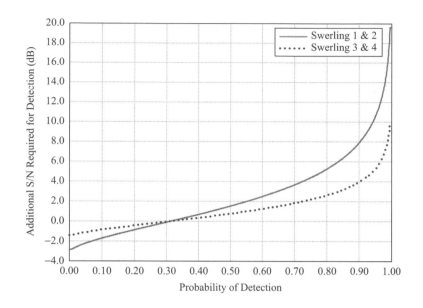

TABLE 3-3 ▪ Additional Signal-to-Noise Ratio (dB) Required for Detection as a
Function of P_d and P_{fa}—Swerling Cases 1 and 2

P_d	P_{fa}										
	10^{-4}	10^{-5}	10^{-6}	10^{-7}	10^{-8}	10^{-9}	10^{-10}	10^{-11}	10^{-12}	10^{-13}	10^{-14}
0.05	−1.6	−2.0	−2.2	−2.4	−2.6	−2.7	−2.8	−2.9	−3.0	−3.0	−3.1
0.10	−1.3	−1.5	−1.7	−1.8	−1.9	−2.0	−2.1	−2.2	−2.2	−2.3	−2.3
0.15	−1.0	−1.1	−1.3	−1.4	−1.5	−1.5	−1.6	−1.6	−1.7	−1.7	−1.8
0.20	−0.6	−0.8	−0.9	−0.9	−1.0	−1.1	−1.1	−1.1	−1.2	−1.2	−1.2
0.25	−0.3	−0.4	−0.5	−0.5	−0.6	−0.6	−0.7	−0.7	−0.7	−0.7	−0.8
0.30	0.0	−0.0	−0.1	−0.1	−0.2	−0.2	−0.2	−0.3	−0.3	−0.3	−0.3
0.35	0.4	0.3	0.3	0.3	0.2	0.2	0.2	0.2	0.2	0.2	0.1
0.40	0.7	0.7	0.7	0.7	0.7	0.6	0.6	0.6	0.6	0.6	0.6
0.45	1.1	1.1	1.1	1.1	1.1	1.1	1.1	1.1	1.1	1.1	1.1
0.50	1.5	1.5	1.5	1.5	1.5	1.6	1.6	1.6	1.6	1.6	1.6
0.55	1.9	2.0	2.0	2.0	2.0	2.0	2.1	2.1	2.1	2.1	2.1
0.60	2.4	2.4	2.5	2.5	2.5	2.6	2.6	2.6	2.6	2.6	2.6
0.65	2.9	3.0	3.0	3.1	3.1	3.2	3.2	3.2	3.2	3.2	3.3
0.70	3.5	3.6	3.7	3.7	3.8	3.8	3.8	3.9	3.9	3.9	3.9
0.75	4.2	4.3	4.4	4.5	4.5	4.6	4.6	4.6	4.7	4.7	4.7
0.80	5.0	5.2	5.3	5.4	5.4	5.5	5.5	5.6	5.6	5.6	5.7
0.85	6.1	6.3	6.4	6.5	6.6	6.6	6.7	6.7	6.8	6.8	6.9
0.90	7.6	7.8	7.9	8.1	8.2	8.2	8.3	8.4	8.4	8.5	8.5
0.95	10.2	10.4	10.6	10.8	10.9	11.0	11.1	11.1	11.2	11.3	11.3
0.99	16.3	16.6	16.9	17.1	17.2	17.4	17.5	17.6	17.7	17.7	17.8

provide the same P_d and P_{fa} for Swerling Cases 3 and 4 requires 4.1 dB more S/N (Table 3-4) for a detection threshold of $14.2 + 4.1 = 18.3$ dB. An exercise at the end of this chapter will reinforce this concept and show the resultant impact on detection range.

An often asked question is, "What Swerling Case should I use?" To decide which Swerling Case to use we need to select between two target RCS types and two RCS fluctuation periods. For some analyses this choice is straightforward, but for most the choice can be daunting. Fortunately, practical and mathematical considerations can simplify the choice of Swerling Case. The fast fluctuations of Cases 2 and 4 tend to average over time to a constant (Case 0) target. When the number of pulses integrated, coming up in the next section, is more than about 10 to 20, Case 2 approaches Case 1, and Case 4 approaches Case 3. Thus, all four cases may not have to be addressed individually. Swerling Case 1 is used more than the other cases because it requires a larger (more conservative) detection threshold (S/N) than the other cases [Skolnik, 2001, p. 68]. Of course, one can always conduct the analysis for the most likely Swerling Cases and thus can bound the detection range results.

TABLE 3-4 ■ Additional Signal-to-Noise Ratio (dB) Required for Detection as a Function of P_d and P_{fa}—Swerling Cases 3 and 4

P_d	P_{fa}										
	10^{-4}	10^{-5}	10^{-6}	10^{-7}	10^{-8}	10^{-9}	10^{-10}	10^{-11}	10^{-12}	10^{-13}	10^{-14}
0.05	−0.8	−1.0	−1.1	−1.2	−1.3	−1.3	−1.4	−1.4	−1.5	−1.5	−1.5
0.10	−0.6	−0.8	−0.9	−0.9	−1.0	−1.0	−1.1	−1.1	−1.1	−1.1	−1.2
0.15	−0.5	−0.6	−0.6	−0.7	−0.7	−0.8	−0.8	−0.8	−0.8	−0.9	−0.9
0.20	−0.3	−0.4	−0.4	−0.5	−0.5	−0.5	−0.6	−0.6	−0.6	−0.6	−0.6
0.25	−0.1	−0.2	−0.2	−0.3	−0.3	−0.3	−0.3	−0.3	−0.4	−0.4	−0.4
0.30	0.0	−0.0	−0.0	−0.1	−0.1	−0.1	−0.1	−0.1	−0.1	−0.1	−0.2
0.35	0.2	0.2	0.1	0.1	0.1	0.1	0.1	0.1	0.1	0.1	0.1
0.40	0.4	0.4	0.3	0.3	0.3	0.3	0.3	0.3	0.3	0.3	0.3
0.45	0.5	0.5	0.5	0.5	0.5	0.5	0.5	0.5	0.5	0.5	0.5
0.50	0.7	0.8	0.8	0.8	0.8	0.8	0.8	0.8	0.8	0.8	0.8
0.55	1.0	1.0	1.0	1.0	1.0	1.0	1.0	1.0	1.0	1.0	1.0
0.60	1.2	1.2	1.2	1.3	1.3	1.3	1.3	1.3	1.3	1.3	1.3
0.65	1.5	1.5	1.5	1.5	1.6	1.6	1.6	1.6	1.6	1.6	1.6
0.70	1.7	1.8	1.8	1.9	1.9	1.9	1.9	1.9	1.9	2.0	2.0
0.75	2.1	2.2	2.2	2.2	2.3	2.3	2.3	2.3	2.3	2.4	2.4
0.80	2.5	2.6	2.6	2.7	2.7	2.7	2.8	2.8	2.8	2.8	2.8
0.85	3.0	3.1	3.2	3.2	3.3	3.3	3.3	3.4	3.4	3.4	3.4
0.90	3.8	3.9	4.0	4.0	4.1	4.1	4.2	4.2	4.2	4.2	4.3
0.95	5.1	5.2	5.3	5.4	5.4	5.5	5.5	5.6	5.6	5.6	5.7
0.99	8.2	8.3	8.4	8.5	8.6	8.7	8.7	8.8	8.8	8.9	8.9

Radar designers strive to provide a radar system that provides the necessary S/N within size, weight, power, and cost constraints. As seen in Figure 3-9, small changes in S/N can greatly increase the P_d and/or reduce the P_{fa}. At some point it becomes extremely difficult and/or expensive to provide any additional S/N. For example, many people often ask why radar systems don't use cooled receivers like some electro-optical or infrared (EO/IR) systems. To reduce the receiver thermal noise power by half (–3 dB), the receiver would have to be cooled to 145 K (–199 °F) (see Equation 2-14), a very challenging amount. However, the radar community realized that, instead of performing detection based on a single pulse, multiple pulses will be received by waiting a short amount of time. Detection could then be performed based on the integration of the multiple received pulses.

3.3 | INTEGRATION OF MULTIPLE PULSES

The primary method of enhancing the S/N and detection performance of a radar system is to use multiple pulses to "dig" the target signal out of the receiver noise [Barton, 1988, Section 2.3]. In this operation, multiple pulses are sent to all range, Doppler, and angle resolution bins (see Chapter 5), and the coincident signals add in their respective bins. The resultant summing is called integration. In implementing integration of multiple received pulses the amplitude and phase of the resultant target signal plus noise is stored. This may be done in a computer with digitized or analog signals, as shown in Figure 3-11 [Richards et al., 2010, Chapter 15; Skolnik, 2001, Section 5.6; Stimson, 1998, pp. 129–133]. As ensuing returns arrive they are added. Accumulated stored returns exceeding the detection threshold are declared as detections.

The recirculating delay line is an elegant form of integrator [Skolnik, 2001, pp. 294–295]. The delay of the feedback loop is made equal to the radar's pulse repetition interval (PRI). The number of pulses integrated is flexible, making

FIGURE 3-11 ■
Simple Integrator
and Delay Line
Integrator

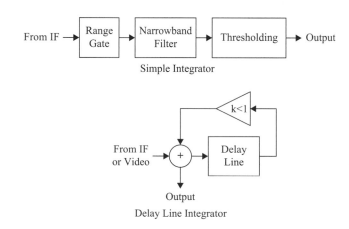

this integrator useful for electronically steered phased array radar systems, which potentially have arbitrary dwell times (see Section 7.4). The drawback is that the loop gain (k) must be kept slightly below one to avoid oscillations, meaning that the integration performance falls short of ideal.

Is the technique of integration any different from sending the same resultant power in a long pulse? Theoretically, it is not, but losses are associated with integration in mismatches due to elapsed time. In general, however, multiple-pulse integration is more economical than generating a single pulse of the same total energy. Additionally, a long pulse results in range measurement problems like poor range resolution and long blind zones (see Section 5.2).

There are two main types of integration: coherent and noncoherent. Coherent integration is also called predetection integration because it takes place before the envelope detector (see Section 2.4) using both the amplitude and phase of the received target signal pulses and receiver noise samples (S+N). Noncoherent integration is also called postdetection integration because it takes place after the envelope detector using only the magnitude (envelope) of the S+N. For coherent integration, a coherent (i.e., all the pulses are in phase with each other or the phase relationship between pulses is known in the receiver) waveform is transmitted, and thus the received pulses are coherent. Hence, the amplitude of each pulse simply adds (in voltage or increases quadratically in power) in phase with the other pulses (Equation 3-31) [Barton, 1988, Section 2.3]. The receiver thermal noise samples are random and thus add (in power) as complex (amplitude and phase) random numbers (Equation 3-32). Combining these two equations shows that coherent integration provides a linear increase in the S/N (Equation (3-33).

$$S_n = n_p^2 \, S \tag{3-31}$$

$$N_n = n_p \, N \tag{3-32}$$

$$\left(\frac{S}{N}\right)_n = \frac{n_p^2 \, S}{n_p \, N} = n_p \left(\frac{S}{N}\right) \tag{3-33}$$

where:

S_n = Received target signal power after integration of n_p coherent pulses, watts

n_p = Number of pulses integrated, no units

S = Received target signal power from one pulse, watts

N_n = Receiver thermal noise power after integration of n_p noise samples, watts

N = One sample of receiver thermal noise power, watts

$(S/N)_n$ = Target signal-to-noise ratio after integration of n_p coherent pulses, no units

(S/N) = Single pulse target signal-to-noise ratio, no units

It is difficult and expensive to build a coherent radar system (transmitter, receiver, and signal processor), especially when using a high-power transmitter. Therefore, the majority of radar systems use a noncoherent waveform. With a noncoherent waveform, the received pulses are also not necessarily in phase with each other, and the phase relationships between pulses are unknown in the receiver. They are correlated (i.e., from the same source and reflected from the same target) signals, not purely random signals. Thus, they add as complex (amplitude and phase) correlated numbers. The improvement in S/N due to noncoherent integration is a complicated function of P_d, P_{fa}, and target RCS fluctuation characteristics [Skolnik, 2001, p. 65].

To account for the improvement in S/N due to integration of multiple pulses, radar engineers often use an integration gain term (Equation 3-34). For coherent integration, the integration gain is given in Equation (3-35). For noncoherent integration, the range of integration gain for most practical cases is given in Equation (3-36) [Barton, 1988, Section 2.3]. Marcum [1960] stated that the optimum noncoherent integration gain is approximately the number of pulses integrated raised to the 0.76 power (Equation 3-37). These integration gain terms are shown in Figure 3-12, with the legend identifying the exponent of the number of pulses integrated with coherent = 1, Marcum = 0.76, etc. We will discuss the specifics of noncoherent integration in detail shortly in Section 3.3.1.

$$\left(\frac{S}{N}\right)_n = G_I \left(\frac{S}{N}\right) \tag{3-34}$$

$$G_I = n_p \quad \text{Coherent integration: Swerling Cases 0, 1, and 3} \tag{3-35}$$

$$n_p^{0.7} < G_I < n_p^{0.9} \quad \begin{array}{l}\text{Noncoherent integration:} \\ \text{Swerling Cases 0, 1, and 3}\end{array} \tag{3-36}$$

$$G_I \approx n_p^{0.76} \quad \begin{array}{l}\text{Noncoherent integration Marcum:} \\ \text{Swerling Cases 0, 1, and 3}\end{array} \tag{3-37}$$

where:

G_I = Integration gain, no units

The signal-to-noise ratio after the integration of multiple pulses is the product of the single-pulse signal-to-noise ratio (Equation 2-25) and the integration gain (Equation 3-38). Oftentimes the logarithmic form of this equation is mathematically more convenient (Equation 3-39). This equation can be simplified by using the carrier frequency instead of wavelength and combining the constants (Equation 3-40). As seen in Figure 3-13, a radar system can greatly improve the S/N through integration of multiple pulses and in turn increase the target detection range.

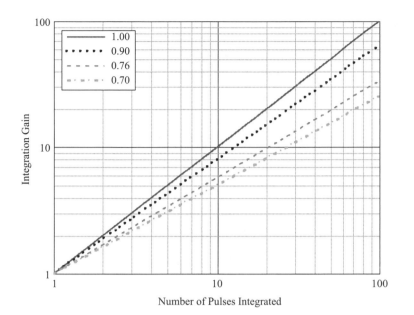

FIGURE 3-12 ■
Integration Gain—
Coherent and
Noncoherent

$$\left(\frac{S}{N}\right)_n = \frac{P_R\ G_{RT}{}^2\ \lambda^2\ \sigma\ G_{sp}\ G_I}{(4\pi)^3\ R_{RT}{}^4\ F_R\ k\ T_0\ B_R\ L_R} \tag{3-38}$$

$$\begin{aligned}
10\log\left[\left(\frac{S}{N}\right)_n\right] =\ & 10\log(P_R)+20\log(G_{RT})+20\log(\lambda)\\
&+10\log(\sigma)+10\log(G_{sp})+10\log(G_I)\\
&-30\log(4\pi)-40\log(R_{RT})-10\log(F_R)\\
&-10\log(k)-10\log(T_0)-10\log(B_R)\\
&-10\log(L_R)
\end{aligned} \tag{3-39}$$

$$\begin{aligned}
10\log\left[\left(\frac{S}{N}\right)_n\right] =\ & 10\log(P_R)+20\log(G_{RT})-20\log(f_c)\\
&+10\log(\sigma)+10\log(G_{sp})+10\log(G_I)\\
&-40\log(R_{RT})-10\log(F_R)-10\log(B_R)\\
&-10\log(L_R)+340.5434
\end{aligned} \tag{3-40}$$

The number of pulses integrated is a function of the integration time and the PRF (Equation 3-41). The integration time is often a fixed value based on radar signal processing conditions. This is the case for pulse-Doppler and tracking radar systems. The integration time can also be the target illumination time (Equation 3-42). This is the case for scanning radar systems. The target illumination time is the time the target is within the radar antenna's half-power (−3 dB) beamwidth (see Chapter 4). The maximum number of pulses available for integration for scanning radar is computed using the target illumination

FIGURE 3-13 ■
Multiple Pulse
Signal-to-Noise
Ratio (S/N)$_n$,
Detection Threshold
(SNR$_{dt}$), and Single
Pulse Signal-to-
Noise Ratio (S/N) as
a Function of Radar-
to-Target Range

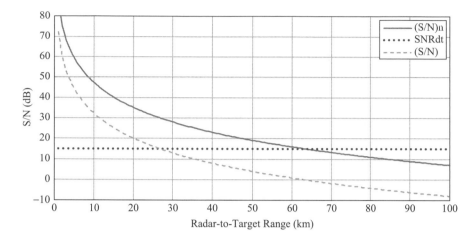

time. Some scanning radar systems integrate all the pulses in one target illumination time, while others integrate only a portion of the pulses.

$$n_p = T_I \, PRF \tag{3-41}$$

$$T_{ill} = \frac{\theta_{3dB}}{\theta_{dot}} \tag{3-42}$$

where:

T_I = Integration time, seconds
PRF = Pulse repetition frequency, hertz
T_{ill} = Target illumination time, seconds
θ_{3dB} = Antenna half-power (–3 dB) beamwidth, degrees or radians
θ_{dot} = Antenna scan rate, degrees/second or radians/second

The multiple-pulse radar equation can be solved for the range at which the radar will detect the presence of a target with a given S/N, that is, the S/N required for detection (detection threshold). The radar detection range is given in Equation (3-43). Often, the logarithmic form of this equation is mathematically more convenient (Equation 3-44). This equation can be simplified by using the carrier frequency instead of wavelength and combining the constants (Equation 3-45).

$$R_{dt} = \sqrt[4]{\frac{P_R \, G_{RT}^2 \, \lambda^2 \, \sigma \, G_{sp} \, G_I}{(4\pi)^3 \, SNR_{dt} \, F_R \, k \, T_0 \, B_R \, L_R}} \tag{3-43}$$

$$
\begin{aligned}
40 \log(R_{dt}) = {} & 10 \log(P_R) + 20 \log(G_{RT}) + 20 \log(\lambda) \\
& + 10 \log(\sigma) + 10 \log(G_{sp}) + 10 \log(G_I) \\
& - 30 \log(4\pi) - 10 \log(SNR_{dt}) - 10 \log(F_R) \\
& - 10 \log(k) - 10 \log(T_0) - 10 \log(B_R) \\
& - 10 \log(L_R)
\end{aligned} \tag{3-44}
$$

$$
\begin{aligned}
40 \log(R_{dt}) = \ &10 \log(P_R) + 20 \log(G_{RT}) - 20 \log(f_c) \\
&+ 10 \log(\sigma) + 10 \log(G_{sp}) + 10 \log(G_I) \\
&- 10 \log(SNR_{dt}) - 10 \log(F_R) - 10 \log(B_R) \\
&- 10 \log(L_R) + 340.5434
\end{aligned} \tag{3-45}
$$

where:

R_{dt} = Radar detection range, meters

SNR_{dt} = Signal-to-noise ratio required for detection (detection threshold), no units

One last but very important remark on integration: there are limits to the number of pulses integrated. A tracking radar system provides a target detection and associated measurements every integration time. A search radar system scans past a target once every scan (or frame) time (time to scan the search volume). Thus, the integration time is constrained by the update rate required by the operational considerations of the radar system. Modern radar systems perform integration in its range gates or Doppler filters (see Chapter 5). Therefore, to maximize the integration gain, all the pulses in one integration time must be in the same range gate and/or Doppler filter. If the integration time is too long, relative to the target states (e.g., speed, heading), the received pulses will migrate across range gates and/or Doppler filters. Also, as we will discuss in Chapter 5, the range rate resolution is inversely proportional to the integration time. The range rate resolution is an operational requirement of the radar system. In Chapter 5 we will also discuss how both range and range rate measurement ambiguities are a function of the PRF. Radar engineers knowledgeably balance these interrelationships to ensure that the radar system has the required performance, such as update rate, detection range, and measurements. Radar is one big systems engineering problem.

3.3.1 Noncoherent Integration Gain: The Details

In the previous section we discussed the basics of noncoherent integration gain using Marcum's [1960] work. However, frequently we need the details of noncoherent integration gain as a function of P_d, P_{fa}, and target RCS fluctuation characteristics. We will start with a long-standing concept for the mathematical description of radar detection performance, the equivalent single-pulse signal-to-noise ratio. The equivalent single-pulse S/N is the S/N each pulse must have so that after the integration of multiple noncoherent pulses the resultant integrated S/N is that required for detection. The use of equivalent single-pulse S/N allows for mathematically simple detection range equations, an advantage in the "dark ages" of calculation. However, many people are confused with the numerical values associated with the equivalent single-pulse S/N—often negative decibels ("how can a radar detect a target with a negative S/N?").

Also, the noncoherent integration gain is not clear because it is included in the equivalent single-pulse S/N.

The detailed approach for determining noncoherent gain as a function of P_d, P_{fa}, and target RCS fluctuation is based on Barton [1988, pp. 65–73] and Barton and Barton [1993, p. 131]. The equivalent single-pulse S/N for a Swerling Case 0 target fluctuation model using the Q probability function is given in Equations (3-46) and (3-47). Albersheim [1981] developed simple empirical formulas for the equivalent single-pulse S/N for a Swerling Case 0 target fluctuation model (Equation 3-48), which we can compute using a scientific calculator. Equation (3-48) is said to be accurate to within 0.2 dB for $10^{-7} \leq P_{fa} \leq 10^{-3}$, $0.1 \leq P_d \leq 0.9$, and $1 \leq n_p \leq 8096$. This accuracy is much better than we could read from a graph. The noncoherent integration gain for a Swerling Case 0 target is given in Equation (3-49). The noncoherent integration gain for the four Swerling target fluctuation cases is given in Equations (3-50) and (3-51), based on the additional S/N relative to an SW0 target required for detection of an SW1 target (Equation 3-29).

$$D_{0n} = \frac{D_c}{2 n_p} \left[1 + \sqrt{1 + \frac{9.2 n_p}{D_c}} \right] \quad \text{Swerling Case 0} \qquad (3\text{-}46)$$

$$D_c = \frac{1}{2} [Q^{-1}(P_{fa}) - Q^{-1}(P_d)]^2$$
$$= [\text{erfc}^{-1}(2 P_{fa}) - \text{erfc}^{-1}(2 P_d)]^2 \qquad (3\text{-}47)$$

$$D_{0ndB} \approx -5 \log (n_p) + \left(6.2 + \frac{4.54}{\sqrt{n_p + 0.44}} \right) \log (A + 0.12 \, A \, B + 1.7 \, B)$$
$$\text{Swerling Case 0}$$
$$(3\text{-}48)$$

$$G_{I0} = \frac{D_0}{D_{0n}} \quad \text{Swerling Case 0} \qquad (3\text{-}49)$$

$$G_{If} = G_{I0} \, (\Delta SNR_{f1})^{\gamma_f} \quad \text{Swerling Cases } 1,2,3,\&4 \qquad (3\text{-}50)$$

$$\gamma_f = -0.03 \log (n_p) \qquad \text{Swerling Case 1}$$
$$= 1 - \frac{1 + 0.03 \log (n_p)}{n_p} \qquad \text{Swerling Case 2}$$
$$= \frac{-0.03 \log(n_p)}{2} \qquad \text{Swerling Case 3} \qquad (3\text{-}51)$$
$$= \frac{1}{2} - \frac{1 + 0.03 \log(n_p)}{2 \, n_p} \qquad \text{Swerling Case 4}$$

where:

D_{0n} = Equivalent single-pulse signal-to-noise ratio required for detection after noncoherent integration of multiple pulses for a Swerling Case 0 target, no units

n_p = Number of pulses integrated, no units

D_c = Ideal signal-to-noise ratio required for coherent detection, no units

D_{0ndB} = Equivalent single-pulse signal-to-noise ratio required for detection after noncoherent integration of multiple pulses for a Swerling Case 0 target using Albersheim's formula, dB

G_{I0} = Noncoherent integration gain for a Swerling Case 0 target, no units

D_0 = Signal-to-noise ratio required for detection (detection threshold) for a constant target signal (Equation 3-22), absolute, or Equation (3-25), dB, no units

G_{If} = Noncoherent integration gain for a Swerling Case 1, 2, 3, or 4 target, no units

ΔSNR_{f1} = Additional signal-to-noise ratio, relative to a Swerling Case 0 target, required for detection for a Swerling Case 1 target, Equation (3-29), no units

γ_f = Swerling fluctuation case exponent, no units

As we can see in the previous equations, the noncoherent integrating gain is a complicated function of P_d, P_{fa}, and target RCS fluctuation characteristics. Figure 3-14 shows the noncoherent integration gain as a function of the number

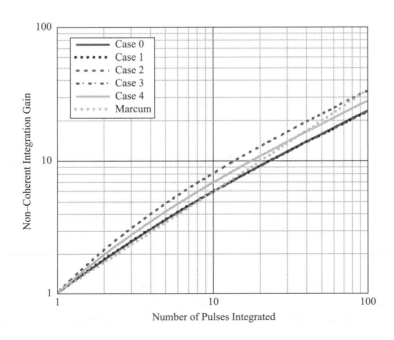

FIGURE 3-14 ■ Non-Coherent Integration Gain: $P_d = 50\%$ and $P_{fa} = 10^{-6}$

FIGURE 3-15 ■
Non-Coherent
Integration Gain:
$P_d = 90\%$ and
$P_{fa} = 10^{-8}$

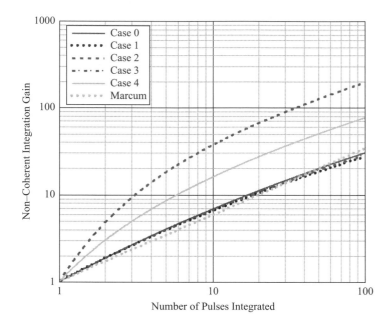

FIGURE 3-15 ■ Non-Coherent Integration Gain: $P_d = 90\%$ and $P_{fa} = 10^{-8}$

of pulses integrated for a nonfluctuating target (Swerling Case 0) and the four Swerling target fluctuation cases when the $P_d = 50\%$ and $P_{fa} = 10^{-6}$. Figure 3-15 shows the same information when $P_d = 90\%$ and $P_{fa} = 10^{-8}$. Both these figures include Marcum's noncoherent integration gain (Equation 3-37), for comparison. As seen in Figure 3-14, the noncoherent integration gain from Marcum's equation and Swerling Cases 0, 1, and 3 are very similar when the number of pulses integrated is less than about 20. We can see a similar relationship in Figure 3-15, except it extends over the entire numerical range of the number of pulses integrated. Both these figures show higher noncoherent integration gain associated with Swerling Cases 2 and 4 (fast) fluctuation targets, especially for the more demanding detection criteria.

3.3.2 Detection Threshold, Noncoherent Integration Gain, and Swerling Case

As we have seen in the previous two sections, there is a complicated relationship between P_d, P_{fa}, detection threshold, and noncoherent integration gain, and we are curious about how all these factors translate into detection range and differences in detection range for each Swerling case. After all, detection range is the main reason we are looking at detection theory in the first place. Table 3-5 contains numerical examples of all these relationships for two common detection criteria (P_d and P_{fa}), demonstrating that the detection range factor (detection range relative to the value for an SW0 target) is bounded by

TABLE 3-5 ■ Numerical Examples: Swerling Case, Noncoherent Integration Gain, Detection Threshold, and Detection Range Factor for 32 Pulses Integrated

Swerling Case	$P_d = 50\%$ and $P_{fa} = 10^{-6}$			$P_d = 90\%$ and $P_{fa} = 10^{-8}$		
	G_I (dB)	SNR_{dt} (dB)	Range Factor	G_I (dB)	SNR_{dt} (dB)	Range Factor
0	10.9	11.2	1.00	11.9	14.2	1.00
1	10.8	$11.2 + 1.6 = 12.8$	0.91	11.5	$14.2 + 8.2 = 22.4$	0.61
2	12.3	$11.2 + 1.6 = 12.8$	1.00	19.8	$14.2 + 8.2 = 22.4$	0.98
3	10.8	$11.2 + 0.8 = 12.0$	0.96	11.7	$14.2 + 4.1 = 18.3$	0.78
4	11.6	$11.2 + 0.8 = 12.0$	1.00	15.8	$14.2 + 4.1 = 18.3$	0.99

Swerling Cases 0 and 1—another reason these cases are most often used. Exercises at the end of this chapter will numerically reinforce noncoherent integration gain and its relationship with detection threshold and detection range, including one for the numerical examples in Table 3-5.

The discussions in the previous sections on detection threshold and non-coherent integration gain for Swerling targets can lead to the assumption that they go hand in hand, that is, that they are based on the same Swerling case. While this may be true for some radar–target combinations, it is not for many others. The detection threshold is set based on the expected Swerling case for the targets the radar system intends to detect. The noncoherent integration gain is based on the Swerling case of the target, which can change with time. For example, an aircraft can be SW0 when flying straight and level and SW4 when dynamically maneuvering. Also, there may be different Swerling cases for the many different targets in the environment. An exercise at the end of this chapter will numerically show noncoherent integration gain for one Swerling case and the detection threshold set for another and the associated impact on detection range.

3.4 | SOME DETECTION TECHNIQUES

Many radar systems improve detection performance by using multiple-event probability theory–based detection techniques such as cumulative probability, sequential detection, and M-out-of-N detection. These detection techniques all use multiple detection attempts to improve detection performance, such as detection range, P_d, and P_{fa}. Thus, they provide an opportunity to trade radar detection performance for time. Another detection technique forgoes a fixed detection threshold for an adaptive threshold that "follows" the perceived interference and varies with the perceived interference to provide a constant false alarm rate (CFAR).

3.4.1 Cumulative Probability

The cumulative probability is the probability of an event occurring at least once out of N attempts. The cumulative probability of detecting the target over multiple attempts is higher than the single attempt probability of detection. The cumulative probability of detection (P_{dc}) is given in Equation (3-52). P_{dc} as a function of the single attempt P_d is shown in Figure 3-16. The cumulative probability of false alarm (P_{fac}) can also be computed using this equation by simply replacing the individual probability of detection with the individual probability of false alarm (Equation 3-53).

$$P_{dc} = 1 - \prod_{i=1}^{N} (1 - P_{di})$$
$$= 1 - (1 - P_d)^N$$

(3-52)

$$P_{fac} = 1 - \prod_{i=1}^{N} (1 - P_{fai})$$
$$= 1 - (1 - P_{fa})^N$$

(3-53)

where:

P_{dc} = Cumulative probability of detection, no units
N = Number of individual detections attempts

FIGURE 3-16 ■
Cumulative
Probability of
Detection

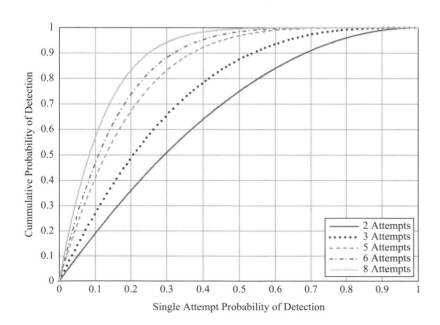

P_{di} = Probability of detection for the i-th individual detection attempt, no units

P_d = Probability of detection (the same for all detection attempts), no units

P_{fac} = Cumulative probability of false alarm, no units

P_{fai} = Probability of false alarm for the i-th individual detection attempt, no units

P_{fa} = Probability of false alarm (the same for all detection attempts), no units

For example, if $P_d = 50\%$, the cumulative probability of detection $P_{dc} = 87.5\%$ after three detection attempts. Likewise, if $P_{fa} = 10^{-6}$, the cumulative probability of false alarm $P_{fac} = 3 \times 10^{-6}$ after three detection attempts. The large increase in probability of detection combined with a small increase in probability of false alarm is an advantage of cumulative probability, especially when considering the associated difference in detection thresholds between the single attempt P_d and P_{fa}, and the cumulative P_{dc} and P_{fac}. An exercise at the end of this chapter will numerically reinforce cumulative probability and its relationship with detection threshold and detection range.

It is common to see detection ranges defined for a certain cumulative probability of detection, especially for airborne radar systems. Care must be taken in interpreting such a detection range because the equation for cumulative probability of detection is based on all individual probabilities of detection being associated with the same target (correlated to use the statistical term). This correlation is easy to ensure in theory but very hard in practice. Tracking radar systems with their high update rate (integration time) or search radar systems with very fast target revisit times can provide the necessary correlation; however, many other radar applications do not.

3.4.2 Sequential Detection

Another method of enhancing detection performance, if system timelines allow it, is to use more than one radar waveform and/or detection decision to determine whether a target is present. This approach is often called sequential detection or alert-confirm [Richards et al., 2010, p. 99; Jeffery, 2009, pp. 149–150; Stimson, 1998, p. 506]. Sequential detection can be performed in many ways, but most such tactics are really probability games. One may set a detection threshold at relatively low S/N, where the P_d is adequate but the P_{fa} is high. Range, Doppler, and angle resolution bins in which threshold crossing occurs are noted, and an additional radar waveform (sometimes called "verification" or "confirmation") is transmitted. If the detection threshold is crossed again in the same range, Doppler, and angle bins, target detection is declared. If not, the original detection is considered a false alarm. The confirmation waveform may be the same as, similar to, or very different from the initial waveform. Note what has happened to the detection statistics because of sending the confirmation waveform. The probability of noise alone exceeding the detection threshold twice is markedly lower than just once (Equation 3-54).

$$P_{fa}(2) = P_{fa}(1)^2$$

$$P_{fa}(N) = P_{fa}(1)^N$$

(3-54)

where:

$P_{fa}(2)$ = Probability of noise alone exceeding the detection threshold twice
$P_{fa}(1)$ = Probability of noise alone exceeding the detection threshold once (P_{fa})
$P_{fa}(N)$ = Probability of noise alone exceeding the detection threshold N times

The utility of this approach is twofold. It permits an improvement in radar detection performance (at a small sacrifice in time and radar complexity) without increasing effective radiated power (ERP), and it makes for more efficient use of existing capability because no confirmation waveform is sent when the detection threshold is not crossed. We must take care, however, in interpreting what the sequential detection thresholding scheme means. The detection threshold was set to make the P_d as high as necessary, at the expense of letting the P_{fa} also be relatively high. The confirmation waveform does not allow us to increase P_d (a posterior it is 1.0, given the signal-plus-noise has already crossed the threshold is not noise alone). What the confirmation waveform does allow us to do is reduce the P_{fa}. For example, if the chance of a noise sample crossed the threshold once were one in a thousand (10^{-3}), the chance it would cross twice would be one in a million (10^{-6}). The sequential P_{fa} as a function of the single-attempt P_{fa} is shown in Figure 3-17.

Since the end result of sequential detection is to reduce the P_{fa}, one must also consider the impact on the FAR, the associated average time between false alarms (T_{fa}), and search time as discussed in Section 3.2.2. There is a balance between having a high initial P_{fa} and the time necessary for all the confirmation waveforms. If too many confirmation waveforms are initiated, the overall search time (nominal search time based on the initial waveform plus the time for all the required confirmation waveforms) of the radar system will at some point become operationally unacceptable. An exercise at the end of this chapter will numerically address this situation.

3.4.3 M-out-of-N Detection

Another multiple detection attempt scheme uses the binomial probability theorem. Instead of declaring target detection based on a single detection attempt, some radar systems require at least M detections in N attempts. M-out-of-N detection is often called binary integration. A threshold is established, and an event is based on whether the threshold is crossed. The probability of exceeding

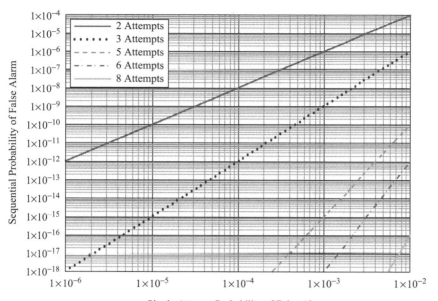

FIGURE 3-17 ■
Sequential
Probability of False
Alarm

the detection threshold (a detection) M times out of N attempts is given in Equation (3-55), the binomial probability theorem.

$$P(M, N, p) = \frac{N\,!}{M\,!(N - M)\,!}\, p^M\,(1 - p)^{(N-M)} \qquad (3\text{-}55)$$

where:

$P(M, N, p)$ = Probability of M detection threshold crossing out of N attempts
M = Number of times the detection threshold is crossed
N = Number of detection attempts
p = Probability of crossing the detection threshold for each attempt

For simplicity, let us consider a two-attempt binomial detection. If the threshold is set to achieve a $P_d = 0.5$ ($p = 0.5$), the possible outcomes are two threshold crossings with probability 0.25, one threshold crossing with probability 0.5, and zero threshold crossings with probability 0.25. The calculations for this example are given in Equation (3-56).

$$P(2,\ 2,\ 0.5) = \frac{2\,!}{2\,!(2 - 2)\,!}\, (0.5)^2\,(1 - 0.5)^{(2-2)} = 0.25$$

$$P(1,\ 2,\ 0.5) = \frac{2\,!}{1\,!(2 - 1)\,!}\, (0.5)^1\,(1 - 0.5)^{(2-1)} = 0.5 \qquad (3\text{-}56)$$

$$P(0,\ 2,\ 0.5) = \frac{2\,!}{0\,!(2 - 0)\,!}\, (0.5)^0\,(1 - 0.5)^{(2-0)} = 0.25$$

The probability of at least M detections out of N attempts is given in Equation (3-57), the resultant P_d. The probability of at least one threshold crossing out of two attempts is $0.25 + 0.50 = 0.75$. Therefore, we have increased the P_d by 50% (from 0.50 to 0.75) using a two-attempt binomial detection. The binomial probability of detection for at least M detections out of N attempts as a function of the single-attempt P_d is shown in Figure 3-18.

$$P = \sum_{k=M}^{N} \frac{N!}{k!(N-k)!}\, p^k\,(1-p)^{(N-k)} \tag{3-57}$$

What about the resultant probability of false alarm? Assume the detection threshold is set to achieve a $P_d = 0.50$ with a $P_{fa} = 10^{-4}$. Again, there are three possible outcomes for the two attempts: two threshold crossings with probability 10^{-8}; one threshold crossing with probability 1.9998×10^{-4}; and zero threshold crossings with probability 0.9998. The probability of at least one threshold crossing is $10^{-8} + 1.9998 \times 10^{-4} = 1.9999 \times 10^{-4}$, which means the P_{fa} has approximately doubled.

Referring to Table 3-1, the detection threshold to achieve a $P_d = 0.50$ with a $P_{fa} = 10^{-4}$ is 9.4 dB. The detection threshold to achieve a $P_d = 0.75$ with a $P_{fa} = 2 \times 10^{-4}$ is 10.4 dB. Thus, the detection criteria (P_d and P_{fa}) can be associated with the higher detection threshold and a detection range associated with the lower detection threshold.

FIGURE 3-18 ■
Binomial Probability
of Detection for at
Least M Detections
Out of N Attempts

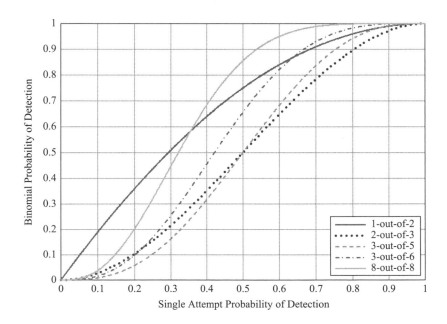

If a probability of at least one threshold crossing out of two attempts is desired to be 0.5, then the probability of detection of each attempt would be approximately 0.3. Likewise, if a probability of at least one false alarm out of two attempts is desired to be 10^{-4}, than the probability of false alarm of each attempt would be approximately 0.5×10^{-4}. The detection threshold to achieve a $P_d = 0.3$ with a $P_{fa} = 0.5 \times 10^{-4}$ is 8.5 dB. This reduction in the required detection threshold (and associated increase in detection range), while maintaining a desired P_d and P_{fa}, is one of the values of binomial detection.

Imagine a five-attempt binomial detection with a three-out-of-five threshold crossing criterion. We retain single-attempt $P_d = 0.5$, $P_{fa} = 10^{-4}$, and S/N = 9.4 dB. The probability of at least three threshold crossings when the single-attempt $P_d = 0.5$ is 0.5. For this criterion, the probability of detection remains unchanged. For a single-attempt $P_{fa} = 10^{-4}$, the probability of at least three threshold crossings is about 10^{-11}. For a three-out-of-five criterion, we have succeeded in lowering the probability of false alarm by seven orders of magnitude while keeping the probability of detection unchanged. This reduction in the probability of false alarm, while maintaining a desired P_d and detection threshold, is one of the advantages of binomial detection. Once again referring to Table 3-1, the detection threshold to achieve $P_d = 0.5$ with a $P_{fa} = 10^{-11}$ is 13.9 dB. Exercises at the end of this chapter will numerically reinforce M-out-of-N detection and its relationship with detection threshold and detection range.

M-out-of-N detection criteria are used in many search radar systems because they are easy to implement in a computer and allow reductions in the probability of false alarm at constant, or even improving, probabilities of detection with modest detection thresholds. The resultant P_d and P_{fa} can be manipulated rather easily as well. To raise P_d, minimize the number of crossings required for detection; to lower P_{fa}, do the opposite. With the right combination of M and N, the resultant binomial P_{fa} can be reduced and the resultant binomial P_d increased compared with the single detection case. The only drawback of M-out-of-N detection is the additional time necessary for N detection attempts instead of only one. Search radar systems typically perform the multiple detection attempts over multiple scans. Richards et al. [2010, Section 3.3.10] and Edde [1993, Section 5-4] both discuss M-out-of-N detection in detail.

3.4.4 Constant False Alarm Rate Detection

Target signals can be accurately and consistently detected in the presence of receiver thermal noise using a predetermined fixed detection threshold. However, situations where clutter (see Chapter 6) and/or noise jamming electronic attack (see Chapter 8) signals are much stronger than the receiver thermal noise can result in detections of clutter and/or noise jamming (clutter

FIGURE 3-19 ▪
Constant False
Alarm Rate
Threshold and
Receiver Output
versus Range

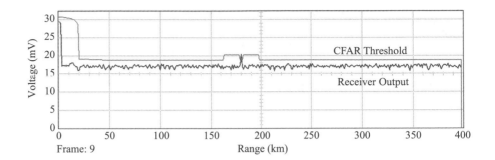

or noise jamming alone exceeds the fixed detection threshold). CFAR algorithms dynamically adjust the detection threshold in response to the perceived interference (thermal noise, clutter or noise jamming) to provide just what its name says—a constant false alarm rate (see Section 3.2.2)—in the presence of the ever-varying receiver contents. CFAR detection helps keep the display clear for the operator and/or helps prevent the data extractor from being overloaded so the radar system can consistently provide quality target detections and measurements. There are numerous CFAR algorithms, including cell averaging, smallest of, greatest of, and ordered statistics. A cell-averaging CFAR detection threshold is shown in Figure 3-19. As seen in this figure the detection threshold is increased in response to the clutter (left side) and a target (center). CFAR detection can be very effective in minimizing detections of clutter and/or the false alarms produced by noise jamming. Jeffery [2009, Section 3.11.1], Levanon [1988, Chapter 12], and Richards et al. [2010, Chapter 16] all closely examine CFAR detection.

3.5 | DETECTION THRESHOLD AND RADAR DETECTION RANGE

Of course the reason we are discussing the detection threshold in the first place is so that we can determine the detection range of the radar system, an essential radar performance metric. How do changes in the detection threshold translate into changes in the detection range? When all the other radar, target, and geometry/environment characteristics are the same, the relationship between detection ranges and detection thresholds is given in Equation (3-58). The change in detection range as a function of the change in the detection threshold is shown in Figure 3-20. As seen in this figure, slight (<0.3 dB) changes in the detection threshold result in slight ($\approx 2\%$) changes in the detection range. Therefore, for many radar–target situations, it is not worth agonizing over slight changes in the detection threshold, whatever their cause.

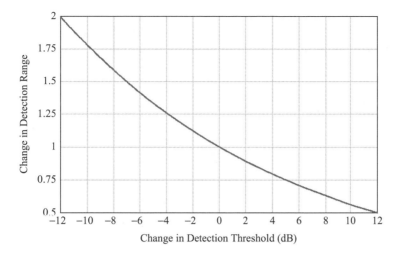

FIGURE 3-20 ▪
Change in Detection
Range as a Function
of the Change in the
Detection Threshold

$$R_{dt2} = \sqrt[4]{\frac{SNR_{dt1}}{SNR_{dt2}}} R_{dt1} \qquad (3\text{-}58)$$

where:

R_{dt2} = New radar detection range, meters
SNR_{dt2} = New detection threshold, no units
R_{dt1} = Old radar detection range, meters
SNR_{dt1} = Old detection threshold, no units

3.6 | SUMMARY

Detection theory is often one of the most confusing radar system topics. Hopefully, this chapter clearly addressed detection theory, threshold detection based on P_d and P_{fa}, multiple-pulse integration, fluctuating target RCS, and additional detection techniques. Target detection is based on probability theory. We developed the concepts and math behind P_{fa} and P_d, starting with the basics of a constant amplitude target signal and continuing to the Swerling fluctuating target RCS cases. We also saw how integrating multiple pulses can greatly improve a radar system's detection performance (i.e., P_d, P_{fa}, and detection range). Radar engineers use detection theory to understand the relationships between the detection threshold, P_{fa}, FAR, T_{fa}, P_d, integration time, number of pulses integrated, coherent pulses, and noncoherent pulses.

There are many ways to obtain a desired detection performance. As we will discuss in Chapter 5, target measurement and tracking capability and performance are also a function of the characteristics of the radar system. Radar engineers knowledgeably balance these interrelationships to ensure the radar system has the required detection, measurement, and tracking capability and performance. Radar is one big systems engineering problem.

■■■■ 3.7 I EXERCISES

3-1. A radar system has the following characteristics: time to make an individual detection decision $T_d = 25$ msec; number of range gates $n_{rg} = 300$; number of Doppler filters $n_{df} = 125$; and number of antenna beam positions $n_b = 640$. The radar system is required to have an average time between false alarms $T_{fa} = 60$ sec. Compute the number of detection decisions, N_d; the number of detection decisions per second; the probability of false alarm, P_{fa}; the average false alarm rate, FAR; the number of false alarms in one time period of interest, N_{fa}; and signal-to-noise ratio threshold, SNR (no units and dB).

3-2. Use detection theory Equations (3-20) and (3-22) to calculate the signal-to-noise ratio required for detection (detection threshold), D_0 (dB), to provide a single-pulse probability of detection $P_d = 0.9$ and probability of false alarm $P_{fa} = 10^{-6}$. Compare with Figure 3-9, Table 3-1, and Albersheim's formula, Equations (3-25) and (3-26).

3-3. A radar system requires the following detection criteria: $P_d = 0.5$ and $P_{fa} = 10^{-6}$. What is the signal-to-noise ratio required for detection (detection threshold) for Swerling Case 0, 1, 2, 3, and 4 targets? Compute the results using the equations in Section 3.2.4, and then compare them to the results obtained in Figure 3-10 and Tables 3-1, 3-3, and 3-4. If the detection range is 100 km when the threshold is set based on a Swerling Case 0 target, what is the detection range for the other Swerling cases?

3-4. Repeat Exercise 3-3 for the detection criteria $P_d = 0.9$ and $P_{fa} = 10^{-8}$.

3-5. A radar system has a single-pulse SNR = –2 dB. If the required probability of false alarm, $P_{fa} = 10^{-6}$, what is the probability of detection, P_d (no units), associated with the single-pulse SNR? If the radar system has an integration time $T_I = 40$ msec and pulse repetition frequency PRF = 800 Hz, how many pulses are integrated, n_p? What is the coherent integration gain, G_I (no units and dB); the signal-to-noise ratio after coherent integration, SNR_n (no units and dB); and the probability of detection, P_d (no units), associated with the SNR_n after coherent integration? What is the noncoherent integration gain, G_I (no units and dB); the signal-to-noise ratio after noncoherent integration, SNR_n (no units and dB); and P_d associated with the SNR_n after noncoherent integration? Use Table 3-1 for the relationship between the S/N required for detection, P_d, and P_{fa}.

3-6.   To increase the performance of a power-limited radar system, you plan to use multiple-pulse integration. You currently achieve a single-pulse $P_d = 0.5$ and $P_{fa} = 10^{-4}$. How many coherent pulses are required to be integrated to obtain $P_d = 0.95$ and $P_{fa} = 10^{-12}$? How many non-coherent pulses are required to be integrated to obtain the new P_d and P_{fa}? Use Marcum's optimum noncoherent integration gain.

3-7.   A radar system requires the following detection criteria: $P_d = 0.5$ and $P_{fa} = 10^{-6}$. The radar system has the following characteristics: antenna half-power (–3 dB) beamwidth $\theta_{3dB} = 0.96$ degrees; antenna scan rate $\theta_{dot} = 36$ degrees/second; and pulse repetition frequency PRF = 1200 Hz. What is the noncoherent integration gain for Swerling Case 0, 1, 2, 3, and 4 targets? What is the detection range factor (detection range relative to that for a Swerling Case 0 target) for the Swerling cases? Compute the results using the equations in Section 3.3.1, and then compare them with the results obtained in Figure 3-14 and Table 3-5. (Hint: use the results of Exercise 3-3.)

3-8.   Repeat Exercise 3-7 for the following detection criteria: $P_d = 0.9$ and $P_{fa} = 10^{-8}$. Compute the results using the equations in Section 3.3.1, and then compare them with the results obtained in Figure 3-15 and Table 3-5. (Hint: use the results of Exercise 3-4.)

3-9.   A Swerling Case 0 target is detected by the radar system in Exercise 8 at a range $R_{dt} = 100$ km. If the detection threshold is unchanged, at what ranges can the radar system detect Swerling Case 1, 2, 3, and 4 targets? (Hint: use the results of Exercise 3-8.)

3-10. A radar has a probability of detection $P_d = 0.4$ and probability of false alarm $P_{fa} = 10^{-6}$ for each detection event. Calculate the cumulative probabilities of detection (P_{dc}) and false alarm (P_{fac}) when 10 individual detections are combined. Use Figure 3-9 or Table 3-1 to determine the detection threshold associated with each detection attempt and the detection threshold associated with the cumulative probabilities of detection and false alarm. What is the change in detection range associated with these two detection thresholds?

3-11. A radar system has the following characteristics: time to make an individual detection decision $T_d = 25$ msec; probability of false alarm $P_{fa} = 10^{-6}$; number of range gates $n_{rg} = 300$; number of Doppler filters $n_{df} = 125$; and number of antenna beam positions $n_b = 640$. What is the probability of false alarm when using a two-step (alert–confirm) sequential detection approach? What is the average number of false alarms in the first step (alert) of the sequential detection approach, N_{fa}?

What is the consequence of the second step (confirm) of the sequential detection approach in terms of the number of antenna beam positions (absolute and percent) and additional time (absolute and percent)? (Hint: use the results of Exercise 3-1.)

3-12.   A four-attempt binomial detection approach is proposed. The detection threshold is set to provide $P_d = 0.4$ and $P_{fa} = 10^{-3}$ for each attempt. Calculate the binomial probabilities of detection and false alarm associated with exactly one, two, three, and four detections out of four attempts. Repeat for at least one, two, three, and four detections out of four attempts.

3-13.   A radar system uses a five-attempt binomial detection with a three-out-of-five threshold-crossing approach: the single-attempt probability of detection $P_d = 0.55$; probability of false alarm $P_{fa} = 10^{-4}$; and associated detection threshold S/N = 9.7 dB (Table 3-1). What is the binomial probability of detection for at least three or more detection attempts? What is the binomial probability of false alarm for at least three or more detection attempts? Use Figure 3-9 or Table 3-1 to determine the detection threshold associated with the binomial probabilities of detection and false alarm after at least three or more detection attempts.

3-14.   When the radar in Exercise 3-13 uses at least three-out-of-five threshold crossing detection approach its detection range is $R_{dt} = 100$ km. What is the detection range if the radar uses a single threshold crossing detection approach? (Hint: use the results of Exercise 3-13.)

3-15.   The radar system from Exercise 2-5 has the following additional characteristics: pulse repetition frequency PRF = 500 hertz; antenna half-power beamwidth $\theta_{3dB} = 1$ degree; antenna scan rate $\theta_{dot} = 25$ deg/sec; noncoherent pulses; and detection threshold $SNR_{dt} = 12$ dB. What is Marcum's optimal noncoherent integration gain, G_I (no units and dB)? What is the target signal-to-noise ratio after integration of multiple pulses, SNR_n (no units and dB)? What is the resultant radar detection range, R_{dt} (meters)? (Hint: use the results of Exercise 2-5.)

▋ 3.8 | REFERENCES

Albersheim, W. J., 1981, "A Closed-Form Approximation to Robertson's Detection Characteristics," *Proceeding IEEE,* vol. 69, no. 7, July, p. 839. Skolnik [2001] and Richards et al. [2010] both contain Albersheim's formula.

Barton, D. K., 1988, *Modern Radar System Analysis,* Norwood, MA: Artech House. Barton's update and expansion of his 1979 book is the source for many of the detailed detection theory equations.

Barton, D. K., and Barton W. F., 1993, *Modern Radar System Analysis Software*, Norwood, MA: Artech House. Includes the details necessary for the software modeling and simulation of radar systems, concepts, and components; source for many of the detailed detection theory equations.

Blake, Lamont V., 1986, *Radar Range-Performance Analysis*, Norwood, MA: Artech House. Covers a wide range of topics, with detailed supporting equations for radar detection and propagation.

DiFranco, J. V., and Rubin, W. L., 2004, *Radar Detection*, Raleigh, NC: SciTech Publishing. Standard for radar detection; however, be prepared for the calculus. The authors are meticulous in their development, solution, and presentation of the results.

Edde, Byron, 1993, *Radar Principles, Technology, Applications*, Upper Saddle River, NJ: Prentice-Hall. Covers a wide range of topics, with supporting equations and numerical examples.

Jeffery, Thomas W., 2009, *Phased-Array Radar Design Application of Radar Fundamentals*, Raleigh NC: SciTech Publishing. Good discussion on sequential detection, false alarm rate, probability of false alarm, and related topics in Section 7.2.

Levanon, Nadav, 1998, *Radar Principles*, New York: John Wiley & Sons. Discusses fundamental detection theory in Chapter 3 and CFAR detection in Chapter 12.

Marcum, J. T., 1960, "A Statistical Theory of Target Detection by Pulsed Radar," *IRE Transactions in Information Theory*, Volume IT-6, April, pp. 145–267. This source has stood the test of time.

North, D. O., 1943, "An Analysis of Factors Which Determine Signal/Noise Discrimination in Pulsed Carrier Systems," RCA Labs Tech Report PTR6C. (Reprinted in *Proceedings of the IRE*, Volume 51, July 1963, pp. 1016–1027.)

Richards, M. A., Sheer, J. A., and Holm, W. A. (editors), 2010, *Principles of Modern Radar*, Volume 1: Basic Principles, Raleigh, NC: SciTech Publishing. The fundamentals of detection theory are in Chapter 3. Chapter 15 covers the details of threshold detection of radar targets. Chapter 16 covers constant false alarm rate detectors.

Skolnik, Merrill I., 2001, *Introduction to Radar Systems*, 3rd Edition, New York: McGraw-Hill. Meticulous in its presentation of all equations and references; answers are emphasized.

Stimson, George W., 1998, *Introduction to Airborne Radar*, 2nd Edition, Raleigh, NC: SciTech Publishing. The fundamentals of detection theory are in Chapters 10 and 11. Chapter 40 covers advanced radar techniques including sequential detection.

Swerling, P., 1954, "Probability of Detection of Fluctuating Targets," Rand Corporation Research Memo RM-1217. (Reprinted *IRE Transactions*, IT-6, April 1960.)

Radar Antennas

HIGHLIGHTS

- Antenna gain and effective area
- Remarkable utility of the paraboloid
- Deriving the far-field antenna gain pattern with calculus
- Design features for mainbeam gain, beamwidths, and sidelobes
- Unique features of array antennas
- Several ways to steer the beams of phased arrays, emphasizing phase shifters

An antenna is the mechanism by which the electromagnetic wave is radiated into and received from the environment. For radar systems, although not necessarily for antennas in other electromagnetic applications, it is essential the antenna enhance radar performance. A radar antenna has three roles: to be a major contributor to the radar's detection performance; to provide the required surveillance coverage; and to allow measurements of angle of sufficient accuracy and precision.

A reasonable place to begin our discussion of radar antennas is with the two expressions used in Chapter 2 to derive the radar equation: antenna mainbeam gain and effective area. For a transmitting antenna, antenna mainbeam gain is simply a measure of how much the transmitted waveform is being focused. This is achieved using a directional antenna, similar to the way a flashlight focuses, or directs, the light in a desired direction at the expense of others. The ratio of this focusing over isotropic (gain of one everywhere) radiation is the mainbeam gain of the antenna. The antenna gain is often represented in decibels relative to isotropic (dBi).

The ability to preferentially add up signals provides focusing. A signal arriving at the antenna from a preferred direction is integrated; those arriving from elsewhere are not. This assumes the signal arriving at the antenna is in the form of plane waves, that is, the phase of the arriving signal is constant

FIGURE 4-1 ▪ A Plane Wave

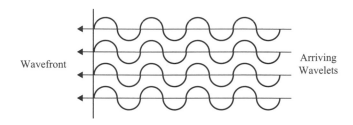

over any plane perpendicular to the direction of arrival, as shown in Figure 4-1. Because most sources of electromagnetic waves are physically small, these wavefronts are really spherical, but at ranges of interest the approximation to a plane wave is very good. For now, we will assume the phase differences are "indistinguishable" from an extended source of in-phase signals, otherwise known as being in the "far field" of the antenna. How to determine what is indistinguishable and the far-field distance is described in Section 13.3.

For a receiving antenna, effective area is simply a measure of ability of the antenna to intercept the incident power density. Electromagnetics theory tells us the relationship between antenna mainbeam gain and the effective area of the antenna (Equation 4-1). Effective area is related to the physical area of the antenna by the antenna efficiency [Blake, 1986, pp. 11, 171, 361; Skolnik, 2001, pp. 543–544]. The antenna efficiency term allows us to account for factors such as antenna manufacturing tolerances, antenna losses, illumination function, feed networks, and array element characteristics.

$$G = \frac{4\pi\, A_e}{\lambda^2} = \frac{4\pi\, \rho\, A}{\lambda^2} \qquad (4\text{-}1)$$

where:

G = Antenna mainbeam gain, no units
A_e = Effective area of the antenna, square meters (m^2)
λ = Wavelength, meters
ρ = Antenna efficiency, $\approx$ 30–70% for most practical radar antennas, no units
A = Physical area of the antenna, square meters (m^2)

▮ 4.1 | THE ANTENNA GAIN PATTERN

It is one thing to have found the antenna mainbeam gain; it is another to know how the antenna gain varies across the spectrum of angles—in other words, to

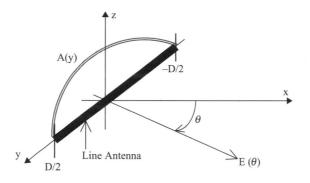

FIGURE 4-2 ■
Simple Line Antenna

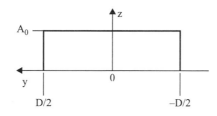

FIGURE 4-3 ■ A
Uniform Current
Distribution

know the directional antenna gain pattern. There are three main approaches for determining the antenna pattern: electromagnetic theory; Fourier transforms; and an incremental algebraic. We will concentrate on the electromagnetic theory approach and will identify how the other approaches can be used.

Electromagnetics theory tells us the antenna gain (power) is the electric field (E-field) squared. Thus, if we can determine the E-field, we can compute the antenna gain. The E-field is often referred to as the antenna voltage pattern. We will start with a simple line antenna with a current distribution across the dimension of the antenna, as shown in Figure 4-2. Often, the current distribution is called an illumination function. Electromagnetic theory states an E-field is produced perpendicular to the current distribution.

The relationship between the current distribution, antenna dimensions, and E-field is given in Equation (4-2). We will solve this equation for the uniform current distribution case (current distribution is constant across the antenna), as shown in Figure 4-3. For a uniform current distribution, Equation (4-2) is rewritten as (4-3). The integral in Equation (4-3) is solved and evaluated at its limits (Equation 4-4). Equation (4-4) looks pretty formidable until we use the trigonometry identity given in Equation (4-5). Using this trigonometry identity, the E-field is now given in Equation (4-6).

$$E(\theta) = \int_{-D/2}^{D/2} A(y)\, e^{\left(\frac{2\pi\, j\, y \sin(\theta)}{\lambda}\right)} dy \qquad (4\text{-}2)$$

where:

$E(\theta)$ = E-field produced by the current distribution, perpendicular to A(y), voltage antenna gain pattern, no units

D = Antenna dimension, meters

$A(y)$ = Current distribution in the y–z plane

λ = Wavelength, meters

$$E(\theta) = \int_{-D/2}^{D/2} A_0\, e^{\left(\frac{2\pi j y \sin(\theta)}{\lambda}\right)}\, dy \tag{4-3}$$

$$E(\theta) = \frac{A_0}{\left(\frac{2\pi j \sin(\theta)}{\lambda}\right)} \left[e^{\left(\frac{2\pi j y \sin(\theta)}{\lambda}\right)} \right]_{-D/2}^{D/2}$$

$$E(\theta) = \frac{A_0}{\left(\frac{2\pi j \sin(\theta)}{\lambda}\right)} \left[e^{\left(\frac{2\pi j \frac{D}{2} \sin(\theta)}{\lambda}\right)} - e^{\left(\frac{-2\pi j \frac{D}{2} \sin(\theta)}{\lambda}\right)} \right] \tag{4-4}$$

$$\sin(\phi) = \frac{e^{j\phi} - e^{-j\phi}}{2j} \tag{4-5}$$

$$E(\theta) = \frac{A_0 D}{\left(\frac{\pi D \sin(\theta)}{\lambda}\right)} \sin\left(\frac{\pi D \sin(\theta)}{\lambda}\right) \tag{4-6}$$

We often normalize the E-field so that $E(0) = 1$. This is accomplished by setting $A_0 = 1/D$. The normalized E-field is given in Equation (4-7). The antenna gain (power) pattern is the square of the E-field (Equation 4-8).

$$E(\theta) = \frac{\sin\left(\frac{\pi D \sin(\theta)}{\lambda}\right)}{\left(\frac{\pi D \sin(\theta)}{\lambda}\right)} \tag{4-7}$$

$$G(\theta) = [E(\theta)]^2 = \left[\frac{\sin\left(\frac{\pi D \sin(\theta)}{\lambda}\right)}{\left(\frac{\pi D \sin(\theta)}{\lambda}\right)} \right]^2 \tag{4-8}$$

where:

$G(\theta)$ = Normalized antenna gain pattern (power) for a uniform current distribution, no units

The antenna gain pattern can be visualized as an x–y plot or a polar plot, as shown in Figure 4-4. The antenna gain pattern is characterized by its mainbeam, sidelobes, nulls, and backlobe, as shown in Figure 4-5. It is common for antenna gain pattern to be plotted in decibels, as in both of these figures. Decibels are used so the large numerical dynamic range of the actual pattern can be easily visualized. The peak of the mainbeam is the mainbeam gain, as computed using Equation (4-1). The majority of the antenna gain is concentrated in the mainbeam, with the remaining antenna gain distributed in the sidelobes and the backlobe.

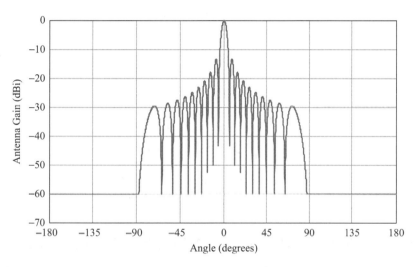

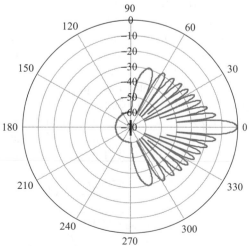

FIGURE 4-4 ■
Antenna Gain Pattern, x–y Plot (top) and Polar Plot (bottom)

FIGURE 4-5 ■
Antenna Pattern
Characteristics

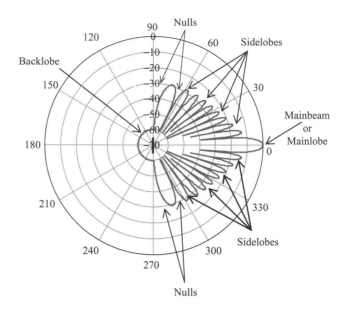

FIGURE 4-6 ■
Antenna Half-Power
Beamwidth

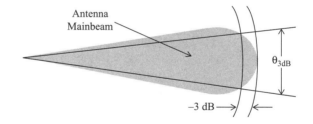

4.1.1 Finding Half-Power and Null-to-Null Beamwidths

We routinely use two different measures of the angular width of the mainbeam: the half-power beamwidth and the null-to-null beamwidth. The half-power (−3 dB) beamwidth is the angular width between the half-power points of the mainbeam, as shown in Figure 4-6. We start the process of determining the half-power beamwidth by solving for the angle at which Equation (4-8) equals 0.5 (Equation 4-9). The $[\sin(x)/x]^2$ function equals 0.5 when $x = 0.443\pi$. We use this relationship in Equation (4-10). Next, we use the small angle approximation for sine [$\sin(\theta) \approx \theta$ for small values of θ in radians] (Equation 4-11). We then write the equation for the half-power beamwidth (Equation 4-12). Converting from radians to degrees gives us the half-power beamwidth in Equation (4-13).

$$G(\theta) = \left[\frac{\sin\left(\frac{\pi D \sin(\theta)}{\lambda}\right)}{\left(\frac{\pi D \sin(\theta)}{\lambda}\right)}\right]^2 = 0.5 \qquad (4\text{-}9)$$

$$\left(\frac{\pi D \sin(\theta)}{\lambda}\right) = 0.443\pi$$

$$\sin(\theta) = \frac{0.443\,\lambda}{D} \tag{4-10}$$

$$\theta \cong \frac{0.443\,\lambda}{D} \tag{4-11}$$

$$\theta_{3dB} = 2\theta = \frac{0.886\,\lambda}{D} \quad \text{radians} \tag{4-12}$$

$$\theta_{3dB} = \frac{51\,\lambda}{D} \quad \text{degrees} \tag{4-13}$$

where:

θ_{3dB} = Antenna half-power (−3 dB) beamwidth for a uniform current distribution, radians or degrees

λ = Wavelength, meters

D = Antenna dimension, meters

We can determine the null-to-null beamwidth in a similar manner as the half-power beamwidth. The first null in the antenna gain pattern occurs when the argument of the sine term equals π [$\sin(\pi) = 0$] (Equation 4-14). Solving for $\sin(\theta)$ we obtain Equation (4-15). Once again, we use the small angle approximation for sine [$\sin(\theta) \approx \theta$ for small values of θ in radians] (Equation 4-16). We then write the equation for the null-to-null beamwidth (Equation 4-17). Converting from radians to degrees gives us the null-to-null beamwidth in Equation (4-18).

$$\left(\frac{\pi D \sin(\theta)}{\lambda}\right) = \pi \tag{4-14}$$

$$\sin(\theta) = \frac{\lambda}{D} \tag{4-15}$$

$$\theta \cong \frac{\lambda}{D} \tag{4-16}$$

$$\theta_{nn} = 2\theta = \frac{2\,\lambda}{D} \quad \text{radians} \tag{4-17}$$

$$\theta_{nn} = \frac{115\,\lambda}{D} \quad \text{degrees} \tag{4-18}$$

where:

θ_{nn} = Antenna null-to-null beamwidth for a uniform current distribution, radians or degrees

We have seen the antenna gain pattern is of the form $[\sin(x)/x]^2$. We can manipulate the integral of this function (Equation 4-19), to produce some interesting findings. Computing the ratio of an integration from 0 to 0.443π (the half-power point) and from 0 to ∞ reveals that just less than 50% of the antenna gain is contained within the half-power beamwidth. Computing the ratio of an integration from 0 to π (the first null) and from 0 to ∞ shows that just less than 90% of the antenna gain is contained within the null-to-null beamwidth.

$$\int_0^x \left(\frac{\sin x}{x}\right)^2 dx \qquad (4\text{-}19)$$

4.1.2 Finding Sidelobe Levels

Equations (4-14) through (4-16) have already demonstrated that the nulls in the antenna gain pattern are at λ/D, $2\lambda/D$, $3\lambda/D$, and so forth. Peaks in the antenna gain pattern occur when the argument of the sine term equals 0 and then at approximately $\dfrac{3\lambda}{2D}, \dfrac{5\lambda}{2D}$, and so forth (argument of the sine term equals 1 or $x = 3\pi/2$, $5\pi/2$, etc.). The first peak is the antenna mainbeam (the mainlobe) occurs at zero. The second peak, the peak of the first sidelobe, occurs at approximately $\dfrac{3\lambda}{2D}$. The antenna gain at the peak of the first sidelobe is computed as given in Equation (4-20). This means the antenna gain of the peak of the first sidelobe of an antenna with a uniform current distribution is down from the mainbeam gain by a factor of approximately 22.2, or 13.46 dBi. The antenna gain of the peak of the n-th sidelobe of an antenna with a uniform current distribution can be readily defined by Equation (4-21). The 10th sidelobe, therefore, is down from the mainbeam gain by a factor of 1/1088, or -30.37 dBi.

$$\left(\frac{\sin\left(\frac{3\pi}{2}\right)}{\frac{3\pi}{2}}\right)^2 = \frac{1}{22.2} = 0.045 \quad 10\log(0.045) = -13.46 \text{ dB} \quad (4\text{-}20)$$

$$\left(\frac{2}{\pi}\right)^2 \frac{1}{(2n+1)^2} \qquad (4\text{-}21)$$

By using a different current distribution, different relationships between antenna beamwidth and sidelobe levels can be obtained. Generally, lower sidelobe levels are desired to minimize clutter (see Chapter 6) and/or sidelobe jamming (see Chapter 8). Thus, current distributions with less amplitude farther out in the distribution (toward the edge of the antenna) are used. This is similar to amplitude weighting, or taper, concepts in digital signal processing. One current distribution appearing naturally is the cosine function, as shown in

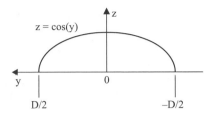

FIGURE 4-7 ■ A Cosine Current Distribution

Figure 4-7. The feedhorns used to radiate the electromagnetic wave into a parabolic reflector (Section 4.2) have antenna patterns shaped like a cosine in amplitude (cosine^2 in power). Dipoles and other elements in array antennas (Section 4.3) also have a similar far-field pattern. What does this do to the far-field antenna gain pattern? We can find out by solving Equation (4-2) when a cosine current distribution is used (Equation 4-22) and then by squaring the result to determine the antenna gain pattern.

$$E(\theta) = \int_{-D/2}^{D/2} \cos(y)\, e^{\left(\frac{2\pi j\, y \sin(\theta)}{\lambda}\right)}\, dy \qquad (4\text{-}22)$$

Solving this integral is left to the reader in Exercise 3 at the end of this chapter. The antenna gain pattern for an antenna with a cosine current distribution is shown in Figure 4-8. The peak of the first sidelobe is 23 dBi down from the mainbeam gain. The lower sidelobe levels come at a price; the half-power beamwidth is about 35% wider, 69 λ/D, than for the uniform current distribution (51 λ/D).

Unfortunately, all sidelobe suppression techniques fatten the mainbeam as well as reduce its peak level. In trying to select a near optimum illumination function, the radar designer must do extensive trade-off analyses, particularly

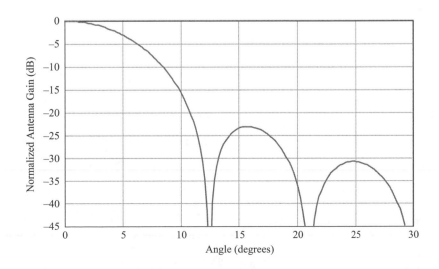

FIGURE 4-8 ■ Normalized Antenna Gain Pattern for a Cosine Current Distribution (One Side)

on the transmit side, because any weighting of the illumination function from uniform reduces the power density (watts/m^2) in the mainbeam by a substantive amount. Table 4-1 shows representative illuminations functions with their mainbeam performance, both half-power beamwidth and relative gain, and the first sidelobe level below peak. A common approximation for the half-power beamwidth in radians is λ/D (57.3 λ/D in degrees), though we can see by Table 4-1 how approximate this is. The beamwidth increase and corresponding gain loss have to be made up elsewhere in the system.

The list in Table 4-1 is not exhaustive; the Dolph-Chebyshev and Taylor illumination functions are two of the other more common ones used for radar antennas [Richards et al., 2010, p. 316; Skolnik, 2001, pp. 621–623]. The main characteristic of these illumination functions is the ability to design to a specific sidelobe level. For a Dolph-Chebyshev illumination function, the peak sidelobe levels are all the same, just like a Chebyshev filter response. For a Taylor illumination function, the peak sidelobe levels are the same for a select number of close-in sidelobes (e.g., the first three or four). The relationships between beamwidth, gain loss, and first sidelobe levels for a Dolph-Chebyshev or Taylor are more complex than those given in Table 4-1.

TABLE 4-1 ■ Characteristics of Different Illumination Functions

Illumination Function	Half-Power Beamwidth (deg)	Gain Loss	First Sidelobe Level (dBi)
Uniform: $A(y) = 1$	51 λ/D	1.000	−13.2
Cosine: $A(y) = \cos^n\left(\frac{\pi}{2}y\right)$			
$\quad n = 0$	51 λ/D	1.000	−13.2
$\quad n = 1$	69 λ/D	0.810	−23.0
$\quad n = 2$	83 λ/D	0.667	−32.0
$\quad n = 3$	95 λ/D	0.575	−40.0
$\quad n = 4$	111 λ/D	0.515	−48.0
Parabolic: $A(y) = 1 - (1 - \Delta)y^2$			
$\quad \Delta = 1$	51 λ/D	1.000	−13.2
$\quad \Delta = 0.8$	53 λ/D	0.994	−15.8
$\quad \Delta = 0.5$	56 λ/D	0.970	−17.1
$\quad \Delta = 0$	66 λ/D	0.833	−20.6
Triangular: $A(y) = 1 - |y|$	73 λ/D	0.750	−26.4
Circular: $A(y) = \sqrt{1 - y^2}$	58.5 λ/D	0.865	−17.6
Cosine^2 on a Pedestal: $0.33 + 0.66 \cos^2\left(\frac{\pi}{2}y\right)$	63 λ/D	0.880	−25.7
$0.08 + 0.92 \cos^2\left(\frac{\pi}{2}y\right)$ Hamming	76.5 λ/D	0.740	−42.8

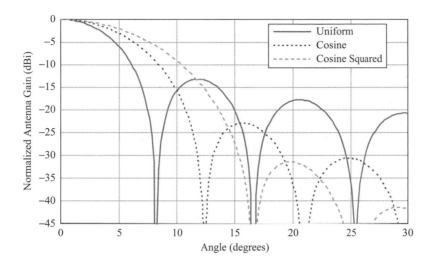

FIGURE 4-9 ◼
Normalized Antenna
Gain Patterns for
Some Illumination
Functions

Figure 4-9 shows the sidelobe behavior of a few of the illumination functions of Table 4-1. Note how the mainbeams of the weighted illumination functions extend into the first sidelobe of the uniform illumination function. Thus, quoting the level of the first sidelobe of various illumination functions alone is misleading; the increase in beamwidth must also be given. This figure clearly shows how the antenna gain pattern achieves a lower sidelobe level at the expense of a wider beamwidth. This essentially follows "conservation of energy" applied to antenna gain patterns. If the sidelobes are reduced, something must increase: the beamwidth.

In general, the mainbeam and first few sidelobes for most actual antenna patterns are very similar to what theory states. However, the farther-out sidelobes become nonsymmetrical and do not fall off in amplitude as quickly due to manufacturing tolerances, installation effects, alignment and calibration issues.

Since an antenna is a resonate device, it has a range of frequencies over which it can transmit and receive signals. There are a few ways of describing the frequency coverage of an antenna: a range, such as 3.0–3.5 GHz; bandwidth, such as 500 MHz; percentage, such as $(3.5 - 3.0)/3.25 = 0.154$ or 15.4%; or ratio, such as $3.5/3.0 = 1.17$ to 1.

4.1.3 Finding the Antenna Gain within the Antenna Gain Pattern

Useful sidelobe-level approximations for radar antennas without detailed calculations have long been sought. The only accurate approximation is that when integrated over 4π steradians the gain of an antenna must be one. One approximation is to define an average sidelobe level for the entire sidelobe region. As we have seen, there are many ways to make sidelobes arbitrarily low by weighting the illumination functions of the antenna; thus, there is no simple way to determine the average sidelobe level other than eyeballing the antenna gain pattern. However, this simple relationship can give insight. Sometimes an approximation will do, but most of the time we need an accurate antenna gain value at a specific angle.

A more detailed, and accurate, approach is used to find the antenna gain at a specific angle in the antenna gain pattern. It is necessary to know only the antenna gain pattern and the mainbeam gain. For example, if the antenna in Figure 4-8 has a mainbeam gain of 32 dBi, the antenna gain at an angle of 10° is 32 dBi – 15 dBi = 17 dBi. Likewise, the antenna gain at an angle of 20° is 32 dBi – 35 dBi = –3 dBi.

4.2 | REFLECTOR ANTENNAS

The classical shape for focusing electromagnetic waves is the parabolic reflector (often referred to as a "dish" antenna). Recall that a parabola, shown in Figure 4-10, has interesting characteristics [Gardner, 1981]. Parallel lines, drawn from a line perpendicular to the axis of the parabola to the parabola and then to its focus, will all be of the same length. Consequently, arriving plane waves will maintain constant phase over this distance. Furthermore, the angle made by the reflected line (with a line tangent to the parabola at that point) is always the same as the angle made by the incident line. If the lines are thought

FIGURE 4-10 ■ A Parabolic Reflector Antenna

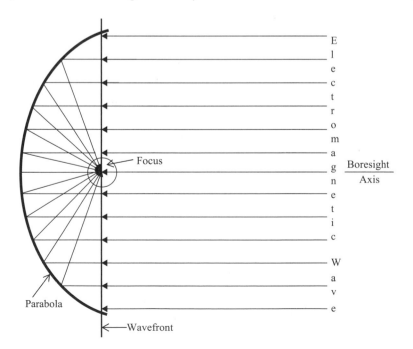

of as rays, then the angle of incidence equals the angle of reflection, which is Snell's law. The application to electromagnetic waves is obvious. All waves arriving parallel to the axis of the parabola and oscillating in phase will add up in phase at the focus; other arriving waves will not add up. In effect, the parabola is focused on infinity. Only waves arriving from a source at infinity (a few miles approximating infinity in practice) are parallel. The same can be said for radiating waves. An electromagnetic wave leaving the focus and reflecting off the parabola will radiate as a plane wave. When extended to a third dimension, the figure becomes a paraboloid. Put a small antenna, such as a feedhorn, at its focus, and we have a parabolic reflector. Mount it on a pedestal allowing it to pivot on two axes, and we have the canonical radar antenna [Richards et al., 2010, Section 9.6; Skolnik, 2001, Section 9.4]. The reflector antenna is mechanically scanned (moved) to allow the antenna pattern to cover the required scan volume—just like we do with a flashlight.

Notice how any segment of the parabola has equivalent characteristics. If only the top one-third of the reflector is illuminated, the feed is offset and does not interfere with the incident radiation. Reflectors with offset feeds are commonplace in radar. The feed might also be embedded in the center of the reflector, radiating its electromagnetic wave outward against a plate at the paraboloid focus, which in turn would reflect it against the reflector. For the geometry to work, the "splash plate" must be a hyperboloid. This arrangement is known as a Cassegrain feed after the optical telescopes of similar design. Of course, these applications require careful design of the feeds, which are antennas themselves.

Reflector antennas have a wide range of shapes including rectangular, circular, and elliptical. Traditionally, reflector antennas have modest antenna efficiencies. Computer-controlled manufacturing has greatly improved the precision of reflector antennas and thus has increased their efficiencies.

4.3 | ARRAY ANTENNAS

An important class of radar systems uses arrays of individual radiating elements instead of reflectors for its antennas [Allen, 1963; Hansen, 1966; Stark, 1974; Richards et al., 2010, Sections 9.7, 9.8; Skolnik, 2001, Section 9.5; Stimson, 1998, Chapters 37, 38]. Array antennas are the distant past, present, and future of radar systems. They are a collection of individual radiating elements: dipoles, slots cut in waveguides, horns, yagis, passive elements, and active elements. Some illumination functions difficult or impossible to achieve with reflector antennas are straightforwardly achieved with an array. The individual elements are usually arranged along a line (linear array), as shown in Figure 4-11, or in a two-dimensional plane (planner array) made by a circle, rectangle, etc. A few arrays have the elements arranged on a geometric surface: cylinder, hemisphere, etc.

FIGURE 4-11 ■ A Linear Array Antenna

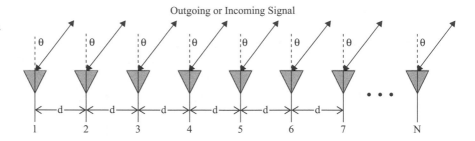

The boresight mainbeam gain for an array antenna is a function of the number of elements, the gain of each element, and the element spacing [Johnson, 1993, Chapter 3]. The variation of the boresight mainbeam antenna gain with element spacing for an array antenna is not expressed by a simple formula. However, when the element spacing is in the region of a half-wavelength or an integer multiple of the wavelength, the boresight mainbeam gain is given in Equation (4-23). Often each element is assumed isotropic (gain of one everywhere). The boresight mainbeam antenna gain is then given in Equation (4-24). Crossed half-wave dipoles, or their equivalent, are often used as array elements. Using the previously developed equation for mainbeam antenna gain (Equation 4-1), the mainbeam antenna gain of a half-wave dipole element on a ground plane is given in Equation (4-25). Thus, the mainbeam antenna gain for an array of N half-wave dipoles on a ground plane is given in Equation (4-26).

$$
\begin{aligned}
&G = 2\,N\,g_e\left(\frac{d}{\lambda}\right) && 1/4 \leq d/\lambda \leq 3/4 \\
&G = N\,g_e && d/\lambda = 1,\,2,\ldots
\end{aligned}
\tag{4-23}
$$

$$
\begin{aligned}
&G = 2\,N\left(\frac{d}{\lambda}\right) && 1/4 \leq d/\lambda \leq 3/4 \\
&G = N && d/\lambda = 1,\,2,\ldots
\end{aligned}
\tag{4-24}
$$

$$
g_x = \frac{4\pi\,A}{\lambda^2} = \frac{4\pi}{\lambda^2}\left(\frac{\lambda}{2}\right)^2 = \pi
\tag{4-25}
$$

$$
\begin{aligned}
&G = 2\,N\,\pi\left(\frac{d}{\lambda}\right) && 1/4 \leq d/\lambda \leq 3/4 \\
&G = N\,\pi && d/\lambda = 1,\,2,\ldots
\end{aligned}
\tag{4-26}
$$

where:

G = Boresight mainbeam antenna gain for an array antenna, no units
N = Number of array elements, no units
g_e = Antenna gain of each element, no units

d = Element spacing, meters

λ = Wavelength, meters

g_x = Antenna gain of a half-wave dipole on a ground plane, no units

The array antenna gain pattern is a function of the pattern of its elements and the array of elements. The pattern of its elements is called the "element factor." The pattern resulting from the array of elements are called the "array factor." The array factor for a uniformly illuminated array of isotropic elements is given in Equation (4-27). The array factor is normalized to have a value of one at zero degrees (along the antenna boresight). When the elements are not isotropic, they have an element pattern or factor; we include it with the array factor to produce the array antenna gain pattern (Equation 4-28). An array antenna gain pattern can be better understood by visualizing it along with the array and element factors, as shown in Figure 4-12 when the element spacing is a half-wavelength.

$$G_a(\theta) = \left[\frac{\sin\left(N\pi\left(\frac{d}{\lambda}\right)\sin(\theta)\right)}{N\sin\left(\pi\left(\frac{d}{\lambda}\right)\sin(\theta)\right)} \right]^2 \tag{4-27}$$

$$G(\theta) = G_e(\theta)\,G_a(\theta) = G_e(\theta) \left[\frac{\sin\left(N\pi\left(\frac{d}{\lambda}\right)\sin(\theta)\right)}{N\sin\left(\pi\left(\frac{d}{\lambda}\right)\sin(\theta)\right)} \right]^2 \tag{4-28}$$

where:

$G_a(\theta)$ = Normalized array factor as a function of angle off boresight (θ) for a uniform current distribution, no units

$G(\theta)$ = Normalized array antenna gain pattern as a function of angle off boresight for a uniform current distribution, no units

$G_e(\theta)$ = Normalized element factor (element antenna gain) as a function of angle off boresight, no units

A key to the array factor is the element spacing with respect to the wavelength (d/λ). Without careful selection of the element spacing relative to the wavelength, maxima in the array factor can be produced at angles other than zero degrees. These other maxima are called grating lobes, and they have the same gain as the mainbeam gain. Grating lobes can be avoided if the element spacing is less than or equal to the wavelength, $d \leq \lambda$. The element factor can be used to minimize the effect of some grating lobes by designing an element factor with a null at the angle of the grating lobe. The relationships between element spacing, element factors, and grating lobes are discussed in more detail in Section 4.3.3.

The half-power and null-to-null beamwidths for an array antenna can be determined using the approach from Section 4.1.1. The half-power and null-to-null beamwidths for an array antenna are given in Equations (4-29) and (4-30), respectively. When the number of elements is large, which it is for a practical

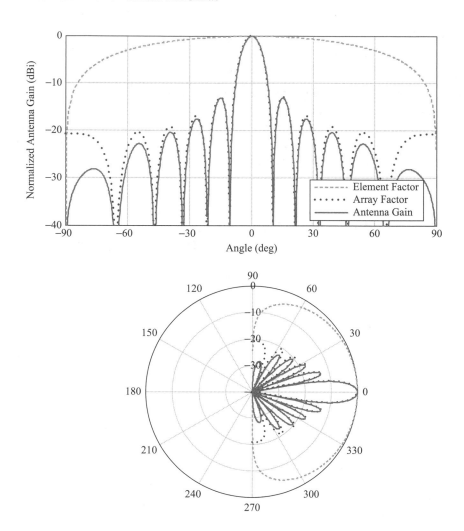

radar array antenna, the first sidelobe is 13.2 dBi below the mainbeam gain (see Section 4.1.2).

$$\theta_{3\,dB} = \frac{51\,\lambda}{N\,d} \quad \text{degrees} \tag{4-29}$$

$$\theta_{nn} = \frac{115\,\lambda}{N\,d} \quad \text{degrees} \tag{4-30}$$

where:

$\theta_{3\,dB}$ = Array half-power (–3 dB) beamwidth for a uniform current distribution, degrees

θ_{nn} = Array null-to-null beamwidth for a uniform current distribution, degrees

4.3.1 Electronically Steered Arrays

Some array antennas can electronically steer the antenna beam in angle. This means the time and mechanical stresses of moving the antenna beam around the scan volume are eliminated. An electronically steered (or scanned) array (ESA), whose beam steering is essentially inertialess, is more complex and capable of less precision than the reflector antenna. Conversely, it is much more cost-effective when the mission requires surveying large solid angles while tracking large numbers of targets and perhaps guiding interceptors as well. Whereas one or more reflector antennas might handle tens of targets simultaneously, if hundreds, or even thousands, of targets are involved, electronic steering is the only practical answer. This is why missions such as space-track, submarine-launched missile warning, multiple target track/engage fire control, and navy battle group defense all use an ESA in various forms.

In somewhat less demanding missions, electronic scanning may be used in one dimension while mechanical scanning is used in the other. Some air traffic control radar systems have this feature, as do some air-to-air and air-to-ground airborne radar systems. Generally, these radar systems electronically scan in the elevation (vertical) plane and mechanically scan in azimuth (horizontal).

Although there are many approaches to achieving electronic beam steering, it must be done by time-delay networks, phasing networks, or a combination of the two. Time-delay steering is the simplest to grasp conceptually. A beam can be formed in almost any direction by adjusting the time at which the electromagnetic wave is permitted to emerge from each different element. In Figure 4-13, a wavefront is formed at an arbitrary angle by delaying by progressive time increments the emission of signals across segments of the one-dimensional array. A beam is formed pointing in the desired angular direction by the relationship between elements given in Equation (4-31).

$$\theta_0 = \sin^{-1}\left(\frac{c\,\Delta t}{d}\right) \qquad (4\text{-}31)$$

where:

θ_0 = Desired mainbeam pointing angle relative to the array boresight, radians or degrees

c = Speed of light, 3×10^8 meters/second

Δt = Time delay between successive array elements, seconds

d = Array element spacing, meters

True time-delay arrays have been built, but consider the complexity: for n_b beam positions, we have $n_b \times \Delta t$ time-delay networks at each element. For an N-element array, we have $N \times n_b \times \Delta t$ time-delay networks. Because for practical radar systems, n_b can be hundreds and N thousands, the result is $>10^5$ time

FIGURE 4-13 ■
Time-Delay Steering

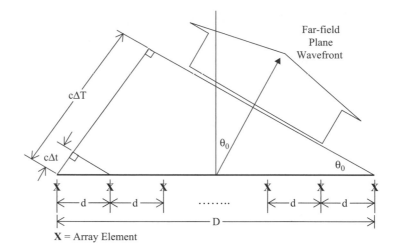

FIGURE 4-14 ■ A
Phase Steered Array

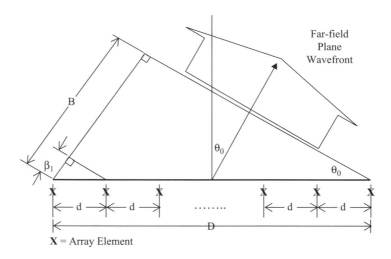

delays. Within any radar pulse, however, only the phase relationships count. Thus, provided the pulse is long compared with the size of the array ($c\tau \gg D$), time-delay networks need not be used, and the phase relationship across the elements determines the beam direction. From Figure 4-14, assuming sinusoidal signals, the i-th element produces a signal $V_i \cos(\omega t + \beta_i)$. A beam is formed pointing in the desired angular direction by the phase relationship given in Equation (4-32). The entire phase progression across the array is given in Equation (4-33).

$$\beta_i = \frac{2\pi \, d_i}{\lambda} \sin\theta_0 \qquad (4\text{-}32)$$

$$B = \sum_{i=1}^{N} \beta_i$$

(4-33)

$$B = \frac{2\pi D}{\lambda} \sin\theta_0$$

where:

β_i = Phase shift at the i-th element, radians
d_i = Distance of the i-th array element from the start of the array, meters
λ = Wavelength, meters
θ_0 = Desired mainbeam pointing angle relative to the array boresight, radians
B = Entire phase progression across the array, radians
N = Number of array elements, no units
D = Dimension of the array, meters

Observe that an equivalent β_i may be achieved for any θ_0 by making λ a variable. The array can now be made to point in an arbitrary direction by changing the frequency. To keep the change in λ to a usable level, say, $\pm5\%$, it is necessary to have the electrical distance between elements greater than 5λ. The ratio of this electrical distance to the physical distance is called the wrap-up factor. Some electronic beam steering is done in just this way and is called frequency scanning. For frequency-scan arrays, the direction the mainbeam points is given in Equation (4-34). If the mainbeam is to be steered over the angular limits $\pm\theta_1$, the necessary wavelength (or frequency excursion) is given in Equation (4-35). This type of electronic beam steering is common in older radar systems, primarily to electronically steer the beam in the elevation (vertical) plane. The advantage of relative simplicity must be weighed against its disadvantage of extremely narrow signal bandwidth (a wideband signal causes steering of the beam). Consequently, several other ways to steer beams electronically have been found.

$$\sin(\theta_0) = \frac{L}{d}\left(1 - \frac{\lambda}{\lambda_0}\right) = \frac{L}{d}\left(1 - \frac{f_0}{f}\right)$$

(4-34)

$$\sin(\theta_1) = \frac{L}{2d}\frac{\Delta\lambda}{\lambda_0} \simeq \frac{L}{2d}\frac{\Delta f}{f_0}$$

(4-35)

where:

θ_0 = Beam steering angle, radians or degrees
L = Electrical distance between array elements, meters
d = Physical distance between array elements, meters
λ = Transmitted wavelength, meters
λ_0 = Wavelength corresponding to the beam pointing at broadside, meters
f_0 = Frequency corresponding to the beam pointing at broadside, hertz

$\mathtt{f} =$ Transmitted frequency, hertz
$\theta_1 =$ Beam steering limits, radians or degrees
$\Delta\lambda =$ Wavelength excursion, meters
$\Delta\mathtt{f} =$ Frequency excursion, hertz

4.3.2 Electronic Beam Steering with Phase Shifters

Currently, the most popular way to do electronic beam steering is with phase shifters. This is why so many electronically steered array antennas are called phased arrays. The phase shift necessary to steer the mainbeam to a desired pointing angle is given by Equation (4-36). With this phase shift applied to each element the signals at the beam steering angle from each element are now in phase rather than coincident in time.

$$\Delta\phi = 2\pi\left(\frac{\mathtt{d}}{\lambda}\right)\sin(\theta_0) \quad \text{radians} \tag{4-36}$$

where:

$\Delta\phi =$ Incremental phase shift between elements necessary to steer the beam to an angle θ_0, radians
$\mathtt{d} =$ Physical distance between array elements, meters
$\lambda =$ Transmitted wavelength, meters
$\theta_0 =$ Beam steering angle, radians or degrees

Diode and ferrite phase shifters are used [Stark, Burns, and Clark, 1970; Temme, 1972; Skolnik, 2001, Section 9.6]. Diode shifters are appealing because they are small, relatively inexpensive, inherently digital, and easy to design and understand. Because so many beam positions and elements contribute to the beam, mixing a relatively few discrete phases at each element can efficiently steer the beam. N two-pole single throw switches (like a diode, which can be turned either on or off) allow for 2^N different phases. This situation is illustrated in Figure 4-15, which shows a two-bit phase-shifting system.

The four different phase shifts available from the two switches are 0°, 90°, 180°, and 270°. The higher the number of different phase shifts, the more

FIGURE 4-15 ■
Concept of a Two-Bit Phase Shifter

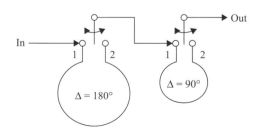

precisely the beam can be steered (Equation 4-37). Two-bit phase shifters are too coarse for practical use; therefore, three-, four-, and five-bit shifters are used. With only three, four, and five discrete delay loops, they can generate 8, 16, and 32 phase variations. Phase shifts of more than 2π radians are redundant. (Mathematicians say that phase shift amounts are "modulo 2π"; that is, a phase shift of $2n\pi + \Delta$ is not different from one of Δ [n an integer].) These combinations are used between 0° and 360° of phase shift. A three-bit phase shifter might have 45°, 90°, and 180° increments, thereby allowing the following combinations: 0°, 45°, 90°, 135°, 180°, 225°, 270°, and 315°.

$$\Delta\theta_0 = \frac{\pi}{4}\frac{1}{2^B}\theta_{3dB} \qquad (4\text{-}37)$$

where:

$\Delta\theta_0$ = Maximum beam steering error, radians or degrees
B = Number of different phase shifts, no units
θ_{3dB} = Array half-power beamwidth, radians or degrees

The same combinations could be achieved less efficiently with seven 45° phase shifts. A four-bit phase shifter, with sixteen available combinations using 22.5° increments, or five-bits with 11.25° increments can be used. If variable phase shifters are used, the beam can be steered as the relative phase between elements is finely changed. Of course, phase-shifter beam positioning will not be done exactly as described. For example, randomization at the points of discontinuity along the phase gradient is essential to keep the sidelobes smooth. Also, each phase shifter has a slight random phase-shift error. For large radar antennas, designers tend to think of three-bit phase shifters as not enough and four-bit shifters as too much.

The antenna beam can be pointed in any direction within the element factor angular coverage. Dynamically steering a large array to thousands of beam positions per second requires the allocation of considerable computing power, which is no longer a problem in the modern world of computational plenty. Adaptively steering the beam based on operational modes/requirements and currently detected and/or tracked targets is very common with radar systems using an active electronically steered array (AESA) (see Section 7.4). For deterministic and/or fixed scan patterns, the phase shifts used for the desired beam positions can be precomputed, stored in memory, and read out as needed.

The normalized (along the antenna boresight) array factor with beam steering for a uniformly illuminated array of isotropic elements is given in Equation (4-38). When the elements are not isotropic, they have an element factor; we include it with the array factor to produce the array antenna gain pattern (Equation 4-39). An array antenna gain pattern with beam steering can be better understood by visualizing it along with the array and element factors, as shown in Figure 4-16 when the element spacing is a half-wavelength.

FIGURE 4-16 ▪
Element Factor,
Array Factor, and
Resultant Antenna
Gain Pattern: Beam
Steered to 50°

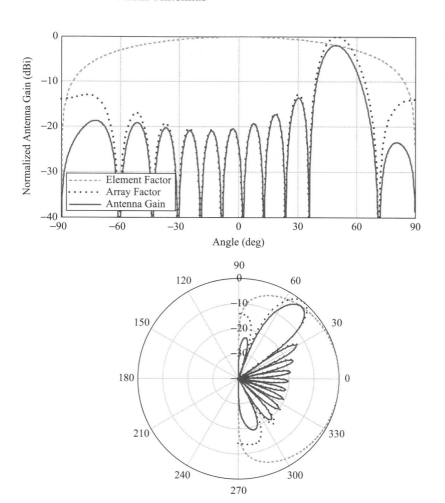

$$G_a(\theta) = \left[\frac{\sin\left(N\pi\left(\frac{d}{\lambda}\right)\left(\sin(\theta) - \sin(\theta_0)\right)\right)}{N\sin\left(\pi\left(\frac{d}{\lambda}\right)\left(\sin(\theta) - \sin(\theta_0)\right)\right)} \right]^2 \qquad (4\text{-}38)$$

$$G(\theta) = G_e(\theta)\,G_a(\theta) = G_e(\theta)\left[\frac{\sin\left(N\pi\left(\frac{d}{\lambda}\right)\left(\sin(\theta) - \sin(\theta_0)\right)\right)}{N\sin\left(\pi\left(\frac{d}{\lambda}\right)\left(\sin(\theta) - \sin(\theta_0)\right)\right)} \right]^2 \qquad (4\text{-}39)$$

where:

$G_a(\theta)$ = Normalized array factor with the beam steered to θ_0 as a function of angle off boresight (θ) for a uniform current distribution, no units

N = Number of elements, no units

      D = Element spacing, meters

      λ = Wavelength, meters

      θ_0 = Beam steering angle, radians or degrees

    G(θ) = Normalized array antenna gain pattern with the beam steered to θ_0 as a function of angle off boresight for a uniform current distribution, no units

  G_e(θ) = Normalized element factor (element antenna gain) as a function of angle off boresight, no units

Just like with element spacing with respect to the wavelength (d/λ), beam steering can also generate grating lobes. A key ESA design trade-off is between element spacing and beam steering limits, all without creating grating lobes. The element factor can be used to minimize the effect of some grating lobes by designing it with a null at the angle of the grating lobe. The relationships between element spacing, element factors, beam steering angle, and grating lobes are discussed in more detail in Section 4.3.3.

Given appropriate element spacing, array antennas obey all the general rules for antenna design I have developed for reflectors. The main difference to bear in mind is that the projected area of the antenna decreases as the beam is steered off boresight. At 60° off boresight the aperture size is down to one-half its maximum, $\cos(60°) = 0.5$. Because the projected area of the array decreases as the beam is steered, the antenna mainbeam gain decreases, and the half-power beamwidth increases. The projected area of the array decreases proportionally to the cosine of the beam steering angle, and thus the mainbeam array antenna gain decreases in the same manner (Equation 4-40). For example, when steered 60°, the mainbeam gain of the array is half (–3 dB) the boresight gain. The beamwidth increases approximately inversely proportional to the cosine of the beam steering angle (Equation 4-41). So when steered 60°, the beamwidth of the array is twice the boresight beamwidth. This approximation is acceptable when the beam is not steered too far off broadside, within several beamwidths of the scan limits. An exact equation for beamwidth as a function of beam steering angle is quite complex.

$$G(\theta_0) = G\cos(\theta_0) \qquad (4\text{-}40)$$

$$\theta_{3\,dB}(\theta_0) \approx \frac{\theta_{3\,dB}}{\cos(\theta_0)} \qquad (4\text{-}41)$$

where:

    G(θ_0) = Mainbeam array antenna gain as a function of beam steering angle θ_0, no units

        G = Mainbeam array antenna gain with the beam pointing at boresight, as in Equations (4-23) through (4-26), no units

$\theta_{3dB}(\theta_0)$ = Array antenna half-power (−3 dB) beamwidth as a function of beam steering angle θ_0, radians or degrees

θ_{3dB} = Array antenna half-power beamwidth with the beam pointing at boresight, radians or degrees

The overall electronically steered angular field of regard (FOR) is limited by array theory and array characteristics. The overall angular FOR is also called the "scan limits." For most practical radar array antennas the FOR is typically ±40° to ±60° off boresight. The array antenna pattern and characteristics: mainbeam gain, beamwidth, sidelobe levels, and symmetry change with the beam steering angle, as shown in Figure 4-17, with an element spacing of a half-wavelength. These changes are most pronounced as the beam steering angle increases. Many modern phased array radar systems have the ability to increase the integration time (see Section 3.3) to compensate for the loss in antenna gain with beam steering angle and associated loss in detection range. To provide a required FOR and array antenna pattern, combinations of mechanical and electronic steering are often used. Ground-based radar systems often use mechanical in azimuth and electronic in elevation, whereas airborne radar systems incorporate a tilted mechanical roll axis with electronic in azimuth and elevation. As I have said a few times before, radar is one big systems engineering problem.

4.3.3 Grating Lobes Based on Element Spacing and Beam Steering

For all the previously discussed array designs, element spacing is assumed to be a half-wavelength ($\lambda/2$). Closer spacing is typically not cost- or performance-efficient (although arrays with close spacing, called "super gain arrays," have been designed). Distant spacing causes grating lobes (high gain lobes) to appear in the antenna pattern at various angles. The expression "grating lobes" comes from the diffraction gratings of physics, which are analogous but at much higher frequencies (6×10^{13} Hz vs. $\sim 10^9$ Hz). For example, for an element spacing of 1λ, when the array beam is pointing on boresight (all elements have the same phase), grating lobes pointing at ±90° appear because phases add up in those directions as well. For spacing of $3\lambda/4$, grating lobes can begin to form as the amplitude of the farther-out sidelobes increases. An example of array antenna patterns for element spacing and the resultant grating lobes are shown in Figure 4-18. To avoid grating lobes, the element spacing for a fixed-beam broadside array is normally less than one wavelength.

For an ESA, grating lobes will appear whenever the denominator of Equation (4-39) equals zero. The grating lobes are formed by a combination of the off-boresight steering angle and the element separation (Equation 4-42); remember that $\sin(n\pi) = 0$. The maximum element physical separation to steer

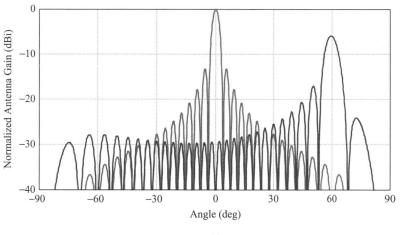

FIGURE 4-17 ▪
Array Antenna Gain
Pattern and
Characteristics
Change with Beam
Steering Angle

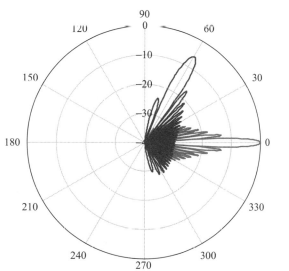

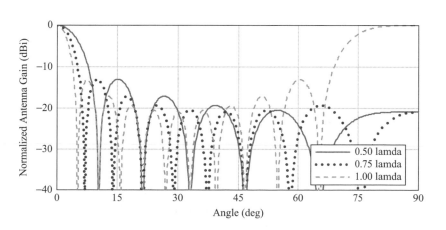

FIGURE 4-18 ▪
Array Factor Grating
Lobes as a Function
of Element Spacing

the beam to a desired angle without generating grating lobes is given in Equation (4-43). As seen in this equation, the greater the maximum beam steering angle the closer together the elements must be. For example, if the maximum beam steering angle is 30° the maximum element spacing is 0.67 λ. While if the maximum beam steering angle is increased to 60°, the maximum element spacing is 0.54 λ.

$$\pi \left(\frac{d}{\lambda}\right)\left(\sin(\theta_g) - \sin(\theta_0)\right) = \pm n\,\pi$$

$$\left|\sin(\theta_g) - \sin(\theta_0)\right| = n\frac{\lambda}{d} \tag{4-42}$$

$$d_{max} = \frac{\lambda}{1 + \sin(\theta_0)} \tag{4-43}$$

where:

   d = Element physical separation, meters
   λ = Wavelength, meters
   θ_g = Angle at which a grating lobe will appear, radians or degrees
   θ_0 = Beam steering angle, radians or degrees
   n = Integer, n = 0, 1, 2, ...
d_{max} = Maximum element physical separation to steer the beam to θ_0 without any grating lobes, meters

Consequently, the radar designer selects the element spacing and maximum beam steering angle so no grating lobes are formed during any part of the radar's beam-steering schedule. An example of array antenna patterns with λ/2 element spacing for different beam steering angles and the resultant grating lobes are shown in Figure 4-19. For most surveillance radar systems, beam

FIGURE 4-19 ■
Array Factor Grating Lobes as a Function of Beam Steering Angle

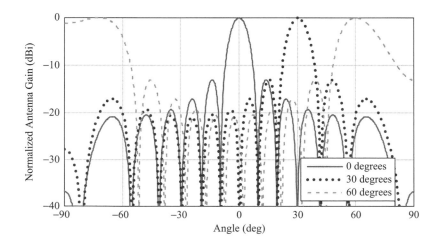

steering of about $\pm60°$ is provided. The accompanying element separation is 0.54 λ. However, the element spacing cannot be too small or the boresight mainbeam antenna gain will suffer (Equation 4-23). Thus, the actual value for the element spacing is usually a careful compromise between controlling grating lobes, boresight mainbeam antenna gain, cost, etc.

Grating lobes can also be controlled by the element gain pattern. By placing nulls of the element gain pattern at the angle of the grating lobe, usually $\pm90°$, the effect of the grating lobe can be minimized. This also allows for correspondingly larger element separations. Although a big deal is often made of grating lobes, a well-designed array does not exhibit grating lobes over its designed scan limits. If grating lobes were a problem, the radar's performance would certainly suffer.

4.3.4 Array Thinning and Sidelobes

A key question about array antennas is, "How does one control sidelobe levels?" Although the same principles apply to arrays as applied to reflectors—amplitude weighting of the illumination function is required—the mechanisms for accomplishing it are entirely different. With reflector antennas, weighting of antenna illumination is hard to avoid, as is weighting of the signal as it scans across the target. Arrays are harder to weight. On transmit, it is generally not feasible to send differing levels of transmit power to the individual elements, changing those levels each few milliseconds as the beam scans over its surveillance and tracking program. On receive, it is wasteful to design some receivers to be less sensitive than others.

However, several practical steps can be taken. On transmit, the amount of power sent to blocks of elements can be adjusted, dividing the transmit array into several subarrays and tapering the power to these. On transmit and receive, the elements can simply be physically thinned; that is, the elements at the outer edges spread out more to give the effect of an amplitude taper. As long as some randomization is done so discontinuities (and the accompanying sidelobe spikes) are avoided, array thinning works. Control of the illumination function is excellent, much better than can be achieved with a reflector (unless an array is used to feed it), and very sophisticated weighting functions can be implemented.

All other things being equal, analysis usually recommends that the transmit array be fully filled and the receive array be tapered. However, when the costs of the separate transmit and receive apertures are taken into consideration, initial costs tend to outweigh improved long-term operation. Even in a single transmit/receive (T/R) aperture, however, it is feasible to fill a smaller transmit aperture fully and thinly fill the receive aperture to give good angle resolution (several receive beams following the transmitter beam around) and good sidelobe characteristics.

What constitutes optimum design also depends strongly on the mission. A surveillance mission allows a wide transmit beam and encourages a large receive aperture (surveillance is primarily driven by adequate signal-to-noise ratio (S/N) for detection at long ranges; see Chapter 3). A tracking mission is primarily driven by narrow beams for fine angle tracking performance more so than S/N. Tracking performance improves directly as antenna mainbeams get narrower, but only by the square root of an improving S/N (see Chapter 5). Moreover, of course, a combined tracker–scanner would be a complicated compromise.

4.3.5 Array Design Considerations

The preceding discussions have emphasized the array of radiating elements, but the techniques for driving those elements with the necessary power are also vital. Figure 4-20 shows several ways of feeding the radiating elements of an array. Each has pros and cons.

The amplifier-per-element approach, often called an active array, has two main advantages: the phase shifting can be performed at low power; and any loss in power in the phase shifters can be made up by the subsequent amplifier. Generally low-power devices tend to have a high bandwidth. The radar system degrades gracefully because random failures of up to half the amplifiers may

FIGURE 4-20 ■
Array Configurations

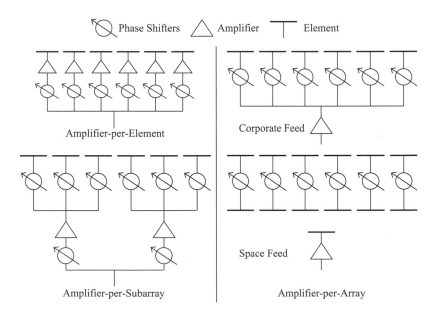

not appreciably affect performance. Solid-state amplifiers operating for 100,000 hours without failing are readily available [Brookner, 1985]. However, the thousands of separate amplifiers needed for even a modestly sized system increase cost, complexity, heat dissipation, etc. Current efforts to build a complete element (amplifier, phase shifters, and antenna) on a single chip—called a monolithic microwave integrated circuit (MMIC)—are providing a revolution in array design. A form of such an array antenna includes a transmit/receive (T/R) module at each element, resulting in an AESA, which can provide additional flexibility for some radar systems (see Section 7.4).

The amplifier-per-array approach, often called a passive array or passive electronically steered array (PESA), allows the entire antenna to be driven by a single high-power tube (klystrons or traveling wave tubes), via a network of waveguides or coaxial cables ("corporate feed") or simply across empty space ("space feed" or "optical feed"). The losses in the feed system and in phase shifting are not made up prior to radiation and the phase shifting is done at high power.

In large arrays, several high-power tubes may be required. These could all be hooked in parallel behind a single feed, or each tube could be associated with a number of elements, creating several subarrays. A system made up of several subarrays is called an array-of-subarrays. The total number of phase shifts can now be reduced by doing coarse steering with the subarrays and fine-grained steering at the elements. If the array dimensions are large compared with the pulse widths to be radiated, the radiated pulses are smeared when the antenna is pointed off boresight. To reduce smearing, the subarrays can be steered with time-delay networks rather than phase shifters.

4.4 | SUMMARY

All radar antenna types—reflectors, arrays, or reflectors illuminated by an array—have the same end result: an antenna gain pattern. The antenna gain pattern is the antenna gain as a function of angle: azimuth, elevation, or off-axis. Thus, the antenna gain is not a single value. All radar antennas can be characterized by their mainbeam (maximum) gain, beamwidth, and sidelobe levels. These antenna characteristics are a function of frequency, current distribution, and antenna physical dimensions. Many antenna rules-of-thumb are left over from the dark ages of calculation. Fortunately, many of them have disappeared over time, but some still exist, either explicitly or embedded into other areas of radar and the associated equations. Therefore, it is important to be careful about applying them in a particular situation.

▇▇▇ 4.5 | EXERCISES

4-1. You are asked to design an antenna for an infrared radar (wavelength about 20 microns [micrometers]) with a gain $G = 10^6$ (60 dBi). Assume the antenna efficiency $\rho = 100\%$. Assuming a square antenna shape, what are the dimensions of the aperture? What size would it be in the microwave region, say, at S band ($f_c = 3000$ MHz)?

4-2. A reflector antenna has an area $A = 100$ m². How many half-wave dipole elements would there be in a fully filled ultra high frequency (UHF) ($\lambda \approx 1$ meter) array of equivalent capability? What would be its area?

4-3. Assume cosine weighting of a linear aperture of dimension $-\pi/2$ to $+\pi/2$. Show $G(\theta)$ is of the form

$$\left[\frac{\sin(1 - \theta)^{\pi}/2}{(1 - \theta)^{\pi}/2} + \frac{\sin(1 + \theta)^{\pi}/2}{(1 + \theta)^{\pi}/2} \right]$$

Calculate sidelobes at $\theta = 4°$, $6°$, $8°$; square them and turn them into decibels to get –23, –30, and –36 dBi. (Hint: Use the identity $\cos x = \frac{1}{2}(e^{jx} + e^{-jx})$ to get the expression into integrable form.)

4-4. Calculate the mainbeam gain and the azimuth and elevation beamwidths for a radar antenna with the following characteristics: Uniform illumination function; 5 meters wide; 2 meters high; antenna efficiency $\rho = 0.5$; and a frequency $f_c = 3$ GHz.

4-5. Calculate the normalized gain of an antenna at an angle of 5° for an antenna with a uniform current distribution, dimension $D = 3$ meters, and a frequency $f_c = 5$ GHz. If the mainbeam gain $G = 35$ dBi, what is the gain of the antenna at an angle of 5°?

4-6. A phase steered array has the following characteristics: frequency $f_c = 6$ GHz; element spacing $d = \lambda/2$; and phase shift between elements $\Delta = 15°$. What angle is the beam steered to?

4-7. A frequency scan array has a transmit frequency $f_c = 2$ GHz, a frequency excursion $\Delta f = 100$ MHz, element spacing $d = \lambda/2$, and electrical distance between elements $L = 5\lambda$. What is the beam steering limit, θ_1 (digress)? What is the wrap-up factor, L/d, needed to provide a beam steering limit $\theta_1 = \pm 60°$?

4-8. An array antenna using three-bit phase shifters and has a beamwidth $\theta_{3dB} = 2°$. How precisely can the beam be steered?

4-9. Calculate the beamwidth of an electronically steered array with a beamwidth $\theta_{3dB} = 2.5°$ when pointed at 0°, when it is steered to 25°, 30°, and 45°.

4-10. Calculate the mainbeam antenna gain (absolute and dBi) for an electronically steered array with the following characteristics: number of elements N = 1000 and element gain g = 1.5 dBi, when it is steered to 25°, 30°, and 45°.

4.6 | REFERENCES

Allen, J. L., 1963, *The Theory of Array Antennas*, Lexington, MA: Lincoln Laboratory Tech. Report #323.

Blake, Lamont V., 1986, *Radar Range-Performance Analysis*, Norwood, MA: Artech House.
Discusses the concept and specifics of antenna efficiency on pp. 11, 171, 361.

Brookner, E., 1985, "Phased Array Radars", *Scientific American*, Volume 252, no. 2, Feb. 1985, p. 100.

Gardner, M., 1981, "Mathematical Games", *Scientific American*, Volume 245, no. 2, Aug. 1981, pp. 16–26.

Hansen, R. C., 1966, *Microwave Scanning Antennas*, Volume 2. New York: Academic Press.

Johnson, Richard C., (editor), 1993, *Antenna Engineering Handbook*, New York: McGraw-Hill.
Chapter 3 – Arrays of Discrete Elements by Mark T. Ma provides a detailed discussion of array antenna concepts and the supporting math.

Richards, M. A., Sheer, J. A., and Holm, W. A. (editors), 2010, *Principles of Modern Radar*, Volume 1: Basic Principles. Raleigh, NC: SciTech Publishing.
The fundamentals of radar antennas are in Chapter 1. Chapter 9 covers antenna details such as patterns, reflectors, and arrays.

Skolnik, Merrill I., 2001, *Introduction to Radar Systems*, Third Edition. New York: McGraw-Hill.
Has a great deal of additional information on both reflector and array antennas in Chapter 9, The Radar Antenna.

Stark, L., 1974, "Microwave Theory of Phased Array Antennas-A Review," *Proc. IEEE*, Volume 62, Dec., pp. 1661–1701.

Stark, L., Burns, R. W., and Clark, W. P., 1970, "Phase Shifters for Arrays," in M. I. Skolnik (ed.), *Radar Handbook*, New York: McGraw-Hill.

Stimson, George W., 1998, *Introduction to Airborne Radar*, 2nd Edition. Raleigh, NC: SciTech Publishing.
Has a great deal of additional information on array antennas in Chapters 37 and 38.

Temme, D. H., 1972, "Diode and Ferrite Phase Shifter Technology," in A. A. Olner and G. H. Knittel (editors), *Phased Array Antennas*. Dedham, MA: Artech House.

Radar Measurements and Target Tracking

HIGHLIGHTS

- Analyzing the radar waveform in both time and frequency domains
- Matched filtering to optimize signal-to-noise
- Analyzing a pulse burst waveform in both time and frequency domains
- Measuring radar-to-target range: how, resolution, ambiguities, and tracking
- Pulse compression: achieving very fine range resolution through modulation of the radar carrier frequency
- Measuring radar-to-target range rate: how, resolution, ambiguities, and tracking
- Measuring radar-to-target angle: how, resolution, and tracking
- Measurement loose ends: accuracy, altitude, resolution cell, range and Doppler sidelobes, and ambiguity functions
- Sophisticated target trackers

A measurement is a characteristic of a detected target as determined by the radar system: range, range rate, azimuth angle, and elevation angle. Radar systems are often categorized based on the number or type of measurements they provide. Two-dimensional (2D) radar systems provide range or range rate and one angle (azimuth or elevation). Three-dimensional (3D) radar systems provide range or range rate, azimuth angle, and elevation angle.

Target tracking is the following of a time sequence of target measurements. The time sequence of measurements can be continuous for a target tracking radar system. Target tracking radar systems use measurement trackers to independently track in range, range rate, and angle. The time sequence of measurements can be near-continuous for a track-while-scan radar system. For a search radar system the scan period determines the time sequence of target measurements.

5.1 | RADAR WAVEFORMS: TIME AND FREQUENCY DOMAINS

Waveforms are the tools providing the radar systems the means for accomplishing its mission. To make the most out of the minute amounts of electromagnetic power returning to a radar system from far-away targets depends on intelligent design of radar waveforms: carrier frequency, pulse width, duration, and pulse repetition interval (PRI) or pulse repetition frequency (PRF) (see Chapters 2 and 3). Of prime importance is to extract the received signal from noise (detection), but the waveform must be carefully designed to yield the information (measurements) needed. An infinite number of waveform designs are possible, although the practical designs number in the thousands. However, the essentials can be grasped with only a few illustrations, beginning simply and building on the specifics.

5.1.1 Characteristics of the Simple Pulse and Matched Filter Theory

A simple radar system may be used to measure only range and angle. A simple short pulse is adequate for this kind of measurement. As the radar becomes more sophisticated, its waveforms will increase in complexity. Federal Aviation Administration air traffic control surveillance and approach control radar systems may have relatively simple waveforms, whereas those used to support complex weapons systems in the defense department have complex waveforms. Fortunately, the principles governing even the most advanced waveforms are relatively simple and can be developed from features of the simple pulse.

Consider a simple pulse composed of a sine wave, interrupted by turning a switch on and off. It has a duration τ and an amplitude V, as shown in Figure 5-1. We use a voltage pulse here as it allows for somewhat simpler math. This pulse propagates out to the target, scatters off it, and returns to the radar system

FIGURE 5-1 ■ A Simple Pulse—Time Domain, Voltage

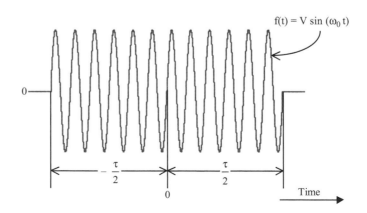

(see Chapter 2). The received signal has been reduced in amplitude enormously, by ten to thirty orders of magnitude (factors of 10) in many cases of interest. Heroic methods are necessary to detect the signal, embedded as it is in the receiver thermal noise (see Chapter 3).

We can evaluate the simple voltage pulse of Figure 5-1 for its power spectral content (spectrum) by squaring its Fourier transform. The rigorous way to do this is to take the Fourier transform of $f(t) = V\sin(\omega_o t)$ between $-\tau/2$ and $\tau/2$. Richards et al. [2010, Chapter 8] and Stimson [1998, Chapter 17] show this Fourier transform in all its rigorous detail. This will yield an expression with $\sin(x)/x$ shapes around both $-\omega_o$ and $+\omega_o$. Because we know in radar applications ω_o is on the order of 10^9 hertz and ω's of interest are 10% or less, we can simply take the Fourier transform of the envelope and remember when $\omega = 0$ in the result, it is really at ω_o. Our problem then reduces to taking the Fourier transform of the envelope (Equation 5-1), which results in Equation (5-2). Using the trig identity for sinc (Equation 5-3), we get the voltage spectrum given in Equation (5-4). Squaring the voltage spectrum results in the power spectrum (Equation 5-5).

$$f(t) = V \qquad -\frac{\tau}{2} < V < \frac{\tau}{2} \qquad (5\text{-}1)$$

$$g(\omega) = \frac{1}{2\pi} \int_{-\frac{\tau}{2}}^{\frac{\tau}{2}} V e^{-j\omega t}\, dt \qquad (5\text{-}2)$$

$$\sin(x) = \frac{e^{jx} - e^{-jx}}{2j} \qquad (5\text{-}3)$$

$$g(\omega) = \frac{V\tau}{2\pi} \frac{\sin\left(\omega\frac{\tau}{2}\right)}{\omega\frac{\tau}{2}} \qquad (5\text{-}4)$$

$$G(\omega) = \left(\frac{V\tau}{2\pi} \frac{\sin\left(\omega\frac{\tau}{2}\right)}{\omega\frac{\tau}{2}}\right)^2 \qquad (5\text{-}5)$$

where:

$f(t)$ = Time domain representation of the voltage pulse envelope, volts
V = Pulse amplitude, volts
τ = Pulse duration, seconds
$g(\omega)$ = Frequency spectrum of the pulse, volts
ω_o = Angular frequency of the pulse, radians/second
$2\pi f_c$, where f_c = carrier frequency, hertz
$G(\omega)$ = Frequency spectrum of the pulse, watts

Figure 5-2 is a plot of the power spectrum of the pulsed signal with the mean receiver thermal noise power (see Section 2.3.1) also shown. As discussed in Section 2.3.1, the question arises as to how to determine the bandwidth of the radar receiver. As seen in Figure 5-2, as the bandwidth of the receiver is increased outward from ω_o to ω, signal is added more rapidly than noise for a short distance, after which noise is added more rapidly. It follows that the radar receiver wants to admit only the portion of the signal in which gains over noise can be made. A filter whose spectral response is identical to the transmitted pulse will pick up the entire signal if it is not distorted in transit and if the noise really has a constant root mean square value across the bandwidth (Gaussian white noise). A filter having this characteristic, and the added one of its phase response being the conjugate of the transmit pulse, is called a "North" [North, 1943] or "matched" filter. A matched filter can be shown to be an optimum filter against the type of noise described; it maximizes the received target signal power relative to the receiver thermal noise power. Mathematical developments of this are frequent [Deley, 1964, 1970; Richards et al., 2010, pp. 538–543]. The bandwidth of the matched filter is given in Equation (5-6).

$$B_R = \frac{1}{\tau} \qquad (5\text{-}6)$$

where:

B_R = Radar receiver matched filter bandwidth, hertz

In practice, a conjugate filter is unnecessary, and simple third- to fourth-order bandpass filters can be used as matched filters. Notice that their ideal cutoff frequency can be found by solving for the point where the rate of change

FIGURE 5-2 ■
Spectral Plot of
Signal and Noise
Power, Log Scale

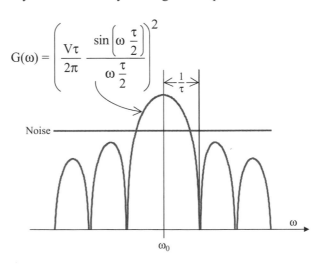

$$G(\omega) = \left(\frac{V\tau}{2\pi} \frac{\sin\left(\omega\frac{\tau}{2}\right)}{\omega\frac{\tau}{2}} \right)^2$$

of integrated signal power crosses over the rate of change of integrated noise power. Because the integrand is the derivative of the integral in this case, the point is trivial to find, as given in Equation (5-7). The value of receiver thermal noise depends on the expected operating point of the radar system. A simple filter such as this has excellent qualities for suppressing noise beyond its passband. As a result, the filter bandwidths of many radar systems range in frequency as given in Equation (5-8) [Barton, 1979, p. 20].

$$\left(\frac{V \tau}{2\pi} \frac{\sin\left(\omega \frac{\tau}{2}\right)}{\omega \frac{\tau}{2}} \right)^2 = N \qquad (5\text{-}7)$$

$$\frac{0.8}{\tau} < B_R < \frac{1.2}{\tau} \qquad (5\text{-}8)$$

where:

N = Radar receiver thermal noise power, watts

5.1.2 Characteristics of a Pulse Burst Waveform

The overwhelming majority of radar systems use multiple-pulse, or pulse burst, waveforms to provide improved detection performance through integration (see Section 3.3). Pulse-Doppler radar systems use pulse burst waveforms to provide target range rate measurements (see Section 5.3). The pulse burst waveform can be tailored to the precise needs of the radar system using them. To see how this is done, it is necessary to understand a pulse burst waveform in both the time and frequency domains.

For radar systems using a noncoherent (the phase relationship varies from pulse to pulse and is unknown in the receiver) pulse burst waveform, the spectrum is the same as a simple pulse [Stimson, 1998, Chapter 17]. This spectrum was previously shown in Figure 5-2.

For radar systems using a coherent (the phase relationship is constant from pulse to pulse or known in the receiver) pulse burst waveform, the spectrum is determined by taking the Fourier transform of the coherent pulse burst waveform. Richards et al. [2010, Chapter 8] and Stimson [1998, Chapter 17] show this Fourier transform in all its rigorous detail. A pulse burst waveform in the time domain and its Fourier transform (spectrum) are shown in Figure 5-3. Often the integration time (T_I), the time over which multiple pulses are integrated from Chapter 3, is called the pulse burst duration or the coherent processing interval (CPI). Notice how the spectrum of a coherent pulse burst waveform is a sequence of spectral lines at positive and negative integer multiples of the PRF centered about the carrier frequency. To further visualize this relationship, Figure 5-4 shows the time and frequency domains for two

FIGURE 5-3 ■ A Coherent Pulse Burst Waveform and Its Spectrum

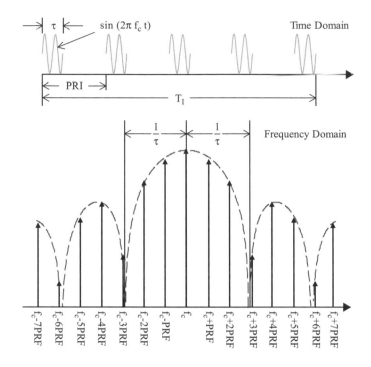

coherent pulse burst waveforms: one with a higher PRF (top) and one with a lower PRF (bottom). The spectrums for these two waveforms are identical except for the spacing of the spectral lines.

A pulse burst waveform is essential for pulse-Doppler, moving target indicator (MTI), radar imaging, and ballistic missile defense radar systems, but they are not an unvarnished blessing. With them come the problems of blind ranges and speeds and range and range rate ambiguities, which will result in some targets not being detected and/or ambiguous measurements made by the radar system.

5.1.3 Other Characteristics of the Radar Waveform

A few other characteristics of the radar waveform are used in alternative forms of the radar equation (see Chapters 2 and 3) and in the mathematical relationships for detection theory (see Chapter 3): average power, energy, and voltage. The average power over one pulse repetition interval is the product of the peak power and ratio of the time the pulse is "on" to the PRI, as given in Equation (5-9) and shown in Figure 5-5. The ratio of the time the pulse is on to the PRI is called the transmit duty cycle (Equation 5-10). The energy in one pulse is the product of its peak power and duration (Equation 5-11). The relationship between power

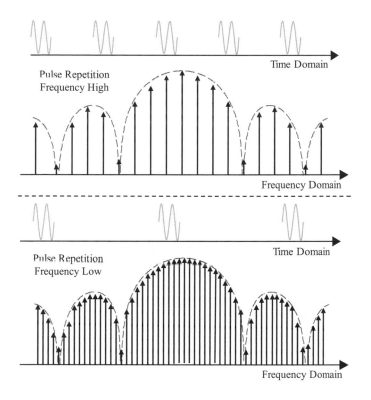

FIGURE 5-4 ■ A Coherent Pulse Burst Waveform: Time and Frequency Domain Examples

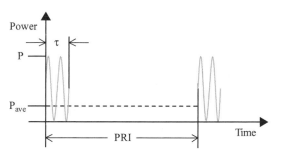

FIGURE 5-5 ■ Average Power over One Pulse Repetition Interval

and voltage is given by Equation (5-12). These power-voltage equations assume impedance matching or an impedance of one (1) ohm.

$$P_{ave} = P \frac{\tau}{PRI} = P \tau \, PRF = P \, d_t \qquad (5\text{-}9)$$

$$d_t = \frac{\tau}{PRI} = \tau \, PRF \qquad (5\text{-}10)$$

$$E = P \tau \qquad (5\text{-}11)$$

$$P = V^2 \qquad V = \sqrt{P} \qquad (5\text{-}12)$$

where:

P_{ave} = Average transmit power, watts
P = Peak transmit power, watts
τ = Transmitted pulse duration, or pulse width, or pulse length, seconds
PRI = Pulse repetition interval, seconds
PRF = Pulse repetition frequency, hertz
d_t = Transmit duty cycle, no units
E = Energy in one pulse, watt-seconds or joules
V = Voltage, volts

5.2 | RANGE MEASUREMENT

A radar system determines the radar-to-target range by measuring the elapsed time, or time delay, from when a pulse was transmitted and when the reflected pulse is received (and of course detected). If you recall from Chapter 2, distance equals speed times time (Equation 5-13). The time delay, and thus range, can be measured with automatic circuits in the radar receiver.

$$R + R = c\,\Delta t \Rightarrow R = \frac{c\,\Delta t}{2} \qquad (5\text{-}13)$$

where:

R = Radar-to-target slant range, meters
Δt = Time delay from when a pulse is transmitted and when the reflected pulse is received, seconds
c = Speed of light, 3×10^8 meters/second

The time delay, and thus range, can also be measured using range gates, or range bins, as shown in Figure 5-6. As shown in this figure, the time bins are usually one pulse width wide. The received pulse will arrive in a specific range gate based on its associated time delay, as shown in Figure 5-7. Rarely will it fall perfectly in one range gate, but usually it straddles two range gates, top and

FIGURE 5-6 ■
Range Gates

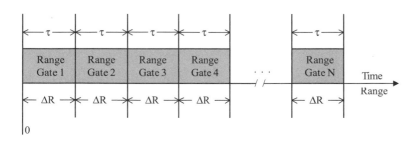

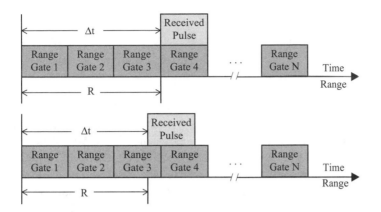

FIGURE 5-7 ■
Received Pulse in
the Range Gates

bottom of Figure 5-7, respectively. The range gate containing the received pulse has the strongest output signal (i.e., highest S/N); the other range gates contain only receiver noise. A threshold detection (see Chapter 3) stage follows each range gate, thereby providing the range measurement associated with the detection.

5.2.1 Range Resolution

A key question about range measurement is: "How far apart in range do targets have to be before they can be resolved as unique targets?" To answer this, it is first necessary to establish some criterion for resolution. When a radar system using a matched filter operates on a single pulse in the time domain, operations occur as shown in Figure 5-8. It is apparent that objects separated by more than two pulse widths are absolutely resolvable. In fact, even if noise contaminates

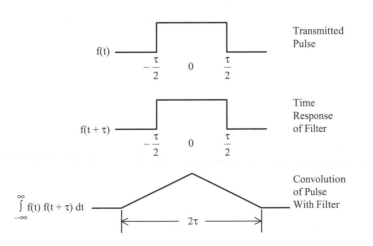

FIGURE 5-8 ■
Convolution of a
Simple Pulse

FIGURE 5-9 ■ Half-Power Points

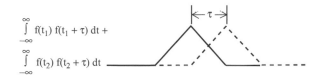

the signal, two targets could be resolved if separated at the half-power points (0.707 voltage level) of the pulses, as shown in Figure 5-9. This criterion of resolution is called the "3 dB" (actually –3 dB) or "Rayleigh" criterion. (Lord Rayleigh suggested that the distance from the peak to the first null of the Fraunhoffer diffraction pattern be used as the criterion for resolving one pattern in the presence of another [Jenkins and White, 1976, p. 327]. The radar community has simply adopted this approach.)

This time separation can be observed as equivalent to the transmitted pulse width. Using the relationship between time and range for radar (Equation 5-13), the resultant range resolution is given in Equation (5-14). Essentially, range resolution is based on the finest time delay that can be resolved, which is equal to the transmitted pulse width. For example, a radar with a 1 μsec pulse width will have a range resolution of $(3 \times 10^8)(1 \times 10^{-6})/2 = 150$ meters.

$$R = \frac{c\,\Delta t}{2} \Rightarrow \Delta R = \frac{c\,\tau}{2} \tag{5-14}$$

where:

ΔR = Range resolution, meters
c = Speed of light, 3×10^8 meters/second
τ = Transmitted pulse width, seconds

Remembering match filter theory from Section 5.1.1, the radar receiver processing bandwidth is one over the pulse width. Thus, the range resolution is often given in terms of the radar receiver processing bandwidth (Equation 5-15). A plot of range resolution as a function of pulse width is shown in Figure 5-10.

FIGURE 5-10 ■ Range Resolution as a Function of Pulse Width

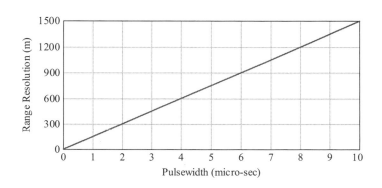

$$\Delta R = \frac{c\,\tau}{2} = \frac{c}{2\,B_R} \qquad (5\text{-}15)$$

where:

B_R = Radar receiver processing bandwidth, hertz

Often there are practical limits to producing the short pulse width required for fine range resolution, especially for high-power transmitters. If finer range resolution than is practical to achieve with a short pulse width is required, all is not lost. The transmitted carrier can be modulated, increasing its bandwidth, and processed in the receiver, as discussed in the next section.

5.2.2 Pulse Compression

For a simple pulse, the pulse width determines its bandwidth and range resolution (Equation 5-15). The technical challenge of producing high-power short duration pulses limits the range resolution for a simple pulse. Frequency or phase modulation of a coherent carrier frequency can be used to increase the bandwidth of the transmitted pulse to provide finer range resolution. The modulation of the transmit carrier frequency and the associated signal processing of the received pulse is called pulse compression [Richards et al., 2010, Chapter 20; Skolnik, 2001, Section 6.5; Stimson, 1998, Chapter 13].

Frequency Modulation

Linear frequency modulation (FM); also called linear frequency modulation on pulse (LFMOP), linearly changes the carrier frequency over the pulse width, as shown in Figure 5-11. It is often called "chirp" because if it were in the audible range it would make a sound similar to a bird's chirp. It is very common and will be discussed in detail in this book. Other FM schemes include sinusoidal and stepped frequency [Stimson, 1998, p. 164].

On receive the modulated pulse undergoes signal processing, which results in shorter pulse width—hence the name "pulse compression." A common time-domain form of pulse compression signal processing is to use a filter whose transit time is proportional to frequency. Such a filter and its ability to compress the pulse width of a linear FM pulse are shown in Figure 5-12. The filter provides a time delay that decreases linearly with frequency at the same

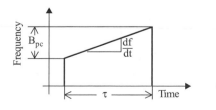
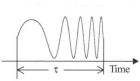

FIGURE 5-11 ■
Linear Frequency
Modulated Pulse

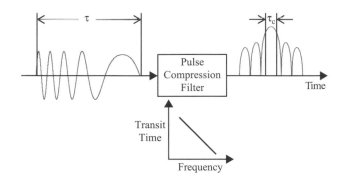

rate as the carrier frequency of the pulse increases. Thus, the higher frequencies associated with the trailing portions of the pulse take less time to pass through the filter than the lower frequencies associated with the leading portion. Successive portions of the pulse will essentially stack up at the output of the filter. Consequentially, when the pulse emerges from the filter, its amplitude is greater and its width is less than when it entered. The pulse still has the same energy. Such a pulse compression filter can be implemented with analog devices such as an acoustical delay line or can be applied digitally. Similarly, frequency domain signal processing, generally the fast Fourier transform (FFT), can be used to compress the received FM pulse.

Phase Modulation

Phase modulation divides the transmitted pulse into coded segments, or "chips," by shifting the phase of the carrier frequency. Biphase (i.e., 180° phase shifts) modulation is the most often used but is by no means the only type. A biphase modulated pulse with three phase changes and associated pulse compression signal processing response is shown in Figure 5-13. When a segment is autocorrelated with a segment 180° out of phase, the two cancel; otherwise they give a response.

By carefully selecting segment phases, the responses can be kept to a height of unity except for the maximum response that is equivalent to the

FIGURE 5-13 ■
Biphase Modulated
Pulse and Pulse
Compression Signal
Processing
Response

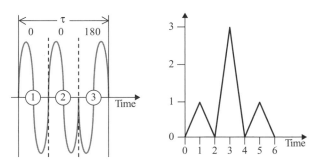

number of segments in the code. In other words, the range sidelobes (see Section 5.5.4) have unit height. Such a code is called a "Barker code" [Richards et al., 2010, pp. 817–818; Skolnik, 2001, pp. 350–351, 361–362; Stimson, 1998, p. 172]. The longest Barker code so far discovered uses 13 chips. While this may not seem to be useful for many radar applications, it can be especially for long-range radar systems. For example, a 13 μsec pulse width provides a range resolution of $(3 \times 10^8)(13 \times 10^{-6})/2 = 1950$ m. While adequate for a radar system with a few 100 km detection range, it's nothing special. Using a 13-chip Barker code provides a range resolution of $1950/13 = 150$ m, which is a very helpful performance improvement for a long-range radar system.

Practical phase codes with larger numbers of segments employ experimentally developed pseudorandom phase selection techniques [Richards et al., 2010, pp. 818–830]. Polyphase ($<180°$ phase shifts, e.g., $90°$, $45°$) codes have come into use with modern radar systems. Examples include binary phase shift keying (BPSK; of which Barker is a subset), minimum peak sidelobe (MPS), maximal length sequences (MLS), Frank, and quadrature phase shift keying (QPSK). Many of these codes are also routinely used as modulation techniques for digital communications. While not providing the uniform range sidelobes of a Barker code, they can provide a much finer range resolution by dividing a long transmitted pulse into many coded segments.

Range Resolution with Pulse Compression

The range resolution for a radar waveform with pulse compression modulation is a function of the modulation bandwidth (Equation 5-16). For FM pulse compression, the modulation bandwidth is the frequency excursion, as shown in Figure 5-11. For phase modulation, the modulation bandwidth is the number of segments or chips divided by the transmitted pulse width (Equation 5-17). For example, a modulation bandwidth of 20 MHz will provide a range resolution of $(3 \times 10^8)/[2(20 \times 10^6)] = 7.5$ m. A plot of range resolution as a function of modulation bandwidth is shown in Figure 5-14.

$$\Delta R_{pc} = \frac{c}{2\,B_{pc}} \qquad (5\text{-}16)$$

$$B_{pc} = \frac{N\phi}{\tau} \qquad (5\text{-}17)$$

where:

ΔR_{pc} = Range resolution with pulse compression, meters
c = Speed of light, 3×10^8 meters/second
B_{pc} = Pulse compression modulation bandwidth, hertz
$N\phi$ = Number of phase coded segments or chips, no units
τ = Radar transmitted pulse width, seconds

FIGURE 5-14 ■
Range Resolution as
a Function of
Modulation
Bandwidth

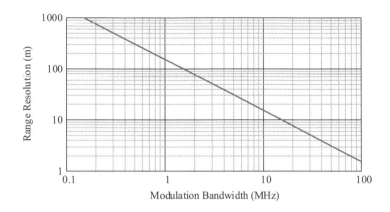

Matched Filter Theory for Pulse Compression

A matched filter (see Sections 2.3.1 and 5.1.1) receiver has a bandwidth that matches the spectral width of the transmitted waveform. Therefore, matched filter theory for pulse compression has the radar receiver bandwidth equal to the pulse compression modulation bandwidth (Equation 5-18). In doing so, a matched filter receiver maximizes the received target signal relative to the receiver thermal noise. Also, from a practical perspective the receiver bandwidth must pass the full pulse compression modulation bandwidth to allow the pulse to compress—which after all is the purpose of pulse compression.

$$B_R = B_{pc} \tag{5-18}$$

where:

B_R = Radar receiver processing bandwidth, hertz
B_{pc} = Pulse compression modulation bandwidth, hertz

Impact of Pulse Compression on Signal-to-Noise Ratio

As seen in Figure 5-12 and Figure 5-13, the amplitude of the received signal is increased as a result of pulse compression signal processing. The increase in amplitude is a signal processing gain, often called the time-bandwidth product, as given in Equation (5-19). Essentially this relationship ensures the conservation of energy (amplitude × time) of the pulse, since as the pulse duration is reduced (compressed) its amplitude must increase. The pulse compression signal processing gain is often called the pulse compression ratio (PCR), the ratio of the transmitted pulse width to the compressed pulse width.

$$G_{sp} = \tau\, B_{pc} \equiv PCR = \frac{\tau}{\tau_c} \tag{5-19}$$

where:

G_{sp} = Radar signal processing gain, no units
τ = Radar transmitted pulse width, seconds
B_{pc} = Pulse compression modulation bandwidth, hertz
PCR = Pulse compression ratio, no units
τ_c = Compressed pulse width, seconds

Since the signal amplitude is increased by the time-bandwidth product, at first glance one would think the target signal-to-noise ratio (S/N) would increase. However, the receiver bandwidth must be wide enough to pass the pulse compression modulation bandwidth of the pulse. Thus, because the receiver thermal noise is a function of the receiver bandwidth, the receiver thermal noise is also increased. The S/N after pulse compression processing is the same as the input S/N [Goj, 1993, Chapter 1]. Conservation of energy holds true. This is shown in an exercise at the end of this chapter.

5.2.3 Range Ambiguity

Since the time delay is measured with respect to the time a pulse was transmitted, only time delays less than the PRI are unambiguous. Time delays greater than the PRI are incorrectly interpreted, as shown in Figure 5-15. The distance to which a pulse can make a round trip before the next pulse is sent is called the "unambiguous" range or "first time around" range. Essentially unambiguous range is the range associated with a time delay equal to the PRI. Using Equation (5-13) we can determine the unambiguous range (Equation 5-20). For example, a radar with a 300 Hz PRF will have an unambiguous range of $(3 \times 10^8)/[2(300)] = 500$ km. A plot of unambiguous range as a function of the PRF is shown in Figure 5-16.

$$R = \frac{c\,\Delta t}{2} \Rightarrow R_u = \frac{c\,PRI}{2} = \frac{c}{2\,PRF} \qquad (5\text{-}20)$$

where:

R_u = Unambiguous range, meters
c = Speed of light, 3×10^8 meters/second
PRI = Pulse repetition interval, seconds
PRF = Pulse repetition frequency, hertz

Range measurements are unambiguous for targets within the unambiguous range. If, however, targets are farther out, inside two, three, or four times the unambiguous range, pulses will return from those targets and, once they start appearing, will show up in every PRI. All targets, regardless of range, will appear within all PRIs. Targets are ambiguous as to which PRI they are in, although their positions within the PRI itself are known accurately.

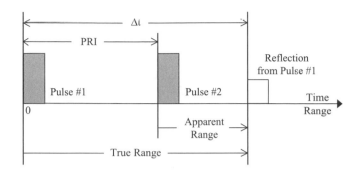

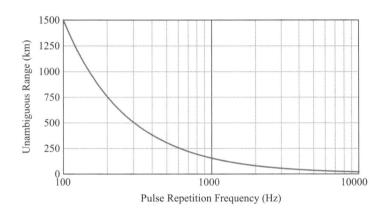

A frequently used way to resolve range ambiguities is to use multiple PRIs [Richards et al., 2010, pp. 660–665; Skolnik, 2001, pp. 175–178; Stimson, 1998, pp. 155, 286]. The overall result is the ambiguous ranges changing position between the different PRIs, and the actual range remains fixed. The procedure is shown in Figure 5-17. The true target range is where the received pulses from each PRI coincide. A coincidence filter can sort out the ambiguities, providing there are not too many. Alternatively, a simple algorithm can be used to compute the true range from the multiple ambiguous ranges. If ambiguities still exist, a third PRI can be used. Even with multiple PRIs it may not be possible to resolve all range ambiguities over the desired maximum range of the radar system.

5.2.4 Dead Zone

Normally, the radar receiver is shut off or isolated from the antenna while the transmitter is on. Therefore, no reflected pulses can be received and detected during this interval. In the meantime, the leading edge of the pulse has traveled out to $c\tau$ from the transmitter. The radius of what is called the "dead zone" is

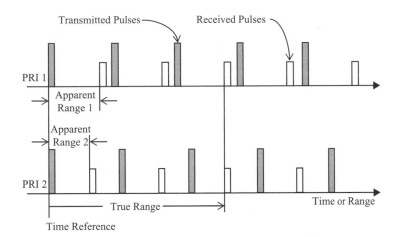

$c\tau/2$. The zone is not really dead because special arrangements can be made so some targets can be detected with the trailing portions of the received pulses.

5.2.5 Eclipsing

Eclipsing occurs when a reflected pulse is received while the transmitter is on and the receiver is not connected to the antenna, as shown in Figure 5-18. Eclipsing exists for targets with a time delay approximately equal to the PRI. Eclipsing results in a loss of received pulse power, which effects detection performance. Full eclipsing occurs when the entire received pulse in eclipsed; when only part of the received pulse is eclipsed, it is called partial eclipsing.

The lower the PRI (the higher the PRF), the more likely eclipsing will impact the radar. The effects of eclipsing can be minimized through the choice of PRI (PRF) and/or radar detection performance (detection range is less than the unambiguous range). Even with higher PRFs, eclipsing can be minimized or

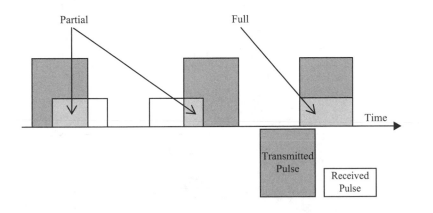

FIGURE 5-18 ■
Eclipsing

controlled. For airborne radar systems, there are dynamic changes in the radar-to-target range. The use of multiple PRIs minimizes the probability of the reflected pulse being eclipsed for all PRIs.

5.2.6 Range Tracking

Range tracking implies following an object in range as a function of time, that is, a time sequence of range measurements. A "tracking gate" is put around an object designated for range track. The radar system may automatically do the tracking with a closed-loop control system, or the operator may do the tracking, or a combination of both can be used. Alternatively, a track file may be activated in the computer. We will discuss both these concepts in the following sections.

Split-Gate Range Tracker

Range measurement tracking is accomplished by following the detected target time delay over time. When a single target is being tracked, a common range tracking technique is a split-gate tracker [Richards et al., 2010, pp. 693–694; Skolnik, 2001, pp. 246–247; Stimson, 1998, p. 384], which divides the received target pulse between two time gates, an early gate and a late gate, as shown in Figure 5-19.

The output of the late gate is inverted. A difference signal is generated by the difference of the outputs of the two gates. A closed-loop servo control system is used to drive (center) the split gates over the received target pulse (i.e., zero the difference of the outputs of the two gates), as shown in Figure 5-20.

The plot of the difference signal is a discriminator curve, as shown in Figure 5-21. The sign (x-axis) of the difference signal indicates the direction, left or right, to move the split gates. The amplitude (y-axis) of the difference signal indicates how far to move the split gates. The design of the range

FIGURE 5-19 ■
Split-Gate Range
Tracker Concept

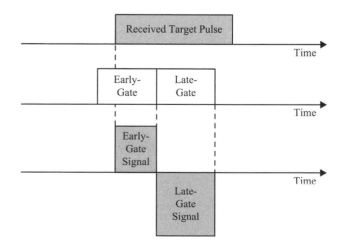

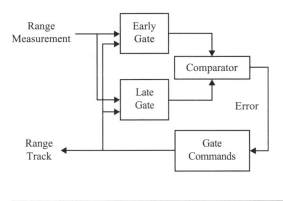

FIGURE 5-20 ■
Split-Gate Range
Tracker

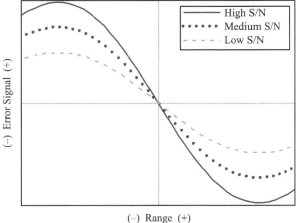

FIGURE 5-21 ■
Split-Gate Range
Tracker
Discriminator Curve

tracking loop is based on the dynamics of the worst expected radar–target engagements. The bandwidth of the range tracking loop determines the amount of tracking lag (error) in a particular radar–target engagement scenario. Some range tracking loops use different bandwidths for acquisition (wide) and tracking (narrow) to improve the overall tracking performance under these two different conditions.

The range tracking loop is normally designed to provide operation in the linear region of the discriminator curve, that is, the part with the linear slope. Operation outside of the linear region leads to nonlinear response from the tracker and to poor tracking performance (e.g., large tracking error, break lock on the target). Some electronic attack (EA) techniques (see Chapter 8) seek to cause operation outside the linear region.

Leading-Edge Range Tracker
If a target tends to be large with respect to a range resolution cell, that is, an "extended" target, or if it has scattering phase centers (see Chapter 6) that jump around, "leading-edge" and/or "trailing-edge" range tracking may be required

FIGURE 5-22 ▪ A
Differentiating Circuit

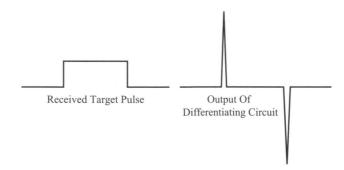

[Barton, 1998, pp. 436–438]. A leading-edge range tracker operates on the pulse rise, while a trailing-edge range tracker works on the pulse fall. The received target pulse is first applied to a differentiating circuit. A differentiating circuit responds only to distinct amplitude changes producing an output at the leading and trailing edges of the pulse, as shown in Figure 5-22.

Either or both of the leading or trailing delta functions may be used for the track operations using a closed-loop control system. Because much of the target signal power is lost in the differentiating process, high S/Ns are necessary. The tendency of the range tracker to wander over the "center of mass" of an extended target is thereby thwarted. A leading-edge tracker can also provide an electronic protection (EP) technique against some EA techniques (see Chapter 8).

Track File from Search Radar Systems

A target track file including the target range and angle (see Section 5.4) is initiated in the radar system, generally in a computer, when a target is detected [Jeffery, 2009, Chapter 5]. Depending on the complexity of the tracking algorithm, the radar system may simply look where last seen, move the tracking gate to the nearest target, and record the results. Alternatively, it may develop updates to the existing track based on the latest measurement. A computer track file usually contains a target path model of relatively simple linear form (Equation 5-21) so that it can provide a predicated target range.

$$R_3 = R_2 + \left(\frac{R_2 - R_1}{t_2 - t_1}\right) \Delta t \qquad (5\text{-}21)$$

where:

R_3 = Predicted radar-to-target range at time Δt in the future, meters
R_2 = Measured radar-to-target range at time t_2, meters
R_1 = Measured radar-to-target range at time t_1, meters

As more data are obtained (i.e., R_3 is measured and R_4 is predicted), the path may be followed very accurately, especially if it is deterministic (no energy is added to or taken from the object). For indeterminate paths such as an airplane might take, it is important to keep the interval between measurements

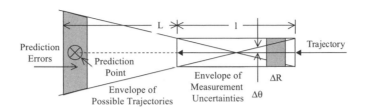

FIGURE 5-23 ■
Gun Barrel Analogy

small compared with the possible changes in range during the time interval. Sophisticated tracking radar systems may contain many complex paths in their computers so that if track is lost the radar will coast, using the computer programs alone while looking in designated regions for the target to reappear.

When an object is on a Newtonian trajectory or following some other reasonably deterministic path, radar tracking data permit predictions of where the object will be at some future time. The exact equations for these calculations are complex, but they can be approximated by using the so-called gun barrel analogy, as shown in Figure 5-23. The radar makes measurements over a distance, and the tracker predicts a future target position. Assuming no improvement by integration, the rate of increase of $\Delta\theta$ after the final measurement is given in Equation (5-22). Converting the prediction distance into a velocity-time product and assuming two equal 1σ measurements results in the angular error at the prediction point (Equation 5-23).

$$\Delta\theta_{dot} = 2\,\Delta\theta \left(\frac{2\,L}{1}\right) \tag{5-22}$$

$$\epsilon_\theta \approx \sqrt{2}\,\Delta\theta \left(1 + \frac{2T}{t}\right) \tag{5-23}$$

where:

$\Delta\theta_{dot}$ = Rate of increase in root mean square angular error after the final measurement, radians/second

$\Delta\theta$ = Root mean square angular error, radians

L = Distance from the last measurement to the prediction point, meters

1 = Distance over which measurements are made, meters

ϵ_θ = Angular error at the prediction point, radians

T = Time ahead the predication is made, seconds

t = Duration of the measurement, seconds

To account for range errors a time factor must initially be introduced. Assuming there is no range rate measurement capability, time is measured perfectly, and the range measurements are 1σ, range errors introduce velocity errors proportionally, Equation (5-24). Assuming the first and last range errors are equal, and time is measured perfectly, results in the range error at the prediction point, Equation (5-25).

$$\Delta v^2 = \frac{\Delta R_i{}^2 + \Delta R_f{}^2}{t} \qquad (5\text{-}24)$$

$$\varepsilon_R \cong \sqrt{2}\,\Delta R \left(1 + \frac{T}{t}\right) \qquad (5\text{-}25)$$

where:

Δv = Error in the velocity estimate, meters/second
ΔR_i = Error in the initial range measurement, meters
ΔR_f = Error in the final range measurement, meters
ε_R = Range error at the prediction point, meters
ΔR = Root mean square range error, meters

Many factors serve to modify these elementary results. The target trajectory will not be a straight line, and smoothing to a polynomial curve will be necessary. The radar errors will decrease substantially with the number of measurements (N) made, perhaps by as much as $1/\sqrt{N}$. Aspect angle will change during the interval of measurement, so there is a cross-correlation between radar range and angle errors. Nevertheless, the simple equations can be used to bound the problem. For example, if the radar has a 50 second measurement duration, a 100 second prediction time, a 10 meter range error, and a 0.5 milliradian angular error, we can determine the angle and range errors at the prediction point for a target 900 km from the radar. Equations (5-23) and (5-25) produce a 1σ error volume of ± 3182 meters cross-range and ± 42.4 meters down range, respectively. This is not far from the numbers obtained by representative radar systems.

$$\varepsilon_\theta \approx \sqrt{2}\,(0.5 \times 10^{-3})\left(1 + \frac{2(100)}{50}\right) = 3.5355 \times 10^{-3} \text{ radians}$$

$$(3.5355 \times 10^{-3})(900 \times 10^3) = 3.182 \times 10^3 \text{ meters}$$

$$\varepsilon_R \cong \sqrt{2}\,(10)\left(1 + \frac{100}{50}\right) = 42.4264 \text{ meters}$$

5.3 | RANGE RATE MEASUREMENTS

A radar system can determine the radar-to-target range rate—time rate of change of range. The range rate is the magnitude of the projection of the object's velocity vector and the radar's velocity vector for a moving radar system on the object-to-radar range vector, as shown in Figure 5-24. Instead of range rate, sometimes the term "relative speed" is used, which is accurate in this context as speed is the magnitude of the velocity vector. Additionally,

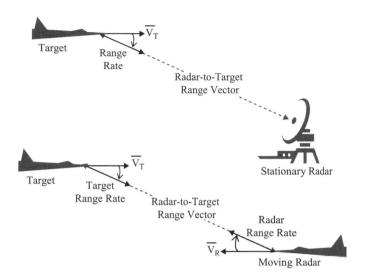

FIGURE 5-24 ■
Range Rate:
Stationary Radar
and Moving Radar
Systems

range rate is often referred to as "radial velocity," but this is inaccurate in this context because velocity is a vector. Since a radar system can measure only the range rate we will use only this term.

Radar systems use several target range rate measurement techniques, the most common of which involves the Doppler shift present in the received target signal. We will discuss range rate measurement using the Doppler shift in detail in this section. Less common range rate measurement techniques, like successive range measurements and using a simple pulse, will be addressed at the end of this section.

The radar system determines the radar-to-target range rate by measuring the Doppler shift of the received target signal relative to the transmitted radar waveform. To measure the Doppler shift, the radar system uses a coherent pulse burst waveform (see Section 5.1.2) and preserves phase throughout the generation, transmission, and reception of the waveform so all the carrier and modulation phases add correctly.

The radar system transmits its carrier frequency: 100's MHz to 10's GHz. The radar-to-target range rate introduces a Doppler shift onto the carrier frequency: ±100's Hz to ±100's kHz. The radar system receives the Doppler shifted carrier frequency (Equation 5-26). The spectrum of the received target signal from a moving target is a Doppler shifted version of the transmitted spectrum, as shown in Figure 5-25. The Doppler shift is given in Equation (5-27). As seen in these equations, closing targets "compress" the radar wave, resulting in a higher received frequency, whereas opening targets "stretch" the radar wave, resulting in a lower received frequency. Negative range rate (i.e., decreasing radar-to-target range) results in a positive Doppler shift, and a

FIGURE 5-25 ◾
Transmitted
Spectrum and
Received Target
Spectrum

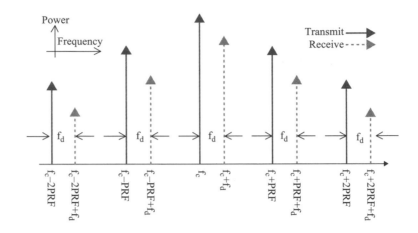

FIGURE 5-25 ◾ Transmitted Spectrum and Received Target Spectrum

positive range rate (i.e., increasing radar-to-target range) results in a negative Doppler shift.

$$f_{cr} = f_c + f_d \tag{5-26}$$

$$f_d = \frac{-2 \, R_{dot}}{\lambda} \tag{5-27}$$

where:

f_{cr} = Received carrier frequency, hertz
f_c = Radar transmitted carrier frequency, hertz
f_d = Doppler shift, hertz
R_{dot} = Object-to-radar range rate; positive for an opening target, radar-to-target range increasing, and negative for a closing target, radar-to-target range decreasing, meters/second
λ = Radar transmitted wavelength (c/f_c), meters
c = Speed of light, 3×10^8 meters/second

For example, the Doppler shift produced by a target closing at a range rate of 150 m/sec for a radar with a wavelength of 0.03 m (10 GHz carrier frequency) is (–2)(–150)/(0.03) = 10 kHz. The negative sign on the range rate value is because the target is closing on the radar, radar-to-target range decreasing. The Doppler shift per unit range rate (1 m/sec) as a function of radar carrier frequency is shown in Figure 5-26. As the figure shows, the higher the radar carrier frequency the higher the Doppler shift for a given range rate. This is why the vast majority of Doppler radars are higher frequency; the radar designer cannot control the range rate of the target, but they can pick the radar frequency.

As evident in the meaning of range rate (see Figure 5-24), the Doppler shift is dependent on the radar-to-target geometry. Figure 5-27 shows examples of

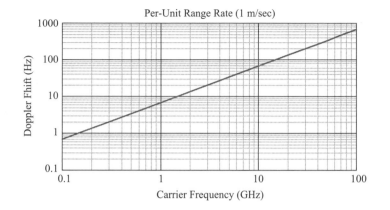

FIGURE 5-26 ■
Doppler Shift Per-
Unit Range Rate
(1 m/sec) as a
Function of Carrier
Frequency

FIGURE 5-26 ■
Doppler Shift Per-
Unit Range Rate
(1 m/sec) as a
Function of Carrier
Frequency

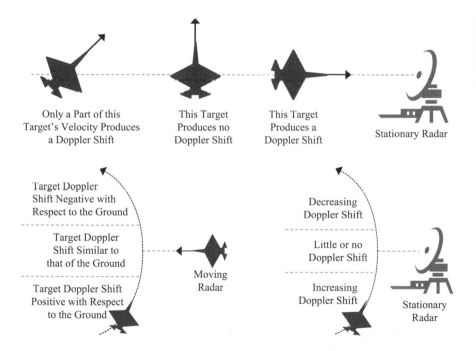

FIGURE 5-27 ■
Doppler Shift
Examples Based on
Radar–Target
Geometry

these dependences for both stationary (generally ground-based), and moving (generally airborne) radar systems.

The Doppler shift is measured in the radar receiver, and the range rate is computed using Equation (5-28). For example, the range rate associated with a Doppler shift of 6000 Hz for the same radar as before is $[-(6000)(0.03)]/2 = -90$ m/sec. The negative sign on the range rate value tells us that this target is closing on the radar; the radar-to-target range is decreasing. The range rate per

FIGURE 5-28 ■
Range Rate Per-Unit
Doppler Shift (1 Hz)
as a Function of
Carrier Frequency

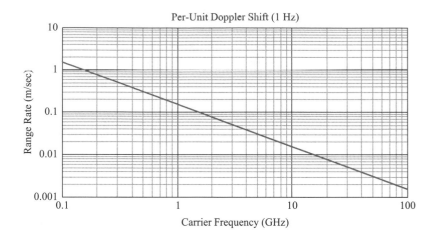

unit Doppler shift (1 Hz) as a function of radar carrier frequency is shown in Figure 5-28.

$$R_{dot} = \frac{-f_d \lambda}{2} \qquad (5\text{-}28)$$

The Doppler shift is measured using a sequence of matched narrow bandpass filters, as shown in Figure 5-29. Actual Doppler filters have a passband and a sequence of frequency domain sidelobes (see Section 5.5.4), not the simple bandpass shown in this figure. The Doppler filter containing the target Doppler shift will have the strongest output signal (i.e., highest S/N); the other Doppler filters contain only receiver noise. A threshold detection (see Chapter 3) stage follows each Doppler filter, thereby providing the range rate measurement associated with the detection.

FIGURE 5-29 ■
Doppler Filter Bank

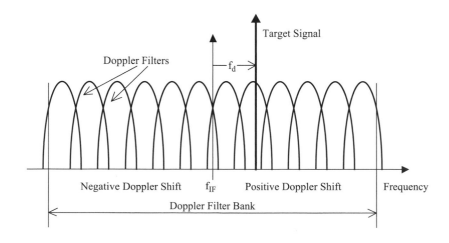

Typically, the Doppler filters overlap the adjacent filter by about their half-power (−3 dB) bandwidth. Often the sequence of Doppler filters is called a Doppler filter bank, which covers the band of all feasible target Doppler shifts (range rates). For a radar dealing with high range rate targets (e.g., aircraft, missiles, satellites), the number of Doppler filters can become large.

In an analog Doppler processor, which is rare but not uncommon, the spectral response required of the Doppler filters is known because the Fourier transform of the transmitted pulse burst waveform is known. In a digital Doppler processor, by far the most common, the Fourier transform of the digitized received signal is computed. Known as the discrete Fourier transform, it can be calculated by solving the number of simultaneous equations equal to the number of points in the digitized version of the signal.

Efficient ways of doing this calculation for a power of two number of pulses (2^{n_p}) are called FFTs [Sheats, 1977]. Therefore, most radar systems with a digital Doppler processor have a pulse burst waveform with a power of two number of pulses: 16, 32, 64, etc. The discrete-time Fourier transform (DTFT) is also a common spectral analysis digital signal processing technique [Richards et al., 2010, Section 14.4]. As a result of computing the Fourier transform, coherent integration is performed. The output signal builds up with the number of pulses processed, ultimately indicating a Doppler signal is present at a specific frequency.

5.3.1 Range Rate Resolution

Narrowband filters achieve their selectivity by integrating the input signal over a period of time (Equation 5-29). Thus, the longer the integration time, the smaller the Doppler filter bandwidth. Often the integration time, the same time over which multiple pulses are integrated from Chapter 3, is called the pulse burst duration (Section 5.1.2) or the coherent processing interval (CPI). Range rate resolution is a function of the Doppler filter bandwidth. In keeping with the criterion established for range resolution (Section 5.2.1), it can be stated that two objects are resolved in Doppler if they are separated in frequency by one Doppler filter bandwidth (the two signals are in adjacent Doppler filters). Converting from the frequency domain to the range rate domain provides the range rate resolution (Equation 5-30). Essentially, range rate resolution is based on the finest Doppler shift that can be resolved, which is equal to the Doppler filter bandwidth. For example, the range rate resolution for a radar system with a 200 Hz Doppler filter bandwidth for the same radar as before is [(200)(0.03)/2] = 3 m/sec.

$$\Delta f_d = \frac{1}{T_I} \tag{5-29}$$

$$R_{dot} = \frac{f_d \, \lambda}{2} \Rightarrow \Delta R_{dot} = \frac{\Delta f_d \, \lambda}{2} = \frac{\lambda}{2 \, T_I} \qquad (5\text{-}30)$$

where:

Δf_d = Doppler filter bandwidth, hertz

T_I = Radar integration time or pulse burst duration or coherent processing interval, seconds

ΔR_{dot} = Range rate resolution, meters/second

λ = Wavelength, meters

The range rate resolution per unit Doppler filter bandwidth as a function of radar carrier frequency is shown in Figure 5-30. From this figure we can observe the higher the radar carrier frequency, the finer the range rate resolution for a given Doppler filter bandwidth, reinforcing why the vast majority of Doppler radars are higher frequency.

A pulse burst waveform permits arbitrarily good Doppler resolution. The pulse burst duration can be as long as necessary (within reason, since it also is the pulse integration time; see Section 3.3). Radar is one big systems engineering problem.

5.3.2 Range Rate Ambiguity

Frequency domain ambiguities exist for a pulse burst waveform. As shown in Figure 5-25, the spectrum of the received signal from a moving target is a Doppler shifted version of the transmitted spectrum. As previously discussed, the radar system measures the Doppler shift using a Doppler filter bank centered about a reference frequency, generally the receiver's intermediate frequency (see Section 2.4). A frequency domain ambiguity exists when the Doppler shift is large relative to the frequency extent of the Doppler filter bank.

FIGURE 5-30 ■
Range Rate
Resolution Per-Unit
Doppler Filter
Bandwidth (1 Hz) as
a Function of Carrier
Frequency

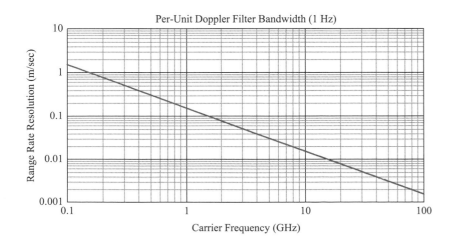

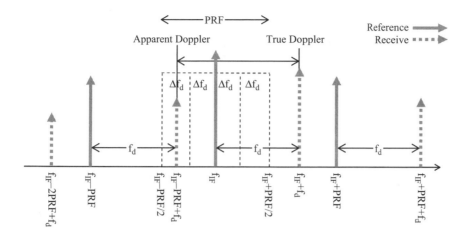

FIGURE 5-31 ▪
Ambiguous Doppler
Shift

The frequency extent of the Doppler filter bank is often the PRF. Therefore, all is well and good for Doppler shifts within ±PRF/2. However, Doppler shifts outside ±PRF/2 will appear between −PRF/2 and +PRF/2 in a position denoted by the excess of their Doppler shift beyond an integer multiple of the PRF, as shown in Figure 5-31.

Target Doppler shifts are ambiguous with respect to the reference frequency, although their position between −PRF/2 and +PRF/2 is known accurately. The unambiguous Doppler shift is ±PRF/2 (Equation 5-31). Converting from the frequency domain to the range rate domain provides the unambiguous range rate (Equation 5-32). Effectively, unambiguous range rate is that associated with a Doppler shift equal to ±PRF/2. For example, the unambiguous range rate for a radar system with a 15 kHz PRF for the same radar as before is ±[(15,000)(0.03)/2] = ±225 m/sec. The unambiguous range rate per unit PRF (1 Hz) as a function of radar carrier frequency is shown in Figure 5-32.

$$f_{du} = \pm \frac{PRF}{2} \qquad (5\text{-}31)$$

$$R_{dot} = \frac{f_d \lambda}{2} \Rightarrow R_{dotu} = \frac{f_{du} \lambda}{2} = \pm \frac{PRF \lambda}{4} \qquad (5\text{-}32)$$

where:

f_{du} = Unambiguous Doppler shift, hertz
R_{dotu} = Unambiguous range rate, meters/second
PRF = Radar pulse repetition frequency, hertz
f_{IF} = Receiver intermediate frequency (see Section 2.4), hertz

In Section 5.2.3 we found that the unambiguous range is inversely proportional to PRF (Equation 5-20). Thus, long unambiguous ranges require

FIGURE 5-32 ▪
Unambiguous
Range Rate Per-Unit
Pulse Repetition
Frequency (1 Hz) as
a Function of Carrier
Frequency

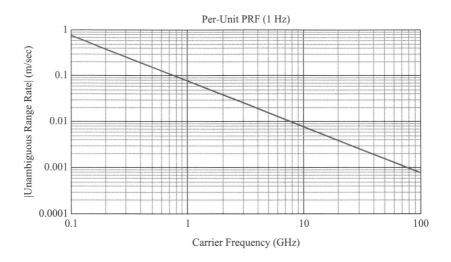

low PRFs. This is opposite to the relationship we just developed for unambiguous range rate. Oftentimes, something has to give: the radar system is either unambiguous in range or range rate but not both at the same time.

An example will make these ambiguities more concrete. Let us say we wish to unambiguously measure Mach 2 targets at a 150 km range with a C-band ($\lambda \approx$ 0.05 m) radar. The perfect waveform would have its range rate ambiguity beyond Mach 2 (660 m/sec) and its range ambiguity beyond 150 km. The PRF necessary to provide the range rate ambiguity is computed using Equation (5-32) and solved in terms of PRF, with the result given in Equation (5-33). The PRF necessary to provide the range ambiguity is computed using Equation (5-20), solved in terms of PRF, with the result given in Equation (5-34). Clearly, the two requirements are incompatible. We must accept either range or Doppler ambiguities.

$$\text{PRF} = \frac{2\,R_{\text{dotu}}}{\lambda} = \frac{2\,(660)}{0.05} = 26.4 \text{ kHz} \tag{5-33}$$

$$\text{PRF} = \frac{c}{2\,R_u} = \frac{3 \times 10^8}{2\,(150 \times 10^3)} = 1000 \text{ Hz} \tag{5-34}$$

In Section 5.2.3 we found that multiple PRFs are a frequently used way around range ambiguities. Doppler ambiguities can be resolved in a like manner [Edde, 1993, pp. 306–310]. Generally, the multiple PRFs are designed so that the return pulses are ambiguous in one measurement domain (i.e., range or range rate) and unambiguous in the other (i.e., range rate or range). Alternatively, some radar systems resolve range and/or range rate ambiguities through the use sequential PRFs: a range PRF followed by a range rate PRF or vice versa (see Section 7.4).

5.3.3 Other Range Rate Measurement Approaches

Successive Range Measurements

This technique allows the range rate to be determined based on the difference between successive range measurements over a known time interval (Equation 5-35) [Skolnik, 2001, p. 315]. The successive range measurements are separated in time by one antenna scan time or measurement interval. This approach for determining the range rate is used by some noncoherent (i.e., no Doppler capability) search radar systems. The range rate resolution is a function of the range resolution; thus, it is not as fine as can be obtained using the Doppler approach for determining range rate we have just discussed. Determining the radar rate using successive range measurements can also provide an electronic protection (EP) technique (see Chapter 8) for Doppler radar systems.

$$R_{dot} = \frac{R_2 - R_1}{t_2 - t_1} \qquad (5\text{-}35)$$

where:

R_{dot} = Range rate, meters/second
R_2 = Range measurement at time t_2, meters
R_1 = Range measurement at time t_1, meters

Single Simple Pulse

This technique allows the range rate to be determined based on the Doppler shifted spectrum of a single simple pulse, as shown in Figure 5-33. This approach for determining the range rate is used by a few noncoherent search radar systems. In keeping with the criterion established for range resolution, two objects can be resolved in Doppler when their peaks are separated by one over the pulse width (see Section 5.1.1 and Figure 5-2). Thus, the Doppler resolution is given in Equation (5-36). This equates to a range rate resolution given in Equation (5-37).

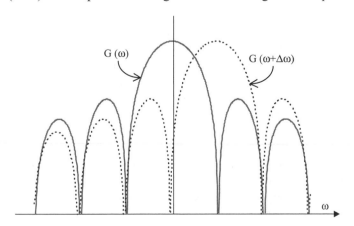

FIGURE 5-33 ■ Doppler Shifted Spectrum and Resolution—Simple Pulse

When targets of unequal radar cross section (RCS) are of interest, much more stringent bases for resolution are adopted; tapering is invoked to decrease the Doppler sidelobes (see Section 5.5.4), with the idea of resolving targets many thousands of times different in amplitude.

$$\Delta f_d = \frac{1}{\tau} \tag{5-36}$$

$$R_{dot} = \frac{f_d \, \lambda}{2} \Rightarrow \Delta R_{dot} = \frac{\lambda}{2 \, \tau} \tag{5-37}$$

where:

Δf_d = Doppler resolution; single simple pulse, hertz
ΔR_{dot} = Range rate resolution; single simple pulse, meters/second

In our previous discussion on range measurement, we found that the range resolution was also a function of the pulse width (see Section 5.2.1). A finer range resolution requires a short pulse width, while a finer range rate resolution requires a long pulse width. Clearly, the radar designer has a trade-off to make when dealing with a simple pulse for range and range rate resolution.

5.3.4 Range Rate Tracking

Range rate tracking implies following an object in range rate as a function of time, a sequence of range rate measurements. A "tracking gate" is put around an object designated for range rate track. The radar system may automatically do the tracking with a closed-loop control system, the operator may do the tracking, or a combination can be used. Alternatively, the output of the radar system's Doppler filter bank can be monitored. We will discuss both these concepts in the following sections.

Split-Gate Doppler Tracker or Speed Gate

Range rate measurement tracking is accomplished by following the target Doppler shift over time. When a single target is being tracked, a common range rate tracking technique is a split-gate range rate tracker [Skolnik, 2001, p. 252; Stimson, 1998, p. 386]. Also called a "speed gate" or Doppler or velocity gate tracker, it divides the received target Doppler between two narrowband Doppler filters—the low-frequency filter and the high-frequency filter—as shown in Figure 5-34.

The output of the high frequency filter is inverted. A difference signal is generated by taking the difference of the outputs of the two filters. A closed-loop servo control system is used to drive (center) the two Doppler filters over the received target Doppler (i.e., zero the difference of the outputs of the two filters), as shown in Figure 5-35.

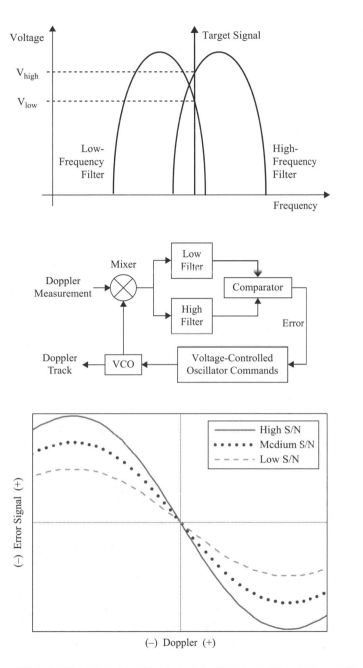

FIGURE 5-34 ▪
Split-Gate Range
Rate Tracker
Concept

FIGURE 5-35 ▪
Split-Gate Range
Rate Tracker

FIGURE 5-36 ▪
Split-Gate Range
Rate Tracker
Discriminator Curve

The plot of the difference signal is a discriminator curve, as shown in Figure 5-36. The sign (x-axis) of the difference signal indicates the direction, left or right, to move the target Doppler shift entering the two filters. The amplitude (y-axis) of the difference signal indicates how far to move the target

Doppler shift. The design of the range rate tracking loop is based on the dynamics of the worst expected radar–target engagements. The bandwidth of the range rate tracking loop determines the amount of tracking lag (error) in a particular radar–target engagement scenario. Some range rate tracking loops use different bandwidths for acquisition (wide) and tracking (narrow) to improve the overall tracking performance under these two different conditions.

The range rate tracking loop is normally designed to provide operation in the linear region of the discriminator curve, that is, the part with the linear slope. Operation outside of the linear region leads to nonlinear response from the tracker and poor tracking performance (e.g., large tracking error, break lock on the target). Some EA techniques (see Chapter 8) seek to cause operation outside the linear region.

Doppler Filter Bank

A simple effective Doppler tracking technique uses banks of narrowband Doppler filters (Doppler filter bank in Section 5.3) and circuitry that provides an indication of which filters are occupied by detected target Doppler signals [Stimson, 1998, pp. 235–240]. The output of each Doppler filter is individually monitored, providing the capability to track multiple targets. Since each Doppler filter has a strong output only during the instant of the target signal's arrival in the radar, each response is associated with a given range and angle bin. The revolution in microelectronics has made it feasible to implement thousands of narrowband Doppler filters along with all the associated logic or to use digital signal processing to implement the FFT in a microprocessor. A target with a diffuse range rate spectrum (e.g., spinning propellers, ionized wakes, pieces falling off; see Section 6.3) may require special treatment.

5.4 | ANGLE MEASUREMENTS

A radar system measures the radar-to-target angle as a result of the mainbeam gain of the antenna (see Chapter 4). The target S/N is highest when the radar antenna mainbeam is pointed at the target (see Chapters 2 and 3). Therefore, when a target is detected it is assumed the antenna mainbeam is pointed at the target. Angle resolvers indicate where a mechanically scanned antenna is pointing. The beam steering computer tells the radar system where an electronically steered mainbeam is pointing. Angle measurements are generally made in two planes: azimuth (horizontal) and elevation (vertical).

5.4.1 Angle Resolution

In keeping with the criterion established for range resolution (Section 5.2.1), two objects can be resolved in angle when they are separated by one antenna half-power (–3 dB) beamwidth (see Chapter 4). Thus, the angular resolution is given in Equation (5-38). There are different angle resolutions for the azimuth and elevation planes when the azimuth and elevation beamwidths are different.

$$\Delta\theta = \theta_{3\,dB} \qquad (5\text{-}38)$$

where:

$\Delta\theta = $ Angle resolution, degrees or radians
$\theta_{3\,dB} = $ Antenna half-power (–3 dB) beamwidth, degrees or radians

5.4.2 Cross-Range Resolution

Frequently, angle resolution is translated into the range dimension, resulting in the cross-range resolution. As shown in Figure 5-37, the cross-range is the arc length perpendicular to the radar-to-target range vector. The cross-range resolution is given in Equation (5-39). Note: in this equation the units of the radar antenna beamwidth are radians. There are different cross-range resolutions for the azimuth and elevations planes if the azimuth and elevation beamwidths are different. The cross-range resolution for a few radar antenna beamwidths as a function of radar-to-target range is shown in Figure 5-38.

$$\Delta CR = R\,\theta_{3\,dB} \qquad (5\text{-}39)$$

where:

$\Delta CR = $ Cross-range resolution, meters
$R = $ Radar-to-target slant range, meters
$\theta_{3\,dB} = $ Antenna half-power (–3 dB) beamwidth, radians

5.4.3 Angle Tracking

Angle measurement tracking is accomplished by following the target angle over time. Its approaches have evolved from primitive methods of having the tracker seek a peak return (the radar antenna mainbeam is pointing at the target)

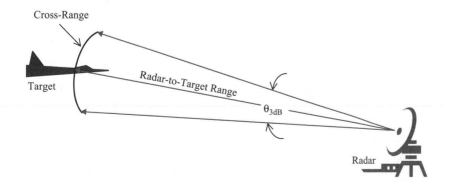

Cross-Range

Target

Radar-to-Target Range

θ_{3dB}

Radar

FIGURE 5-37 ■
Cross-Range
Resolution

FIGURE 5-38 ■
Cross-Range
Resolution as a
Function of Radar-
to-Target Range

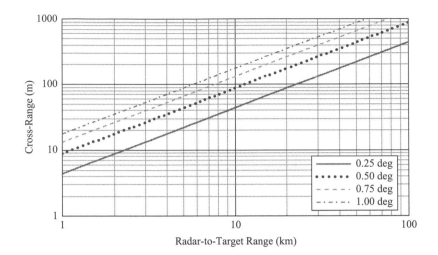

FIGURE 5-39 ■
Angle Tracking
Using Offset Beams

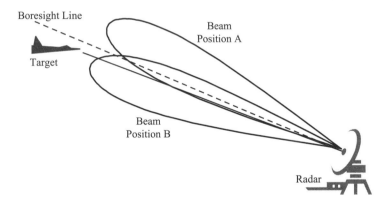

to conical scan and sequential lobing techniques and finally to monopulse. The vast majority of angle tracking approaches use offset antenna beams or lobes, as shown in Figure 5-39.

The outputs of the two beams are added to produce a sum (Σ) signal used for range or range rate tracking. A difference (Δ) signal is generated by taking the difference of the outputs of the two beams. A closed-loop servo control system is used to drive (center) the antenna boresight line to the target, as shown in Figure 5-40.

The plot of the difference signal is a discriminator curve, as shown in Figure 5-41. The sign (x-axis) of the difference signal indicates the direction, left or right, to move the antenna boresight line. The amplitude (y-axis) of the difference signal indicates how far to move the antenna boresight line. The design of the angle tracking loop is based on the dynamics of the worst expected radar–target engagements. The bandwidth of the angle tracking loop determines the amount of tracking lag (error) in a particular radar–target engagement scenario.

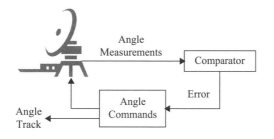

FIGURE 5-40 ▪
Angle Tracker

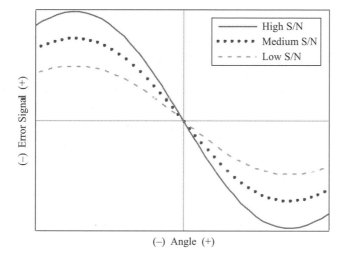

FIGURE 5-41 ▪
Angle Tracker
Discriminator Curve

The angle tracking loop is normally designed to provide operation in the linear region of the discriminator curve, that is, the part with the linear slope. Operation outside of the linear region leads to nonlinear response from the tracker and to poor tracking performance (e.g., large tracking error, break lock on the target). Some EA techniques (see Chapter 8) seek to cause operation outside the linear region.

Angle tracking approaches using a time history of multiple measurements are straightforward to implement. They are often called conical scan and sequential lobing trackers. Angle tracking using a single measurement, call monopulse, is difficult to implement.

Conical Scan and Sequential Lobing Angle Trackers

In conical scan, a squinted antenna mainbeam is rotated around the boresight line at a prescribed rate, which is slower than the PRF but faster than the antenna's angular scan rate [Richards et al., 2010, pp. 700–702; Skolnik, 2001, pp. 225–229; Stimson, 1998, p. 9, 102]. Sequential lobing is similar to conical scan, only the squinted beam is sequentially stepped in angle [ibid.]. The

FIGURE 5-42 ∎
Far-Field Conical
Scan (left) and
Sequential Lobing
(right) Patterns

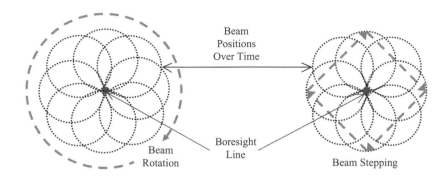

moving of the squinted beam causes it to scribe a small-angle cone in space. Figure 5-42 is a cross section of the far-field beams for conical scan and sequential lobing over time.

When the target is off the boresight line, the radar receives an amplitude modulated sequence of returns, higher when the beam is pointing at the target and lower when it is pointing away. The returns during a scan are compared in amplitude, and the boresight line is moved toward the position where they would balance. In the meantime, returns from a new scan are coming in.

Conical scan and sequential lobing angle trackers are still used because of their simplicity. However, they have a few limitations. Because a time history of multiple returns are used to close the angle tracking loop, short-term changes in the amplitude of returns can induce angle tracking errors. The two main sources of short-term amplitude changes are from the target and EA techniques. The target RCS can fluctuate (see Chapter 6) during the scan interval. An electronic warfare (EW) system can readily detect the scan rate and it can be used to develop an EA technique customized to the angle tracking control loop (see Chapter 8). Also, amplitude comparisons are not particularly accurate, which in turn limits angle tracking accuracy.

Monopulse Angle Tracker

Precision angle tracking in modern radar systems is usually done with a scheme known as monopulse angle tracking [Barton, 1988, Sections 8.4, 8.5; Rhodes, 1959; Richards et al., 2010, pp. 320–322, 701–703; Skolnik, 2001, pp. 213–224; Stimson, 1998, pp. 103–104]. All the limitations of lobing trackers are eliminated in monopulse. The term is used because the information obtained from a single pulse tells the radar where the object is in angle, how far the mainbeam is from pointing exactly at the target in angle, and provides an error signal to drive the mainbeam toward the target. Good monopulse trackers can easily get accuracies of one-tenth of the radar beamwidth. In fact, they can come close to achieving the theoretical angular accuracy capability of a radar system.

Monopulse tracking is essentially simple. It uses multiple simultaneous squinted beams, usually two, three, or four. The multiple simultaneous squinted beams of a monopulse tracker are slightly displaced so that each receives the signal from a slightly different angle. The return in the received beams can be added to form a sum (Σ) signal (which is used for gross pointing) and can be subtracted to form a difference (Δ) signal. The sum signal produces a maximum response, and the difference signal will produce a zero response when the target is on the antenna boresight line. If the difference signal is not zero, the deviation is used to generate a voltage to drive the difference signal toward zero or toward the so-called monopulse null. The technique is portrayed in two dimensions in Figure 5-43.

The plot of the difference signal is a discriminator curve, as shown in Figure 5-41. The angle (x-axis) of the difference signal indicates the direction, left or right, to move the antenna. The amplitude (y-axis) of the difference signal indicates how far to move the antenna. The design of the angle tracking loop is based on the dynamics of the worst expected radar–target engagements.

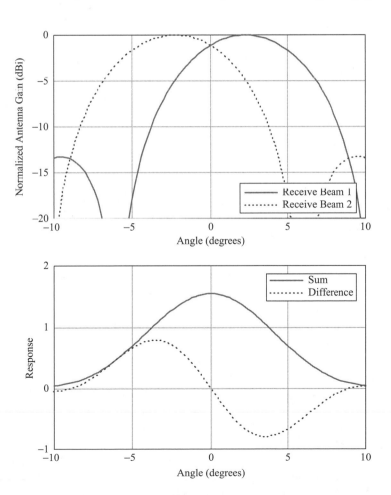

FIGURE 5-43 ▪
Monopulse Tracking: Multiple Beams and Sum and Difference Signals

The bandwidth of the angle tracking loop determines the amount of tracking lag (error) in a particular radar–target engagement scenario. The angle tracking loop is normally designed to provide operation in the linear region of the discriminator curve, that is, the part with the linear slope. Operation outside of the linear region leads to nonlinear response from the tracker and poor tracking performance (e.g., large tracking error, break lock on the target). Some EA techniques (see Chapter 8) seek to cause operation outside the linear region.

Either amplitude or phase comparison approaches can be used for monopulse angle trackers. Amplitude comparison monopulse generates the error signal based on amplitude differences and is often an upgrade from lobing trackers since both are based on signal amplitude. Phase comparison monopulse error signals are based on phase differences. It is easier to make precise phase comparisons than the corresponding amplitude comparisons.

A monopulse angle tracker can theoretically obtain all the information necessary to track a target in angle from each return pulse. Thus, short-term amplitude or phase changes do not induce significant angle errors. A monopulse angle tracker is immune to target RCS fluctuations and can track EA techniques (see Chapter 8) designed to attack lobing trackers, thus making angle tracking of the target (the source of the EA technique) easier [Barton, 1988, pp. 497–503; Chrzanowski, 1990, Section 6.3]. However, most radar systems integrate multiple pulses to provide the S/N required for detection (see Chapter 3). Therefore, most monopulse angle trackers obtain the information necessary to track a target from an integrated pulse train.

5.5 | MEASUREMENT LOOSE ENDS

5.5.1 Accuracy

Another sensible question to ask about radar measurements is, "How accurately they can be made?" An answer is obtained from a great body of work on measurements, measurement resolution, and probability theory. Accuracy is essentially the statistical measure of how well a measurement can be estimated within its resolution. For example, a meter stick marked centimeters has a resolution of 1 centimeter. We can estimate better than 1 centimeter; however, it is just an estimate, and it is a function of other factors. Accuracy is the measure of the quality of the estimate after multiple measurements. In the radar world, this translates to multiple target measurements, each having the same resolution. Therefore, accuracy is often an important target tracking quality.

The usual equation for the potential root mean square (RMS) accuracy is given in Equation (5-40). This accuracy expression is within a few percent of the maximum likelihood estimates of statistics, which are optimum. Manasse [1960] gives the exact equations, while Barton [1979, pp. 38–43] explains the differences. When derived in more sophisticated ways and optimized by various

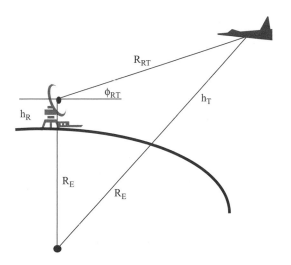

FIGURE 5-44 ■
Target Altitude

methods, the resulting equation changes by small amounts. For example, for $(S/N)_n = 17$ dB, the RMS accuracy is about 1/10 of the measurement resolution.

$$\delta_M = \frac{\Delta M}{\sqrt{2 \left(\dfrac{S}{N} \right)_n}} \tag{5-40}$$

where:

δ_M = Root mean square measurement accuracy, same units as ΔM

ΔM = Measurement resolution

$(S/N)_n$ = Target signal-to-noise ratio (power) after the integration of multiple pulses (see Chapter 3), no units

5.5.2 Target Altitude and Range-Height-Altitude

While not a direct measurement, some radar systems can determine the target altitude. The spherical trigonometry associated with determining the target altitude is shown in Figure 5-44 and includes the measured radar-to-target range, measured radar-to-target elevation angle, radar height, radius of the earth, and radio frequency wave refraction (see Section 13.1) (Equation 5-41). The target altitude resolution is a complicated relationship, with the range resolution, elevation angle resolution, and the radar–target geometry being the main drivers.

$$h_T = \sqrt{(k_r\,R_E + h_R)^2 + R_{RT}{}^2 + 2\,(k_r\,R_E + h_R)\,R_{RT}\,\sin(\phi_{RT})} - k_r\,R_E \tag{5-41}$$

where:

h_T = Target height, meters

k_r = Refraction factor, no units

R_E = Radius of the earth, 6371 kilometers

h_R = Radar height, meters

R_{RT} = Radar-to-target slant range, meters

ϕ_{RT} = Radar-to-target elevation angle, radians or degrees

5.5.3 Radar Resolution Cell

The radar resolution cell is the "intersection" of the range resolution (see Section 5.2.1) and angular resolution (see Section 5.4.1), as shown in Figure 5-45. The resolution cell can also include the range rate resolution (see Section 5.3.1). Often, target tracking is said to work to center the resolution cell about the target.

The vast majority of radar systems have numerous, often millions of, resolution cells. The number of resolution cells is important for establishing the required probability of false alarm (P_{fa}) (see Section 3.2.2). The number of range resolution cells is a function of the unambiguous range (see Section 5.2.3) and the range resolution (see Sections 5.2.1 and 5.2.2) (Equation 5-42). The number of Doppler filters is a function of the PRF and Doppler filter bandwidth (see Section 5.3.1 and Figure 5-31) (Equation 5-43). The number of angular resolution cells is the number of antenna beam positions necessary to provide a desired angular coverage. A mechanically scanned antenna (see Section 4.2) continuously moves through the antenna beam positions. While an electronically steered array (ESA; see Section 4.3.1) discretely steps through the antenna beam positions, the number of antenna beam positions includes both the azimuth (horizontal) plane coverage and the elevation (vertical) plane

FIGURE 5-45 ▪
Radar Resolution
Cell

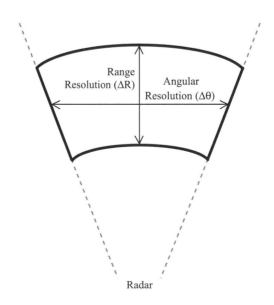

coverage (Equation 5-44). For a radar system performing a sector search, the angular scan coverage is twice the search limits. For example, if the search limits are $\pm 60°$, the angular scan coverage is $2 \times 60° = 120°$.

$$n_{rg} = \frac{R_u}{\Delta R} \tag{5-42}$$

$$n_{df} = \frac{PRF}{\Delta f_d} \tag{5-43}$$

$$n_b = \frac{\theta_S}{\theta_{3dB}} \frac{\phi_S}{\phi_{3dB}} = \frac{\Omega}{\theta_{3dB} \phi_{3dB}} \tag{5-44}$$

where:

n_{rg} = Number of range gates in which detection decisions are made (≥ 1), no units

R_u = Unambiguous range, meters

ΔR = Range resolution, meters

n_{df} = Number of Doppler filters in which detection decisions are made (≥ 1), no units

PRF = Pulse repetition frequency, hertz

Δf_d = Doppler filter bandwidth, hertz

n_b = Number of antenna beam positions in which detection decisions are made (≥ 1), no units

θ_S = Angular scan coverage in the azimuth (horizontal) plane, radians or degrees

ϕ_S = Angular scan coverage in the elevation (vertical) plane, radians or degrees

Ω = Total solid angle scan coverage, radians squared or degrees squared

θ_{3dB} = Antenna half-power (–3 dB) beamwidth in the azimuth (horizontal) plane, radians or degrees

ϕ_{3dB} = Antenna half-power (–3 dB) beamwidth in the elevation (vertical) plane, radians or degrees

5.5.4 Range and Doppler Sidelobes

Pulse compression (see Section 5.2.2) is a very powerful signal processing concept to provide fine range resolution. Likewise, a pulse burst waveform (see Section 5.3) is a very powerful approach to provide fine range rate resolution. However, these have certain deficiencies that make them less than ideal when used in a cluttered environment (see Section 7.1) encountered in many radar–target geometries. These deficiencies arise because the response to the received signal has "sidelobes" in both the range (time) and Doppler (frequency)

FIGURE 5-46 ■
Power Response of
Sidelobes

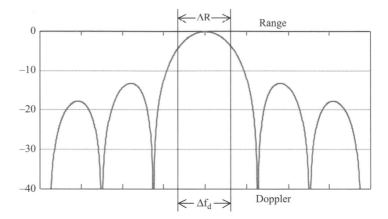

dimensions. The time response of a compressed pulse has the form of $(\sin x/x)^2$, which has monotonically decreasing sidelobes beginning at about 13.4 dB down from the peak response, as shown in Figure 5-46. This is the same as we saw for the antenna gain pattern (see Chapter 4). If the target of interest has a small S/N and another target, say 30 or 40 dB larger, is in a nearby range gate, the spillover of the range sidelobes from the stronger target into the range gate of interest masks the small S/N target. Because the Doppler response is also of the $(\sin x/x)^2$ form, the sidelobes in Doppler are equally troublesome when a large target exists in a nearby Doppler filter.

     To counteract these effects, clever "weighting functions" are often applied to the received signal, just as they are used as antenna illumination functions (see Chapter 4) [Richards et al., 2010, Chapters 8, 14, 17, 20]. For example, when a cosine weighting function, as shown in Figure 5-47, is used, the first sidelobe is down 23 dB from the peak response. The mainlobe, however, is 37% wider than for unweighted (uniform) signal. As suggested in Chapter 4, one can use different weighting functions to achieve certain sidelobe levels. Some of these weighting functions have very low sidelobes, achieved at the expense of worse resolution (due to the widening of the mainlobe). In experimenting, one may run across weighting functions with already famous names. For example, with a Hamming weighting the first sidelobe is about 43 dB down [Temes, 1962].

FIGURE 5-47 ■
Uniform (left) and
Cosine (right)
Weighting Functions

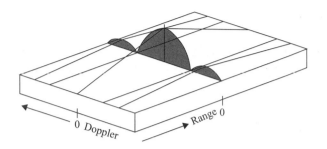

FIGURE 5-48 ■
Ambiguity Function
for a Simple Pulse

5.5.5 Ambiguity Functions

When the range axis and the Doppler axis are combined, they form a plane out of which the responses of various waveforms rise. The resulting three-dimensional surface is called the "Woodward ambiguity function," after Woodward [1955], whose slender volume is an elegant and classical work. The ambiguity function of a simple pulse is shown in Figure 5-48. A whole sector of radar engineering, involving many people working over several decades, has been devoted to manipulating the ambiguity function to obtain an infinitude of desired results. Richards et al. [2010, pp. 800 808] and Rihaczek [1977] contain ambiguity functions for many radar waveforms.

5.6 | SOPHISTICATED TARGET TRACKERS

The previous target tracking sections discussed measurement trackers where, once a target has been detected and a sequence of measurements are made, it can be independently followed in range, range rate (Doppler), and angle. Either the target can be actively followed by the radar system in a special tracking mode, or its whereabouts can simply be noted, by an operator or computer each time it is detected on a radar scan. The latter technique is called "track-while-scan" [Barton, 1979, pp. 82–86; Skolnik, 2001, pp. 212, 252–254; Stimson, 1998, pp. 388–390]. This operation involves correlations and extensive calculations but virtually no radar operations. I shall not discuss it further and refer the reader to the aforementioned references.

A phased array radar system with inertialess, electronically steered beams (see Section 4.3.1) can maintain a routine scan while actively tracking many targets in a mode of operation known as "track-and-scan." We will discuss additional radar modes of operation provided by ESA radar systems in Section 7.4.

Because their ability to track targets at varying angles is physically limited, most single-beam reflector radar systems have only a few tracking channels (one to four). A phased array may have very many (several hundred). The operator of a flight control radar (track-while-scan) may carry tens of tracks in his head. Radar tracking capabilities vary widely and depend on the mission to be accomplished and the radar design. We will discuss two sophisticated target trackers in the following sections: correlated measurement trackers and target state trackers.

5.6.1 Correlated Measurement Trackers

Instead of independently tracking each measurement, a correlated measurement tracker considers multiple measurement domains together. A correlated measurement tracker making use of range and range rate together is shown in Figure 5-49. Sometimes angle measurements are also included for correlation. As the tracker's name implies, the individual measurements must correlate with each other. For example, if the range measurements are changing quickly, the range rate measurements must have the corresponding values. If they do not, the tracker will know something is inconsistent with the range and range rate measurements. This attribute of a correlation tracker provides a powerful EP technique (see Chapter 8). To deceive a correlation tracker, the EA techniques must ensure that the range, range rate, and angle jamming waveforms all correspond, which is a tall order.

5.6.2 Target State Trackers: Alpha-Beta and Kalman Filters

Target state trackers are used to enhance the performance of radar systems. Target state trackers use current target measurements and previous target information to develop a target state estimate, much as we do when we are driving. A target state tracker is shown in Figure 5-50. A target state tracker develops a smoothed, or filtered, estimate of target state (e.g., position, velocity, acceleration) vectors. It can also develop a predicted target state for some time in the future. Target tracking concepts and algorithms have evolved rapidly with the exponential evolution of computers' increase in speed and capacity and decrease in cost and size. The best known target state trackers are the alpha-beta tracker and the Kalman filter, after Kalman and Bucy, who introduced it in 1960.

FIGURE 5-49 ■
Correlated
Measurement
Trackers

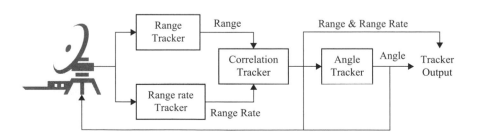

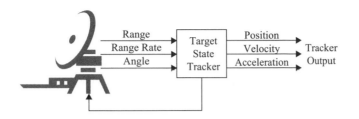

FIGURE 5-50 ■
Target State Tracker

An alpha-beta tracker uses a relatively straightforward set of equations to develop a smoothed target position, smoothed target velocity, and predicted target position (Equation 5-45) [Jeffery, 2009, Chapter 5; Richards et al., 2010, Chapter 19]. If target acceleration is significant, an alpha-beta-gamma tracker is used. Gamma is the acceleration smoothing parameter. There are many ways to select the values for alpha and beta, but unfortunately there are no optimum values. A compromise is made between good smoothing of random measurement errors (narrow bandwidth) and rapid response to maneuvering targets (high bandwidth). An alpha-beta tracker is relatively easy to implement but does not handle maneuvering targets. It can be enhanced with additional algorithms to handle maneuvering targets.

$$\widehat{x}_n = x_{pn} + \alpha(x_n - x_{pn})$$

$$\widehat{v}_n = \widehat{v}_{n-1} + \frac{\beta}{T_s}(x_n - x_{pn}) \qquad (5\text{-}45)$$

$$x_{p(n+1)} = \widehat{x}_n + \widehat{v}_n\, T_s$$

where:

$\widehat{x}_n$ = Smoothed target position, meters
X_{pn} = Predicted target position, meters
α = Position smoothing parameter, no units
X_n = Measured target position, meters
$\widehat{v}_n$ = Smoothed target velocity, meters/second
β = Velocity smoothing parameter, no units
T_s = Time between measurements, seconds

A Kalman filter can inherently handle a maneuvering target through the use of a model of target measurement errors and a model for the target trajectory and its associated uncertainty [Jeffery, 2009, Chapter 5; Richards et al., 2010, Chapter 19]. The Kalman filter is not a filter at all in the electronic circuit sense. Its solutions to radar estimating problems require extensive and tedious calculations including matrix inversions, but the method converges rapidly to useful estimators with a minimum of observations. A feature of the Kalman filter is that it operates only in the present, sometimes giving only the distillate of all history equal billing with the latest observation.

A quantitative understanding of the Kalman filter is available from the idea of a "running average" combined with Bayes formula (published over 200 years ago) for conditional probabilities [Sheats, 1977]. A running average is calculated by subtracting the n-th observation from the previous average and dividing it by the number of observations (Equation 5-46). Noting that the running average is an after-the-event measure and that the n-th observation is weighted the same as all previous observations, two questions arise: Could the running average be used to estimate future events; and is there an optimum weighting for the most recent observation?

$$\overline{X}_n = \overline{X}_{n-1} + \frac{(O_n - \overline{X}_{n-1})}{n} \qquad (5\text{-}46)$$

where:

$\overline{X}_n$ = Average after n observations
$\overline{X}_{n-1}$ = Average after n – 1 observations
O_n = Value of the n-th observation

Bayes's formula gives the probability of an event occurring, conditional on another event already having occurred, and it provides for the weighting of the most recent observation based on the quality of the observation compared with the distilled quality of all the previous ones. For radar measurements, the quality of an observation is given in terms of a standard deviation, which is the square root of the variance, and the standard deviation decreases as the number of observations increases, being inversely proportional to the square root of the number of observations. An intuitively appealing weighting would be one that gave heavier weighting to the "best" measurements, which in this case would mean those with the smallest variance. A weighting function with this appeal (it can also be shown to be optimum) is given in Equation (5-47). Note if the variance of the most current measurement is large, the weighting function tends toward zero. If the variance is small, the weighting function is nearly one.

$$W = \frac{\sigma^2_{/n}}{\sigma^2_{/n} + \sigma_0^{\ 2}} \qquad (5\text{-}47)$$

where:

W = Weighting function
$\sigma^2_{/n}$ = Variance of all measurements that have gone before
$\sigma_0^{\ 2}$ = Variance of the most current measurement

Returning to the equation for a running average, changing the average to an estimate of the mean, and adding the weighting factor results in an equation that illustrates the strategy of the Kalman filter (Equation 5-48). Note how a poor observation, large $\sigma_0{}^2$, will change the estimate little, while a good one will be fully effective. As soon as an additional observation is available, the previous estimate becomes part of the running mean, a new weighting function is entered, and a new estimate is calculated.

$$\overline{X}' = \overline{O}_0 + \overline{W}\,(\overline{X} - \overline{O}_0) \qquad (5\text{-}48)$$

where:

$\overline{X}'$ = Estimate
$\overline{O}_0$ = New observation
$\overline{X}$ = Current mean

The Kalman filter estimates are limited to processes written as differential or difference equations. Obviously, discontinuous future events, such as drastic maneuvering by a tracked target, cannot be accommodated. Nevertheless, for predictive tracking problems, and particularly for navigation and guidance applications, the Kalman filter has proven invaluable. An extended Kalman filter adds a model of target dynamics to overcome the limitations of a Kalman filter with respect to drastic target maneuvers. Extended Kalman filters are essential when dealing with missing measurements, variable measurement errors, and maneuvering targets. Extended Kalman filters are widely used in airborne fire control radar systems.

Kalman filters provide a powerful EP technique (see Chapter 8). To deceive a Kalman, the EA techniques must ensure that the range, range rate, and angle jamming waveforms all provide quality "measurements," match the target dynamics model, and reinforce existing target estimates and predictions, a very tall challenge.

5.7 | SUMMARY

The ability of a radar system to measure characteristics of a target, and performance in doing so, is directly related to the radar system characteristics: carrier frequency, coherence, pulse width, pulse repetition interval, pulse repetition frequency, modulation bandwidth, waveform duration, and antenna beamwidth. The radar system characteristics are optimized for selected measurements and performance. A low PRF provides long unambiguous range. A high PRF provides large unambiguous range rate. A short pulse width or modulated carrier frequency provides fine range resolution. A long waveform duration, or integration time, provides fine range rate resolution. A high carrier frequency provides high Doppler shifts per unit target range rate. And a narrow antenna beamwidth provides fine angular resolution.

Because range and range rate ambiguities are both a function of the PRF, radar systems and/or radar modes are often categorized by PRF: low, high, and medium. Low PRF, also called pulsed, provides range measurements out to a long unambiguous range. High PRF, also called pulsed Doppler, provides range rate measurements over a large unambiguous range rate. Medium PRF has two contrary definitions: for airborne medium PRF radar systems both range and range rate measurements are ambiguous, whereas for ground-based medium PRF radar systems both range and range rate measurements are unambiguous. Low, medium, and high are measurement perspectives, not numerical values. What is considered a "high" numerical PRF for a ground-based radar system would be considered a "low" numerical PRF for an airborne radar system.

Most search radar systems used for target surveillance over considerable ranges are ambiguous in either range or range rate. Both range and range rate ambiguities can be resolved by using multiple PRIs. Alternatively, the radar systems can use waveforms of unambiguous range in one of its modes and waveforms of unambiguous range rate in another.

Target tracking is performed using a sequence of multiple measurements and a closed-loop control system and/or algorithm implemented in a computer. The time between multiple measurements can vary from the PRI to the integration time to the antenna scan time. The bandwidth of the tracker determines the tracking lag (error). Trackers are normally designed to provide operation in the linear region of their discriminator curve. Operation outside the linear region leads to poor tracking performance (e.g., large tracking error, break lock on the target).

There are many ways to obtain a desired target measurement and tracking capability and performance. Radar detection performance is also a function of the characteristics of the radar system and its waveform (see Chapter 3). Radar engineers knowledgeably balance these interrelationships to ensure that the radar system has the required detection, measurement, and tracking capability and performance. Radar is one big systems engineering problem.

5.8 | EXERCISES

5-1.   A radar transmits a $\tau = 0.5$ microsecond pulse at a pulse repetition frequency PRF = 300 Hz. What range resolution and unambiguous range can this radar achieve?

5-2.   What is the dead zone of the radar whose waveform was designed in the preceding question?

5-3.   A space surveillance radar system has the requirement to resolve spacecraft in range if they are 30 meters apart ($\Delta R = 30$ m) and in Doppler if they are moving at range rates that differ by 15 meters/second ($\Delta R_{dot} = 15$ m/sec). Design a compressed pulse burst waveform at X band ($f_c = 10$ GHz) that will meet the requirements.

5-4.   An aircraft approach control radar system with a maximum unambiguous range $R_u = 40$ kilometers has the requirements to resolve aircraft 30 meters apart in range ($\Delta R = 30$ m) and 15 meters/second separation in range rate ($\Delta R_{dot} = 15$ m/sec). Design a pulse burst waveform for this radar at C band ($f_c = 6$ GHz).

5-5.   The waveform of the preceding question is unambiguous in range. The aircraft to be handled have range rates over ± 300 meters/second. (a) Design a waveform that is unambiguous in range rate. (b) What is the unambiguous range of this waveform?

5-6.   A high-precision S-band ($f_c = 3000$ MHz) tracker needs to resolve targets 1 km apart in cross-range ($\Delta CR = 1$ km), 15 meters in range ($\Delta R - 15$ m), and 3 meters/second in range rate ($\Delta R_{dot} = 3$ m/sec) at ranges of 200 km. (a) What diameter is the antenna if it has a uniform current distribution D (meters)? (b) What are the characteristics of the waveform?

5-7.   A radar system has a frequency modulated (FM) chirp waveform with a bandwidth of $B_{pc} = 100$ MHz over a transmitted pulse width $\tau = 200$ microseconds. (a) What is the time-bandwidth product? (b) What is the range resolution, ΔR_{pc} (meters)?

5-8.   Show that the signal-to-noise ratio (S/N) with pulse compression signal processing equals the S/N without. Assume FM or chirp pulse compression.

5-9.   A radar system with a transmitted pulse width of $\tau = 2$ μsec is required to have a range resolution $\Delta R - 5$ meters. Can the radar system provide the required range resolution with its transmitted pulse width? Why or why not? If not, phase modulation pulse compression will be used. How many phase coded segments or chips are necessary to provide the required range resolution $N\phi$ (no units)?

5-10. A radar system has the following characteristics: PRF $= 1000$ Hz, transmitted pulse width $\tau = 1$ μsec, pulse burst duration $T_I = 40$ msec, azimuth beamwidth $\theta_{3dB} = 2$ degrees, and elevation beamwidth $\phi_{3dB} = 4$ degrees. The radar system has an angular scan coverage in the azimuth plane of $\theta_S = 360$ degrees and in the elevation plane of $\phi_S = 8$ degrees. Determine the number of range gates n_{rg} (no units), the number of Doppler filters n_{df} (no units), and the number of antenna beam positions n_b (no units).

5-11. (a) Design a pulse-Doppler (pulse burst) waveform for a transmitted frequency $f_c = 3000$ MHz that is unambiguous in range rate for aircraft flying at $R_{dotu} = \pm 900$ meters/second and can resolve aircraft separated by 30 meters in range (ΔR) and 15 meters/second in range rate (ΔR_{dot}). (b) What is the associated unambiguous range, R_u (meters)? (c) Will the waveform from part (a) support the required measurement performance at ultra high frequency (UHF) ($\lambda = 0.6$ meters)?

■■■ 5.9 ■ | REFERENCES

Barton, D. K., 1979, *Radar System Analysis*, Dedham, MA: Artech House.
  Abstruse theory is made understandable in this detailed work, oriented toward analysis and test of radar systems.
Barton, D. K., 1988, *Modern Radar System Analysis*, Norwood, MA: Artech House.
Chrzanowski, Edward, 1990, *Active Radar Electronic Countermeasures*, Norwood MA: Artech House.
Deley, G. W., 1964, *The Representation, Detection Estimation and Design of Radar Signals*, Santa Barbara, CA: Defense Research Corporation.
Deley, G. W., 1970, "Waveform Design," in M. I. Skolnik, (ed.), *Radar Handbook*, New York: McGraw-Hill.
Edde, Byron, 1993, *Radar Principles, Technology, Applications*, Upper Saddle River, NJ: Prentice-Hall.
Goj, Walter W., 1993, *Synthetic Aperture Radar and Electronic Warfare*, Norwood, MA: Artech House.
  One of the best, and concise, discussions of the effect of pulse compression and Doppler processing (as one needs for a synthetic aperture radar) on the signal-to-noise ratio.
Jeffery, Thomas W., 2009, *Phased-Array Radar Design Application of Radar Fundamentals*, Raleigh, NC: SciTech Publishing.
Jenkins, F. A., and H. E. White, 1976, *Fundamentals of Optics*, New York: McGraw-Hill.
North, D. O., 1943, "An Analysis of Factors Which Determine Signal/Noise Discrimination in Pulsed Carrier Systems," RCA Labs Technical Report PTR6C. (Reprinted in *Proceedings of the IRE*, Volume 51, July 1963, pp. 1016–1027.)
Rhodes, D. R., 1959, *Introduction to Monopulse*, New York: McGraw-Hill.
Richards, M., Scheer, J., and Holm, W. (editors), 2010, *Principles of Modern Radar*, Volume I: Basic Principles, Rayleigh, NC: SciTech Publishing.
Rihaczek, A. W., 1997, *Principles of High Resolution Radar*, Palo Alto, CA: Peninsula Press.
Sheats, L., 1977, "Fast Fourier Transform," in E. Brookner (ed.), *Radar Technology*, Norwood, MA: Artech House.
Sheats, L., 1977, "The Kalman Filter," Chap. 26 in E. Brookner (ed.), *Radar Technology*, Norwood, MA: Artech House.
Skolnik, Merrill I., 2001, *Introduction to Radar Systems*, 3rd Edition, New York: McGraw-Hill.
Stimson, George W., 1998, *Introduction to Airborne Radar*, 2nd Edition, Raleigh, NC: SciTech Publishing.
Temes, C. L., 1962, "Sidelobe Suppression in a Range-Channel Pulse Compression Radar," *IRE Transactions*, Volume MIL-6, April, pp. 162–167.
Woodward, P. M., 1955, *Probability and Information Theory with Applications to Radar*, New York: McGraw-Hill.

Target Signature

HIGHLIGHTS

- Definition of radar cross section
- Start with the radar cross section of simple geometric shapes
- Radar cross section of a sphere, the king of the simple shapes
- Radar cross section of complex, nonsimple shapes, objects
- Target spectrum
- Effect of radar cross section on radar detection range
- Effect of polarization
- Radar cross section of dipoles and clouds of dipoles (chaff)
- The radar cross section of various types of clutter (ground and water)
- Inferring physical characteristics from radar cross section (radar signature)

Radar cross section (RCS) is a measure of the electromagnetic wave intercepted and reradiated at the same wavelength by any object. The dimensions are those of area, usually square meters (m^2), or decibels relative to a square meter (dBsm). The relationship between square meters and dBsm is given in Equation (6-1). The RCS of an object is a complicated function of multiple factors: size, shape, material, edges, wavelength (frequency), and polarization. The RCS of most objects change over time, as was discussed with the impact of target RCS fluctuation or scintillation on radar detection theory in Chapter 3. The changing RCS is a function of changes in the radar waveform, frequency or polarization, and changes in object characteristics and/or orientation relative to the radar system.

$$dBsm = 10\log(m^2) \qquad m^2 = 10^{\left(dBsm/10\right)} \qquad (6\text{-}1)$$

Electromagnetic waves are normally diffracted, absorbed, and/or scattered when incident on an object. Electromagnetic waves reflected back to the radar

FIGURE 6-1 ■
Forward and Back
Scattering from
a Target

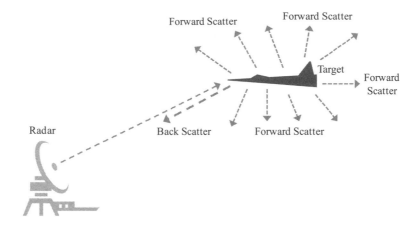

FIGURE 6-1 ■ Forward and Back Scattering from a Target

system are called "backscatter," while electromagnetic waves reflected in all other directions are called "forward scatter." The forward and backscattering of electromagnetic waves off a target is shown in Figure 6-1.

Simple objects tend to have a single, or few, scattering sources. However, complex objects (e.g., airplanes, ships, vehicles) tend to have multiple scattering sources (e.g., nose, fuselage, inlet, wing root, wing, sensors). Thus, for complex objects the RCS is the complex (amplitude and phase) combination of contributions from each scattering source. As the characteristics of the object vary with aspect angle, so does the RCS, often times fluctuating rapidly. Consequently, the RCS of a complex object is not a single value but a pattern of RCS values as a function of wavelength, polarization, azimuth, and elevation angle. For now, we will assume the phase differences of the potentially numerous scattering sources are "indistinguishable" and we are in what is called the "far field" of the target. How to determine what is indistinguishable and the far-field distance is described in Section 13.3.

The RCS of an object can be determined by solving Maxwell's equations. The RCS of an object can also be established by measuring it, either the actual object or a scale model, and comparing it to a reference object. For complex objects attempts to solve Maxwell's equations for the various boundary conditions have not been very successful. RCS prediction computer codes are able to calculate the RCS for some rather complicated objects. Computer programs essentially break down a complicated target into many simple surfaces and superpose their radar cross sections (amplitude and phase) to compute the overall radar cross section. RCS prediction codes have been very widely used in the past few decades.

▌ 6.1 ▐ RADAR CROSS SECTION OF A SPHERE

The understanding of RCS concepts generally starts with an idealized object that is large with respect to a wavelength, has an intercept area of one square unit, is perfectly conducting, and reradiates isotropically. It is easy to build an object with these characteristics: a copper sphere is an example. Providing it is large with respect to the wavelength of the incident electromagnetic wave, a copper sphere of projected area 1 m² has radar cross section of 1 m² (Equation 6-2). For example, the RCS of a 1 meter radius sphere is $\pi(1)^2 = 3.1416$ m² $\equiv 4.9715$ dBsm.

$$\sigma_s = \pi\, r^2 \qquad \frac{2\pi\, r}{\lambda} > 10 \qquad (6\text{-}2)$$

where:

σ_s = Radar cross section of sphere, square meters (m²)
r = Radius of the sphere, meters
λ = Wavelength, meters

For objects with at least one dimension that is small with respect to a wavelength, the simple relationships no longer hold. Again, the conducting sphere is an excellent example. When the circumference of the sphere becomes about equal to a wavelength, the currents induced on the surface of the sphere radiate and add to the reflected wave. When the circumference, $2\pi r$, is somewhat greater than $3\lambda/2$, the radiation from induced currents tends to cancel the direct radiation. At $2\pi r > 2\lambda$, there is enhancement again, at $2\pi r > 5\lambda/2$, there is cancellation, and so forth. This zone of enhancement and cancellation is called the Mie or Resonance region. As the circumference becomes several λ, there are bands of current traveling in opposite directions on the body so that the effects of addition and subtraction are damped. The result of all these phase additions and subtractions is the famous curve for the RCS of a conducting sphere (Figure 6-2) [Crispin and Maffett, 1968, pp. 86–87; Rheinstein, 1968, p. 5].

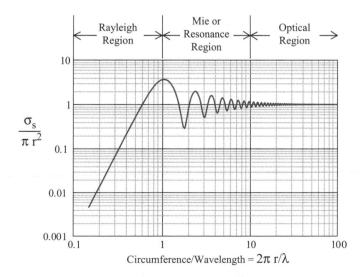

FIGURE 6-2 ■
Radar Cross Section of a Sphere

Although this curve is strictly applicable only to spheres, it is approximate for many objects with roughly equal measurements in all three dimensions, and in the Rayleigh region it becomes a fairly accurate measure of the volume of many objects. The Rayleigh region was defined in classical physics, when Lord Rayleigh first worked out the λ^4 dependency of the amount of light scattered by molecules in the earth's atmosphere. The relationship is analogous to the scattering of electromagnetic waves from objects that are small compared to a wavelength. Although straightforward, the derivation is lengthy but is included in many university texts of basic optics [Jenkins and White, 1976, pp. 467, 471–473]. RCS in the Rayleigh region is given in Equation (6-3).

$$\sigma_{\text{Rayleigh}} = 8\pi^3 \frac{V^2}{\lambda^4} \qquad (6\text{-}3)$$

where:

σ_{Rayleigh} = Radar cross section of a sphere in the Rayleigh region, square meters (m^2)

V = Volume of the sphere, cubic meters (m^3)

▌ 6.2 | RADAR CROSS SECTION OF SIMPLE GEOMETRIC OBJECTS

For several classes of simple geometric objects, RCS can be easily calculated. When the target does not reradiate isotropically, it may have gain in the direction of the radar (Equation 6-4).

$$\sigma = G \, A_e \qquad (6\text{-}4)$$

where:

σ = Radar cross section for nonisotropic object, square meters (m^2)

G = Re-radiation gain in the direction of the radar, no units

A_e = Electromagnetic area of the object as seen by the radar, square meters (m^2)

In Chapter 2, I showed that the reradiation gain is a function of the electromagnetic area and wavelength (Equation 6-5). It should always be remembered that the area is the electromagnetic area as seen by the radar at the instant the radar cross section is being induced. From this, the calculation of the RCS of a flat plate is trivial (Equation 6-6).

$$G = \frac{4\pi \, A_e}{\lambda^2} \qquad (6\text{-}5)$$

$$\sigma_{fp} = \frac{4\pi\,A^2}{\lambda^2} \tag{6-6}$$

where:

σ_{fp} = Radar cross section for a flat plate, square meters (m^2)
A = Area of the flat plate, square meters (m^2)

For example, the RCS of a car license plate (0.3×0.15 m) at 24 GHz (a common police radar frequency) is given in Equation (6-7). People are often amazed at this large numerical value and wonder if they cannot accurately operate their calculator. However, we need to realize the RCS of an object is an electromagnetic relationship what we use units of area to describe, not the physical area we are much more familiar with.

$$\sigma_{fp} = \frac{4\pi\,(0.3 \times 0.15)^2}{\left(\dfrac{3 \times 10^8}{24 \times 10^9}\right)^2} = 162.86\ \text{m}^2 \equiv 22.12\ \text{dBsm} \tag{6-7}$$

Calculating retrodirective targets, such as corner reflectors where the radar is seeing the equivalent of a flat plate, is straightforward. Figure 6-3 shows a two-face (dihedral) and a three-face (trihedral) corner reflector. The faces are at right angles to each other. Some basic trigonometry reveals that many plane waves arriving at these reflectors are directed back toward the radiating source in phase, making the maximum RCS equal to the projected area of the corner reflector [Knott, Shaeffer, and Tuley, 1985, p. 178]. If all facets of these reflectors are equal-size squares, then the RCS of the dihedral and trihedral are given in Equations (6-8) and (6-9), respectively.

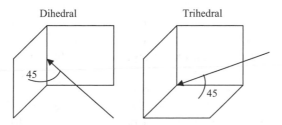

Dihedral   Trihedral

45   45

Note: Arrows Indicated Entry Angle for Maximum RCS

FIGURE 6-3 ■
Two-Corner
Reflector Designs

$$\sigma_d = \frac{8\pi\, a^4}{\lambda^2} \tag{6-8}$$

$$\sigma_t = \frac{12\pi\, a^4}{\lambda^2} \tag{6-9}$$

where:

σ_d = Radar cross section of a dihedral corner reflector, square meters (m^2)
a = Dimension of the square, meters
σ_t = Radar cross section of a trihedral corner reflector, square meters (m^2)

From the idea of intercepted area and reradiated gain, the broadside RCS of a cylinder is easily deduced. A cylinder has projected area of the product of its length and diameter. It reradiates isotropically in the plane perpendicular to its axis of symmetry. Its RCS is given in Equation (6-10). End-on, it is a flat plate.

$$\sigma_{cly} = \frac{\pi\, L^2\, d}{\lambda} \tag{6-10}$$

where:

σ_{cly} = Radar cross section of a cylinder, square meters (m^2)
L = Length of the cylinder, meters
d = Diameter of the cylinder, meters

If we let the diameter of the cylinder get small, we have a long wire. Electrically (with respect to the wavelength), the diameter does not become vanishingly small. However, its contribution decays exponentially with the ratio of the wire diameter to the wavelength. Crispin and Maffett [1968] worked out a relationship for wires many wavelengths long with diameters of two wavelengths or less (Equation 6-11).

$$\sigma = \pi\, L^2 \left(\frac{d}{2\lambda}\right)^{0.57} \tag{6-11}$$

where:

σ_w = Radar cross section of a long wire, square meters (m^2)

The radar cross sections of objects with characteristic dimension 2a tend to be like πa^2 as long as $2\pi a \approx \lambda$. Thus, reentry vehicles (RV) with base diameters of about 0.6 meters, at very high frequencies (VHF) where $\lambda = 1.8$ m, tend to have a nose-on RCS of approximately 1 m^2. At higher frequencies, where geometric optics prevail and the induced currents cancel or are attenuated, only the nose tip and discontinuities such as joins cause reflections back to the radar from a nose-on aspect. A smooth ice cream cone RV will tend to have $\sigma_{ave} = 0.1\lambda^2$ until nose-tip scattering becomes a factor. The RCS of the nose tip is given in Equation (6-12).

$$\sigma_{nosetip} = \pi\, a_{nt}^{2} \qquad a_{nt} \geq \frac{\lambda}{2\pi} \qquad (6\text{-}12)$$

where:

$\sigma_{nosetip}$ = Radar cross section of nose tip, square meters (m^2)
   a_{nt} = Radius of the nose tip, meters

For a 25 mm nose radius at S band, $\sigma = 0.002$ m^2 ≡ –27 dBsm. An RV with a flat back will have an RCS varying between the sum and difference of the $\sigma_{nosetip}$ and the $\sigma_{flatback}$, the latter approximating the RCS of a wire loop of the same circumference and therefore being the dominant scatterer; its RCS is given in Equation (6-13).

$$\sigma_{flatback} = \pi\, r^{2} \qquad r \geq \frac{\lambda}{2\pi} \qquad (6\text{-}13)$$

where:

$\sigma_{flatback}$ = Radar cross section of the flat back, square meters (m^2)
   r = Radius of the flat back, meters

The RCS of these and other simple perfectly conducting shapes are shown in Table 6-1. Jenn [2005, Section 1.6] discusses many of these simple shapes and complex shapes in detail; supporting RCS calculation software can be found on the publisher's website.

6.3 | TARGET SPECTRUM

The spectrum of a simple moving target has the same spectrum as the transmitted radar waveform, only shifted by the Doppler shift (see Section 5.3). The spectrum of a complex moving target also has the contributions from the different components of the target moving at diverse speeds relative to the radar system. Examples of these different target components include rotating propellers, rotor blades, and jet engine compressor and turbine blades. The spectral components from a jet engine are often called jet engine modulation (JEM). JEM spectral lines are primarily a function of radar-to-target aspect angle, engine number, type, speed, and target illumination time. As discussed with target range rate-Doppler measurements in Section 5.3, the target spectrum will be resolved only if the target illumination time is long enough. Therefore, even though the target has a spectrum, the radar system may not be able to resolve it. Skolnik [2001, pp. 380–383] shows example target spectrums associated with JEM and propeller modulation.

TABLE 6-1 ■ Radar Cross Sections of Simple Shapes

Shape	Aspect Angle	RCS	Constraints	Symbols
Sphere	Any	πr^2	$\dfrac{2\pi r}{\lambda} > 10$	r = radius
Sphere	Any	$8\pi^3\left(\dfrac{V^2}{\lambda^4}\right)$	$\dfrac{2\pi r}{\lambda} < 1$	V = volume
Flat Plate	Broadside	$\dfrac{4\pi A^2}{\lambda^2}$	$\dfrac{2\pi a}{\lambda} > 1$	A = area a = dimension
Cone	Axial	$\dfrac{\lambda^2}{16\pi}\tan^4(\theta)$		θ = cone half angle
Cone Sphere	Nose-on	$0.1\lambda^2$	$\dfrac{2\pi r_{\text{nosetip}}}{\lambda} < 1$	r_{nosetip} = nose tip radius
Cone Sphere	Nose-on	$\approx \pi r_{\text{nosetip}}^2$	$\dfrac{2\pi r_{\text{nosetip}}}{\lambda} > 1$	
Truncated Cone	Nose-on	$\approx \pi r_{\text{flatback}}^2$ $\pm\sigma_{\text{nosetip}}$	$\dfrac{2\pi r_{\text{flatback}}}{\lambda} > 1$	r_{flatback} = flat back radius
Cylinder	Broadside	$\dfrac{\pi d L^2}{\lambda}$	$\dfrac{2\pi d}{\lambda} > 1$	d = diameter L = length
Long Wire	Broadside	$\pi L^2\left(\dfrac{d}{2\lambda}\right)^{0.57}$	$\dfrac{L}{\lambda} \gg 1$	d = diameter L = length
Resonant Dipole	Broadside	$\dfrac{\pi}{4}\lambda^2$	$L = \dfrac{\lambda}{2}$	L = length
Resonant Dipole	Spherically Random	$0.15\lambda^2$	$L = \dfrac{\lambda}{2}$	
Cloud of Resonant Dipoles	Random	$0.15\,N\lambda^2$	$L = \dfrac{\lambda}{2}$	N = number of dipoles
Convex Surfaces	Random	$\pi r_x r_y$	$\lambda \ll r_x$ and r_y	r_x = radius of curvature, x-plane r_y = radius of curvature, y-plane
Dihedral Corner Reflector	Axis of	$\dfrac{8\pi a^4}{\lambda^2}$	$\lambda < a$	a = dimension
Trihedral Corner Reflector	Axis of	$\dfrac{12\pi a^4}{\lambda^2}$	$\lambda < a$	a = dimension
Triangular Corner Reflector	Axis of	$\dfrac{4\pi a^4}{3\lambda^2}$		a = dimension

6.4 | TARGET RADAR CROSS SECTION AND RADAR DETECTION RANGE

It is useful to know how changes in the target RCS translate into changes in the detection range, which of course is an essential radar performance metric (see Section 3.3). When all the radar and geometry/environment characteristics are the same, the relationship between detection ranges and target RCS is given in Equation (6-14). The change in detection range as a function of the change in the target RCS is shown in Figure 6-4. In this figure, the radar detection range is normalized to the value at a target RCS of 0 dBsm (1 m^2). As seen in this figure, the detection range doubles when the RCS increases by 12 dB (16 times) and halves when the RCS decreases by 12 dB (1/16).

$$R_{dt2} = \sqrt[4]{\frac{\sigma_2}{\sigma_1}} R_{dt1} \qquad (6\text{-}14)$$

where:

R_{dt2} = Radar detection range for radar cross section σ_2, meters

σ_2 = Target radar cross section for radar detection range R_2, square meters (m^2)

R_{dt1} = Radar detection range for radar cross section σ_1, meters

σ_1 = Target radar cross section for radar detection range R_1, square meters (m^2)

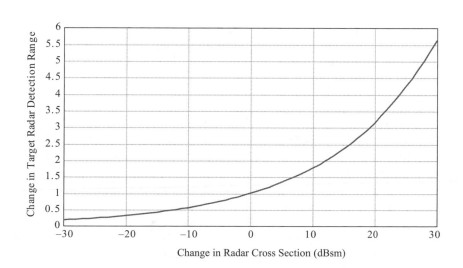

FIGURE 6-4 ■
Change in Radar Detection Range as a Function of the Change in the Target Radar Cross Section

■■■ 6.5 | POLARIZATION

The effects of polarization (see Section 2.2) on RCS can be dramatic. Smooth targets do not depolarize the incident electromagnetic wave. In turn, the reflected wave has the same polarization as the incident wave, but in the opposite sense (the electric vector has flipped over). A weather radar designer who wanted to view droplets and clouds would not worry excessively about polarization. However, a radar system meant to view complex (and depolarizing) targets through weather might transmit horizontal polarization and receive vertical. Or (as is usual) the radar might transmit right-hand circular and receive left-hand circular polarization. Thus, raindrop returns would be filtered out. Aircraft control radar systems employ this technique.

In general, the effects of polarization can be estimated by knowing the angles made with the incident electric vector by edges and long, thin pieces of the target (Equation 6-15). The average RCS is given in Equation (6-16). This result says, on the average, a randomly depolarizing target returns only half the power of a nondepolarizing target. Of course, the target may have preferential polarization angles, which greatly reduces the return. When the target's polarization characteristics are unknown, the return of at least $\sigma_0/2$ can be assured by transmitting circular polarization.

$$\sigma = \sigma_0 \cos^2(\theta) \tag{6-15}$$

$$\overline{\sigma} = \frac{2\sigma_0}{\pi} \int_0^{\pi/2} \cos^2(\theta)\, d\theta = \frac{\sigma_0}{2} \tag{6-16}$$

where:

σ = Radar cross section at a polarization angle, square meters (m^2)
σ_0 = Radar cross section of a nondepolarizing target, square meters (m^2)
θ = Polarization angle, radians
$\overline{\sigma}$ = Average radar cross section across all polarization angles, square meters (m^2)

The return signal is further modified by the characteristics of the receive system. It may receive the same polarization it transmits, or it may receive on two orthogonal polarizations. It is easy to speak in quantitative terms about the various alternatives.

If the intervening medium rotates the polarization angle in a random way, a horizontally polarized transmitter, a randomly depolarized target, and a horizontally polarized receiver will deliver an average RCS as given in Equation (6-17). If the intervening medium does not further depolarize or rotate the polarization angle, then the average RCS is given in Equation (6-18).

$$\overline{\sigma}_{HH} = \frac{2\sigma_0}{\pi} \int\limits_0^{\pi/2} \cos^2(\theta)\, d\theta \cdot \frac{2}{\pi} \int\limits_0^{\pi/2} \cos^2(\phi)\, d\phi = \frac{\sigma_0}{4} \qquad (6\text{-}17)$$

$$\overline{\sigma}_{HH} = \frac{2\sigma_0}{\pi} \int\limits_0^{\pi/2} \cos^4(\theta)\, d\theta = \frac{3\sigma_0}{8} \qquad (6\text{-}18)$$

where:

$\overline{\sigma}_{HH}$ = Average radar cross section (horizontal transmit–horizontal receive), square meters (m^2)

ϕ = Random polarization angle, radians

If a vertically polarized receiver is added, the average RCS is given in Equation (6-19). The sum of the power in the two (horizontal and vertical) channels is $\sigma_0/2$, which is all that is available from the target. Thus, a radar system dealing with targets and media with largely uncertain polarization characteristics minimizes the possible impact of polarization effects by transmitting circular polarization and receiving on two orthogonal linear polarizations. This arrangement guarantees that if the receive channels are summed then the average RCS will equal $\sigma_0/2$.

$$\overline{\sigma}_{HV} = \frac{2\sigma_0}{\pi} \int\limits_0^{\pi/2} \cos^2(\theta)\sin^2(\theta)\, d\theta = \frac{\sigma_0}{8} \qquad (6\text{-}19)$$

where:

$\overline{\sigma}_{HV}$ = Average radar cross section (horizontal transmit–vertical receive), square meters (m^2)

6.6 | CHAFF CHARACTERISTICS

One class of objects whose performance is greatly polarization sensitive is chaff. These are small, light pieces of material with collectively high radar cross section. The use of chaff as an electronic attack (EA) concept is discussed in Chapter 10. The RCS of a single half-wave resonant dipole can be calculated analytically in several ways [Van Vleck, Block, and Hammermesh, 1947]. Experimental results, however, do not agree absolutely with the detailed analysis. It is therefore more convenient to use simple theory to derive radar cross section, especially because the experimental measurements agree relatively closely with simple theory.

Returning to the ideas developed in Chapter 2, we can find the maximum gain of a half-wave dipole in free space. Recall that when the far field of an array (see Section 4.3) contains no grating lobes its near field is fully filled. This occurs when the element separation does not exceed $\lambda/2$ and implies the intercept area of an element in a fully filled array cannot exceed $\lambda^2/4$. Applying the equation $G = 4\pi A/\lambda^2$ to a dipole element gives $G = \pi$. However, this is a dipole backed by a conducting surface, doubling the gain. Radiation into free space reduces G to $\pi/2$. The same result can also be arrived at by noting the dipole pattern is like a donut with an infinitesimal hole. Around its axis, the gain is one. Across its axis (as with the cylinder), $G = \pi L/\lambda$. For the half-wave dipole, $L = \lambda/2$, making $G = \pi/2$.

The peak RCS of a dipole is given as $0.88\lambda^2$ [Barton, 1988, p. 103]. But this is a peak value, not an average within the dipole's defined beamwidth. If the effective dimensions of the intercept area of a dipole in free space is assumed to be enhanced to λ along the dipole axis and $\lambda/2$ across it, making $A = \lambda^2/2$, we can apply $\sigma = GA$ to get Equation (6-20) (which is close to Barton's value).

$$\sigma_p = \left(\frac{\pi}{2}\right) \left(\frac{\lambda^2}{2}\right) = \frac{\pi\lambda^2}{4} \approx 0.8\lambda^2 \qquad (6\text{-}20)$$

where:

σ_p = Peak chaff radar cross section, square meters (m^2)

The average RCS of a half-wave dipole can now be found by assuming a spherically random distribution, applying the two-way $\sin\theta$ amplitude effects ($\sin^2\theta$ power) and $\cos^2\phi$ polarization. The average chaff RCS is given in Equation (6-21). The integration is available from integral tables, and the result is given in Equation (6-22). Others have used from $0.13\lambda^2$ to $0.2\lambda^2$ for this value.

$$\bar{\sigma} = \sigma_p \frac{4}{\pi^2} \int\int_0^{\pi/2} \sin^4(\theta) \cos^2(\phi) \, d\theta \, d\phi \qquad (6\text{-}21)$$

$$\bar{\sigma} = \frac{3\sigma}{16} \approx 0.15\lambda^2 \qquad (6\text{-}22)$$

where:

$\bar{\sigma}$ = Average chaff radar cross section, square meters (m^2)

It is now a simple matter to determine the peak and average RCS of a cloud of randomly oriented perfectly conducting dipoles (Equations 6-23 and 6-24, respectively). Some care must be taken when dealing with these results to differentiate between the radar resolution volume and the cloud volume. If the former is smaller than the latter in any dimension, the RCS per resolution cell must be summed to find the cloud RCS.

$$\sigma_p \cong 0.8\, N\, \lambda^2 \qquad\qquad (6\text{-}23)$$

$$\overline{\sigma} \cong 0.15\, N\, \lambda^2 \qquad\qquad (6\text{-}24)$$

where:

N = Number of chaff dipoles in the cloud, no units

The number of chaff particles in a radar resolution cell has a distinct effect on how the return is viewed by the radar. The RCS of several chaff particles in a single radar resolution cell is noise-like and can be treated like noise for the purposes of detection and signal-processing theory. The RCS of numerous chaff particles in a single radar resolution cell is a target-like return and can be treated like a target for the purposes of detection and signal-processing theory.

6.7 | CLUTTER CHARACTERISTICS

Clutter is reflected signals from objects other than targets, including the earth's surface, vegetation, atmosphere (e.g., rain, sleet, snow), insects, and birds. Scattering from these various types of diffuse clutter sources is an important topic for discussion. For radar altimeters and imaging radar systems (see Chapter 7), the clutter literally becomes the target and is the major portion of the data gathered.

All the considerations about scattering I have touched on so far are applicable when developing scattering models for various types of ground targets. The RCS of clutter is a function of clutter reflectivity and area illuminated by the radar antenna pattern. Clutter reflectivity is a function of radar frequency, polarization, clutter source, and geometry. The wavelength dependence is strong. The leaves of trees may have large, noisy RCS at X band ($\lambda \approx 0.03$ m) but may be virtually invisible at ultra high frequencies (UHF) where $\lambda \approx 0.6$ m. The ground itself tends to be diffuse except in areas of rocky terrain where objects encountered are usually large with respect to a wavelength, and in sandy regions where the roughness scale is much, much smaller than a wavelength. Calm water is a specular reflector whose RCS is very sensitive to the incidence angle. Rough water may be diffuse or specular and periodic, depending on the scale size and makeup of the waves. Man-made objects, such as buildings, power poles, and roads, tend to be specular. Clutter sources are often categorized as distributed or point. Distributed clutter sources have a spatial extent much larger than the radar resolution cell (see Section 5.5.3), such as most terrain and water clutter sources. Point clutter has a spatial extent similar to the radar resolution cell, such as most man-made objects.

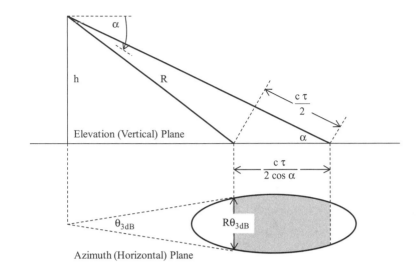

The RCS of the clutter is simply the clutter reflectivity multiplied by the area of clutter intercepted by the radar resolution cell (Equation 6-25). The area of the clutter intercepted by the radar pulse is the product of the projection of the cross-range and range resolutions (i.e., resolution cell size; see Section 5.5.3) on the ground, as given in Equation (6-26) and shown in Figure 6-5.

$$\sigma_c = \sigma_0 \, A_c \tag{6-25}$$

$$A_c = (R \, \theta_{3dB}) \left(\frac{c \, \tau}{2}\right) \left(\frac{1}{\cos(\alpha)}\right) \tag{6-26}$$

where:

σ_c = Clutter radar cross section, square meters (m²)

σ_0 = Clutter reflectivity, square meters/square meters (m²/m²)

A_c = Area of the clutter intercepted by the radar resolution cell, square meters (m²)

R = Radar-to-clutter slant range, meters

θ_{3dB} = Radar antenna azimuth half-power (−3 dB) beamwidth (see Chapter 4), radians

c = Speed of light, 3×10^8 meters/second

τ = Radar transmitted pulse width, seconds

α = Grazing angle of the radar wave relative to the ground, radians or degrees

A number of investigators have gathered a great deal of detailed data on clutter [Barton, 1975; Long, 2001; Billingsley, 2002]. Such studies, which are absolutely essential for the design of a particular radar system for a specific

TABLE 6-2 ▪ Typical Backscattering Coefficients

Terrain	Clutter Reflectivity (dBsm)	Reflection Coefficient (dBsm)
Smooth Water	−53	−45.4
Desert	−20	−12.4
Wooded Area	−15	−7.4
City	−7	0.6
	10° grazing angle and 10 GHz	

mission, tend to obscure important general principles. Sea and ground clutter are usually treated separately. Although their clutter effects have many similarities, the excursions of sea clutter are more extreme. To facilitate incorporation into the radar equation, the clutter reflectivity (radar cross section per unit area) is usually used in the literature (Equation 6-27). Example clutter reflectivity and reflection coefficient values are shown in Table 6-2 [Stimson, 1998, p. 296]. Extensive tables and figures of clutter reflectivity and/or reflection coefficients are given in Long [2001] and Billingsley [2002].

$$\sigma_0 = \Psi \sin(\alpha) \qquad (6\text{-}27)$$

where:

σ_0 = Clutter reflectivity, square meters/square meters (m^2/m^2)
ψ = Reflection coefficient, square meters/square meters (m^2/m^2)
α = Grazing angle of the radar wave relative to the ground, radians or degrees
λ − Wavelength, meters

When the grazing angle is between, say, 10° and 60°, when the radar is looking neither at the horizon nor straight down, for many classes of terrain and wavelengths grazing angle dependence is small and may not be used. The reason for this is for smooth surfaces (granules $\ll \lambda$) there is virtually no clutter return up to very high angles of incidence; for rough surfaces (granules $\leq \lambda$, whose individual radar cross sections are not very dependent on angle) the clutter return will tend to be constant in the region between a few degrees above 0° and a few degrees below 90°. Barton [1985, pp. 198–204] noted the lack of a model for low grazing angles and provided one that includes both reflectivity and propagation characteristics.

Of the models of radar cross section developed for various simple objects in the preceding sections, some depend on λ^2, some depend on λ, and some are wavelength independent. In the Rayleigh region, they are λ^4 dependent, but their absolute level tends to be sufficiently low as to not be a factor. One could model terrain by using various combinations of plates, spheres, cylinders, and wires, making them conducting and nonconducting as appropriate. It is common to find such a terrain model dependent on wavelength. Also to be expected is for ψ to vary over several orders of magnitude, and it does. For water, it may be between 0.100 and 0.001; for flat, desert-like country, 0.010; and for dense undergrowth, 0.100. The fact that even water is not a very good conductor makes the returns lossy, so ψ does not rise near 1.000 except at very high incidence angles, such as with radar altimeters, which look straight down. Here ψ may exceed 10.000 over smooth water. In a super simplification of a large body of data on land clutter (arguing the dependence on λ and α are both small), Nathanson [1969, p. 272] calculates the rough equivalent of our ψ to be between 0.500 and 0.001 m^2/m^2, with a median at 0.040 m^2/m^2.

Wind speed and direction relative to the radar system are factors in determining ψ, especially at sea and also for some terrain types such as forests and crops. The wind also results in a spectrum to the clutter signal [Skolnik, 2001, pp. 152–158]. For ground-based radar systems attempting to detect and track airborne targets, the ground itself is classified as clutter, which must be rejected in some way. Many clutter rejection concepts take advantage of the spectral (frequency domain) difference between the target and clutter signals. Thus, clutter spectrums are very important in assessing the efficiency of clutter rejection concepts (see Section 7.1).

6.8 | RADAR SIGNATURES

I have alluded to the antenna-like characteristics of various objects whose RCS I have discussed. I have also speculated about the physical features resulting in certain RCS behavior. The next step is to ask what might be determined about objects' physical characteristics by studying their radar cross sections. This is the field of radar signatures. It is substantial, and it goes beyond the study of far-field RCS patterns. Nevertheless, some things can be learned about an object from observing its far-field patterns rolling past a radar system's line of sight.

Referring to the elemental antenna of Chapter 2, we can make it equivalent to an RCS far field by noting the energy sources arrive from the radar. The effect is that the pattern changes twice as fast as we move away from perpendicular, making the lobes half as wide. The peak-to-first-null width is now λ/2D instead of λ/D.

Figure 6-6 shows the RCS patterns of some simple objects rotated through 180°. The potential inferences that might be made as to target characteristics are obvious [Crispin and Maffett, 1968, pp. 83–153]. If we can find out the angular rate at which the objects are turning, we can get a measurement of their width. The "beamwidth" of the lobe in the RCS pattern of a rotating object is given in Equation (6-28).

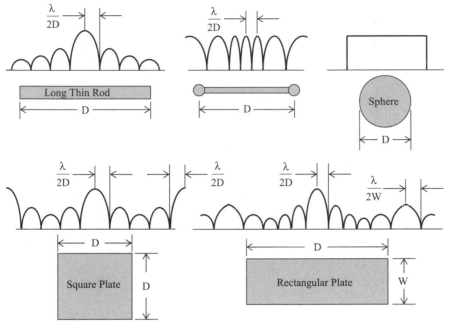

FIGURE 6-6 ■
Some Simple Radar
Cross Section
Patterns

Note: Not to Scale, Approximately −90° to 90° Aspect, D and W ≫ λ

$$\theta = 2\pi \left(\frac{\Delta t}{T}\right) = \frac{\lambda}{2D} \quad \Rightarrow \quad D = \frac{\lambda}{4\pi}\left(\frac{T}{\Delta t}\right) \tag{6-28}$$

where:

θ = Peak to first null in the RCS pattern, radians

Δt = Time it takes for the lobe in the RCS pattern to cross the radar line of sight, seconds

T = Time it takes for the object to rotate one time, seconds

λ = Wavelength, meters

D = Dimension of the object, meters

Much can be done with this kind of analysis using reasonable assumptions, particularly with space objects, where the motion is Keplerian and the observation periods can be long. However, we must always remember that there is no one-to-one correspondence between an illumination function and its far field and, most emphatically, between an illumination function and the external physical characteristics of an object. Finally, the external characteristics of the object may tell us nothing of its internal characteristics.

▌ 6.9 ▐ | SUMMARY

The signature of a target is defined by its radar cross section and spectrum. The RCS determines the amplitude component, and the spectrum determines the frequency domain component of the received target signal power. Target RCS is complex, both electromagnetically and mathematically. The RCS of a target is a function of both target characteristics (i.e., size, shape, materials) and radar characteristics (i.e., frequency, polarization). We often simplify target RCS by looking at simple geometric shapes when possible. The RCS of complex shapes, the targets in which we are most interested, is the complex (amplitude and phase) combination of the scattering off all parts of the target. We often visualize the RCS of complex targets with plots as a function of azimuth angle, elevation angle, frequency, and polarization. The RCS of complex targets is not a single value. It has a very wide range of values—often several orders of magnitude (powers of 10). For Doppler radar systems the target's spectrum needs to be considered along with the RCS.

▌ 6.10 ▐ | EXERCISES

6-1.   You are estimating the RCS of a particular type of aircraft that will frequently fly through the coverage of a surveillance radar system you are designing. You find that it has an L-band antenna of area, A, that will be directly in the radar line of sight much of the time. Your radar is at X band. Assume the antenna is a flat plate. (a) What will the RCS of that antenna be? (b) If it were 50% efficient and matched to your radar, what would its RCS be?

6-2.   You are supervising the designers of a radar system for finding and mapping rainfall. In the region where it will operate, raindrops range in diameter from 2.5 millimeters to 13 millimeters. What frequencies should be used?

6-3.   An aircraft of 30 meter wingspan is flying toward a radar of frequency $f_c = 900$ MHz. Over what changes in aspect angle would the aircraft have to move to make a measurement of wingspan?

6-4.   A sailboat uses a dihedral corner reflector with a dimension a = 0.3 meters to provide a high radar cross section to marine radar systems. A marine radar system with a carrier frequency $f_c = 10$ GHz can detect the dihedral corner reflector at a range $R_{dt} = 10$ km. A sign at the marine supply store states replacing a dihedral corner reflector with a trihedral corner reflector of the same size results in a 50% increase in radar cross section and radar detection range. Determine if these statements are true or false.

6-5. 100,000 aluminum chaff dipoles might weigh a few pounds at L band ($\lambda = 0.3$ meters). What would be the cloud's average RCS?

6-6. A radar system has the following characteristics: half-power beamwidth $\theta_{3dB} = 1.5$ degrees; and transmitted pulse width $\tau = 1$ μsec. The radar system is attempting to detect a target at a range R = 15 km in clutter with a reflectivity $\sigma_0 = -20$ dBsm and a grazing angle of $\alpha = 2$ degrees. What is the clutter radar cross section, σ_c (m^2 and dBsm)?

▌6.11 ▐ REFERENCES

Barton, D. K., 1975, *Radars*, Volume 5, Radar Clutter, Dedham, MA: Artech House. Thirty-eight articles on the subject of land, sea, and aerospace clutter.

Barton, D. K., 1985, "Land Clutter Models for Radar Design and Analysis," *Proceedings of the IEEE*, Feb., pp. 198–204.

Barton, D. K., 1988, *Modern Radar System Analysis*, Norwood, MA: Artech House.

Billingsley, J. B., 2002, *Low-Angle Radar Land Clutter Measurements and Empirical Models*, Raleigh, NC: SciTech Publishing. Just about anything one would want to know about land clutter.

Crispin, J. W., and Maffett, A. L., 1968, "RCS Calculation of Simple Shapes-Monostatic," in J. W. Crispin and K. M. Siegel (editors), *Methods of Radar Cross Section Analysis*, New York: Academic Press. Extensive theory and detailed calculations for many shapes of varying orientations and polarization angles.

Jenkins, F. A., and White, H. E., 1976, *Fundamentals of Optics*, New York: McGraw-Hill.

Jenn, D. C., 2005, *Radar and Laser Cross Section Engineering*, 2nd Edition, Reston, VA: American Institute of Aeronautics and Astronautics, Inc. Very detailed discussion of simple geometric shapes, complex objects, radar cross section predication approaches, and supporting MATLAB software via the publisher's website.

Knott, F. F., Schaeffer, J. F., and Turley, M. T., 2004, *Radar Cross Section*, Second Edition, Raleigh, NC: SciTech Publishing.

Long, M. W., 2001, *Radar Reflectivity of Land and Sea*, 3rd Edition, Boston, MA: D.C. Artech House. Always one of the most comprehensive books; the third edition includes bistatic (physically separate transmit and receive antennas) land and sea clutter.

Nathanson, F. E., 1998, *Radar Design Principles*, Raleigh, NC: SciTech Publishing.

Rheinstein, J., 1968, "Backscatter from Spheres, a Short-Pulse View," *IEEE Transactions*, Volume AP-16, Jan., pp. 89–97. (Also 1966, Lincoln Laboratory Technical Report 414, Lexington, MA.) Exhaustive treatment.

Skolnik, Merrill I., 2001, *Introduction to Radar Systems*, 3rd Edition, New York: McGraw-Hill.

Stimson, George W., 1998, *Introduction to Airborne Radar*, 2nd Edition, Raleigh, NC: SciTech Publishing.

Van Vleck, J. H., Block, F., and Hammermesh, M., 1947, "Theory of Radar Reflection from Wires or Thin Metallic Strips," *Journal of Applied Physics*, Volume 18, p. 274.

Advanced Radar Concepts

HIGHLIGHTS

- Some of the most modern radar system applications and some unusual applications
- Clutter rejection using moving target indicators, pulse-Doppler, and moving target detection
- Synthetic aperture radar systems, the culmination of over forty years of advancing radar and signal processing capability
- Bistatic radar systems, "back to the future" using modern signal processing with the earliest form of radar systems
- The current state-of-the-art, multifunction, mode, mission radar systems
- Over-the-horizon radar and its unique characteristics
- Radar altimeters
- Arecibo, a gigantic radar for ionospheric study and planet gazing
- Laser radar systems or Lidars

Radar systems are pervasive in modern life. Military applications include all of those previously mentioned plus enormously varied additional uses associated with the military missions, such as surveillance, reconnaissance, targeting, and weapons delivery—on land, at sea, and in aerospace. Although military applications dominate, at least as far as big complex radar systems are concerned, nonmilitary applications are legion. Many medium-sized private boats and airplanes are radar equipped, the former for navigation and the latter for obtaining accurate altitude. Commercial airlines and ships are almost all radar equipped. In law enforcement, traffic and border patrol officers often employ radar systems. The Federal Aviation Agency (FAA) operates at least 40 large radar systems and numerous small ones to keep track of the nation's air traffic, with the higher altitudes being under radar control at all times. Many companies and individuals with high-value facilities use radar systems for security.

The availability of coherent circuits, the move to digital electronics, and the rapid evolution of computers and signal processing algorithms have greatly expanded radar systems capabilities. A "new generation" of radar systems have taken advantage of the new technology. To recite a litany of these specific mission-oriented radar systems is probably not as useful as picking out a few illustrative systems to discuss. With the groundwork already laid, we can understand and analyze virtually any radar system. In this chapter, I discuss two classes of advanced radar systems. First, I will give attention to a few applications marking the thrust of modern radar technology: clutter rejection; synthetic aperture radar (SAR); bistatic radar; and multiple function, mode, mission radar systems. Second, I will examine somewhat unconventional systems when compared with the ones I have been discussing: over-the-horizon radar, radar altimeters, ionospheric radar, and laser radar. Although all these radar systems represent extremely complex and intricate applications, their sophistication is achieved by manipulating the fundamental radar properties already discussed in the preceding chapters and/or electronic warfare properties in following chapters.

7.1 | CLUTTER REJECTION

Many ground-based and airborne radar systems must be able to detect and track airborne and/or ground-based targets against a background of diffuse clutter (see Section 6.7). For radar systems operating against diffuse clutter, the radar equation (see Chapters 2 and 3) is modified. The received target signal is embedded in receiver thermal noise (see Section 2.3.1) and the received clutter power. Generally, the received clutter signal power has much greater amplitude than the receiver thermal noise. Thus, we often ignore the receiver thermal noise when discussing clutter at a fundamental level.

The basic approach for discussing clutter is often called "competing clutter" because it is in the same resolution cell (see Section 5.5.3) as the target and thus the target competes with the clutter for detection. Since both the target and clutter are in the same resolution cell, they are at the same range and angle relative to the radar system. The competing target signal-to-clutter ratio (S/C) is the ratio of the received target signal power (see Chapter 2) to the received clutter power. All the terms in the target signal and clutter equations are the same except for the target radar cross section (RCS) and clutter RCS. Therefore, the competing S/C is the ratio of the target RCS to the clutter RCS (see Section 6.7) (Equation 7-1). An exercise at the end of the chapter asks the reader to derive the competing S/C equation.

$$\frac{S}{C} = \frac{\sigma}{\sigma_c} = \frac{2 \cos(\alpha) \sigma}{\sigma_0 \, R \, \theta_{3dB} \, c \, \tau} \qquad (7\text{-}1)$$

where:

S/C = Competing target signal-to-clutter ratio, no units
  σ = Target radar cross section, square meters (m^2)
  σ_c = Clutter radar cross section, square meters (m^2)
  α = Grazing angle of the radar wave relative to the ground (see Section 6.7), radians or degrees
  σ_0 = Clutter reflectivity, square meters/square meters (m^2/m^2)
  R = Radar-to-target/clutter slant range, meters
θ_{3dB} = Radar antenna half-power (−3 dB) beamwidth (see Chapter 4), radians
  c = Speed of light, 3×10^8 meters/second
  τ = Radar transmitted pulse width, seconds

If the clutter returns are high, which they usually are, no amount of increased radar power will redress the situation. Note, too, the S/C becomes worse as the range increases. A radar system in this condition is said to be clutter limited, as its detection range is limited by clutter instead of receiver thermal noise. As one of the exercises demonstrates, a radar system not designed for the task simply cannot detect targets in clutter. Signals from targets below airborne radar systems are obscured, and low-flying targets may be undetectable by ground-based radar systems until ranges are sufficiently long to place the clutter source below the horizon (see Section 13.1).

At the fundamental level, a few ways of improving the S/C are apparent from the equation. One can make the radar antenna beamwidth narrower, make the transmit pulses shorter, and operate with a frequency or polarization that results in a smaller clutter reflectivity. However, all these ways have other consequences for the radar system capability and performance—both target detection and measurements: radar is one big systems engineering problem.

To provide target detection performance in the presence of diffuse clutter, the clutter must be rejected in some way. One approach is to use clutter fences around a ground-based radar site; they are an efficient solution for some clutter problems. Clutter fences merely prevent the radar from illuminating relatively distant objects (e.g., mountains) whose enormous RCS is at the range where targets must be observed. Instead, the signal reflects from the clutter fence at a range so short the unwanted reflections are easily rejected or ignored. For near-in clutter, the radar receiver is simply not turned on until after the received pulse has returned from the clutter source. Where there are points of very high clutter returns, say, buildings or roads, in the radar coverage, these may be blanked in the radar receiver, but at the cost of also blanking targets at the same range and angle.

The previous discussion ignores sidelobe clutter at the same range as the target. In radar systems with good (low) sidelobes, the additional effect of clutter in all the sidelobes can be minimized. In radar system with poor (high) sidelobes, the additional effect of clutter in the sidelobes should be considered.

By far the most common clutter rejection approach is to separate the target from the clutter by the Doppler differences between them: moving target indicator (MTI), pulse-Doppler processing, and moving target detector (MTD). These clutter rejection approaches and associated clutter rejection metrics are discussed in the following sections.

7.1.1 Moving Target Indicator

MTI is really a rudimentary form of pulse-Doppler [Richards et al., 2010, Section 17.4; Schleher, 1991, Chapter 1; Skolnik, 2001, Chapter 3]. Because it has been in operation in various forms since shortly after World War II, it hardly qualifies as "advanced." Yet MTI radar systems of the quality now being developed require all the new technology that can be provided. Moving target indication does just what its name says, it indicates moving targets while suppressing nonmoving targets.

Noncoherent Moving Target Indicator

An early form of MTI was noncoherent. It was called "area MTI" [Barton, 1988, p. 243]. The returns from one scan were simply subtracted from those of the next. All targets that had moved at least one resolution cell and/or whose amplitudes fluctuated in the time between scans were preserved. All unmoving and/or steady amplitude objects, including fixed clutter, were canceled. The method was practical with the available technology of early radar systems and is still used by some radar systems. However, a noncoherent MTI is severely limited in clutter rejection performance since it uses the relatively noisy clutter signal as a reference, as opposed to the very stable internal reference (local oscillator) used by a coherent MTI [Schleher, 1991, pp. 74–76, 345–348].

Coherent Moving Target Indicator

Coherent MTI takes advantage of the different frequency domain (power spectral density) characteristics of the target and clutter signals. Coherent MTI can use either short duration (a few pulses) coherent waveforms or clutter in the target region as the phase reference to cancel clutter in the target region [Schleher, 1991, Section 2.3.1]. The different target and clutter power spectral densities (PSD) are shown in Figure 7-1 for a ground-based radar system. The clutter PSD is narrow, since the clutter source has only a slight range rate (see Section 5.3), and is repeated at integer multiples of the pulse repetition frequency (PRF) (see Section 5.1.2). The width of the clutter PSD is a function of the range rate of the clutter source: windblown trees or ocean waves. Coherent MTI does very well at rejecting clutter at a single Doppler shift or with a narrow PSD.

A coherent MTI has a frequency response with nulls at the clutter PSD and passbands at the target PSD, thereby rejecting the clutter signal and passing the

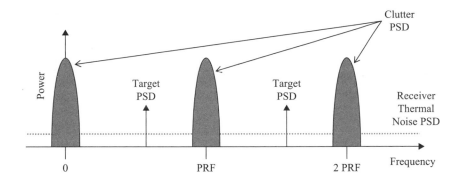

FIGURE 7-1 ■
Target, Clutter, and
Noise Power
Spectral Densities

target signal. Coherent MTI is implemented in delay line cancelers, recursive filters, and digital filters. Due to their legacy as the early implementation of MTI, the term "delay line canceler" is commonly used for this whole group. There are single-delay (two pulse), double-delay (three pulse), and even triple-delay (four pulse) cancelers. The frequency response of a single-delay canceler is given in Equation (7-2) and shown in Figure 7-2 (PRF = 500 Hz).

$$G_p(f) = \left[2 \sin\left(\frac{\pi f}{PRF}\right) \right]^2 \tag{7-2}$$

where:

$G_p(f)$ = Single-delay canceler power frequency response, no units
  f = Frequency, hertz
 PRF — Radar pulse repetition frequency, PRF

As seen in this figure, the nulls of the MTI response are a zero Doppler and positive and negative integer multiples of the PRF. The result is that the clutter PSD is significantly rejected, as shown in Figure 7-3.

Ideally the rejection nulls in the MTI frequency response would completely cover the clutter, but due to the Doppler extent of the clutter PSD there is a

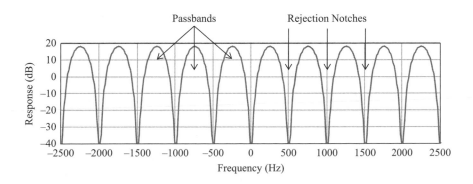

FIGURE 7-2 ■
Single-Delay (Two-
Pulse) Canceler
Power Frequency
Response

FIGURE 7-3 ■
Clutter Rejection
Due to Moving
Target Indicator
Frequency
Response (a Close-
Up of Figure 7-2 at
Zero Doppler
Showing Half the
Passband)

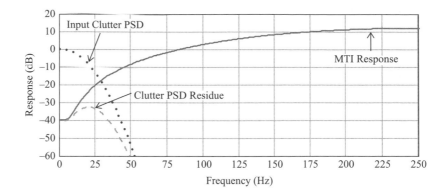

residual clutter PSD. As the clutter PSD becomes wider, additional delays are needed to provide wider rejection nulls in the MTI response. The power response of single- and double-delay cancelers is shown in Figure 7-4. The power response of an N-delay line canceler is given in Equation (7-3).

$$G_p(f) = \left[2 \sin\left(\frac{\pi f}{PRF} \right) \right]^{2N} \tag{7-3}$$

where:

$G_p(f)$ = Multiple-delay canceler power frequency response, no units
  N = Number of delay lines, no units

A delay line canceler does an excellent job of suppressing the clutter at zero Doppler and integer multiples of the PRF. Targets with Doppler shifts within one of the MTI passbands receive modest gain or attenuation. Unfortunately, targets with a Doppler shift at zero or integer multiples of the PRF are rejected along with the clutter PSD. Often the term "blind speed" is used to describe the target range rate which results in a Doppler shift at zero or integer

FIGURE 7-4 ■
Single- and Double-
Delay Cancelers
Power Frequency
Response

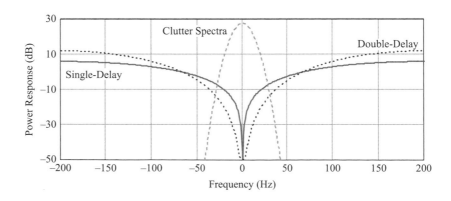

multiples of the PRF. The blind speed is the target range rate resulting in a Doppler shift equal to integer multiples of the PRF (Equation 7-4). For example, the first blind speed for a radar system with a PRF of 300 Hz and a wavelength of 0.1 m ($f_c = 3$ GHz) is $v_1 = \pm(1)(0.1)(300)/2 = \pm15$ m/sec. Likewise, the 10th blind speed is ±150 m/sec.

$$f_d = \frac{2\,v_n}{\lambda} = \pm n\ \text{PRF} \qquad\qquad n = 1,\, 2,\, 3\ldots$$

$$v_n = \pm\frac{n\,\lambda\ \text{PRF}}{2}$$

(7-4)

where:

f_d = Target Doppler shift, hertz
v_n = n-th blind speed, meters/second
λ = Wavelength, meters
n = Integer number of the pulse repetition frequency, no units
PRF = Pulse repetition frequency, hertz

A common solution to the problem of blind speeds is to use staggered PRF MTI processing, in which an alternating sequence of two or more PRIs is transmitted and processed. The result is the nulls in the MTI response occur at the common frequencies of the PRFs, as shown in Figure 7-5. A staggered PRF MTI allows radar designers to put the first blind speed at the speed associated with the fastest target the radar is designed to detect. The deficiencies of MTI have kept Doppler processing evolving as designers strive to make more complete use of what the physics has to offer.

7.1.2 Pulse-Doppler that Processing

The family of pulse-Doppler radar systems is literally pulse radar systems using Doppler information (see Sections 5.1.2 and 5.3). Doppler processing provides

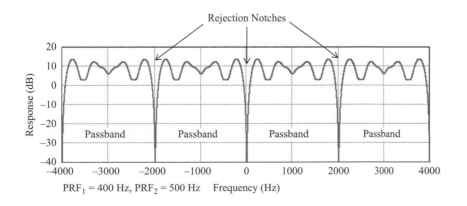

FIGURE 7-5 ■ Staggered Pulse Repetition Frequency Moving Target Indicator

FIGURE 7-6 ■
Frequency
Response of the
Target, Clutter, and
Noise Signals—
Ground-Based
Radar System

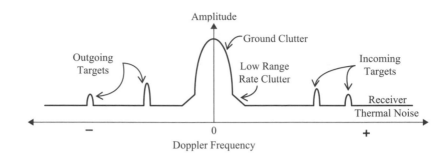

a fine frequency domain characterization of the return signals. The details of the frequency response of the return signals when employed in a ground-based radar system are shown in Figure 7-6. This pattern repeats at each pulse repetition frequency. Immobile or slow-moving targets may be buried in the clutter. However, fast-moving targets are away from the clutter, in the "clutter-free zone."

The most basic form of clutter rejection for ground-based pulsed-Doppler radar systems is to "ignore" the output of the Doppler filters near zero Doppler, the ones containing clutter. Pulse-Doppler radar systems may also have clutter cancelers in their circuitry if the zero-range rate clutter is difficult to accommodate. When the radar system is required to detect slow-moving, close-in targets, the designer will select a waveform with very good range rate resolution, a long pulse burst duration (see Section 5.3.1) which provides a narrow Doppler filter bandwidth. The sidelobe response of the Doppler filters (see Section 5.5.4) is of special concern to minimize the clutter entering the Doppler filter via its sidelobes [Richards et al., 2010, Section 14.4.2; Stimson, 1998, pp. 263–264].

Airborne Pulse-Doppler Radar Systems

The clutter power spectral density for airborne pulse-Doppler radar systems is much more complex than for ground-based radar systems [Stimson, 1998, Part V "Return from the Ground"]. Consider an aircraft flying over terrain, as shown in Figure 7-7. The region of zero-range rate sidelobe clutter will begin at a spot directly under the aircraft and extend to infinity on each beam of the aircraft. Directly to the front of the aircraft are clutter returns at the speed of the aircraft, and directly to the rear are clutter returns at minus the aircraft speed. In the intervening angles over the compass rose are clutter returns ranging from zero to plus and minus the aircraft speed.

Clutter range rate as a function of aircraft speed, magnitude of the velocity vector, and angle to the clutter is given in Equation (7-5). Thus, in range rate space, sidelobe clutter is present at all times in a band twice as wide as the aircraft speed. No matter how much or how little of the region the radar mainbeam scans, the sidelobe clutter is always present. Because it is a specular

FIGURE 7-7 ■
Airborne Pulse-
Doppler Radar
System

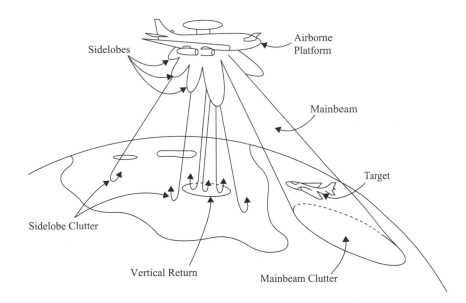

reflection, the clutter directly beneath the aircraft is particularly troublesome due to its potentially high amplitude. It may be dealt with by means other than Doppler processing—for example, range blanking or extension of receiver dead time until these returns have passed the aircraft. Both of these techniques will eliminate coverage of close-in targets. Range and range rate ambiguities further complicate the ground clutter for airborne radar systems [Stimson, 1998, Chapter 23].

$$R_{dotC} = V_{ac} \cos\theta \cos\phi \qquad (7\text{-}5)$$

where:

R_{dotC} = Clutter range rate, meters/second

V_{ac} = Aircraft speed, magnitude of the velocity vector, meters/second

θ = Azimuth, horizontal plane, angle from the aircraft velocity vector to the clutter, radians or degrees

ϕ = Elevation, vertical plane, angle from the aircraft velocity vector to the clutter, radians or degrees

The angular dependency of the Doppler shift defines the azimuth (horizontal plane) and elevation (vertical plane) coverage available from an airborne pulse-Doppler radar system. When the mainbeam is aligned with the velocity vector in azimuth, the clutter for an airborne radar system extends from negative range rates through positive range rates due to antenna sidelobe returns. The altitude return is around zero range rate, and the mainbeam return moves with the elevation beam-pointing angle, as shown in Figure 7-8.

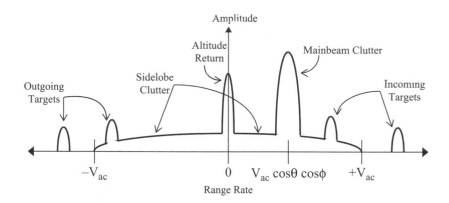

The amplitude of the mainbeam return is large due to to the mainbeam antenna gain. However, if the antenna mainbeam is not illuminating the ground, that is, "picked up" off the ground, there is no mainbeam clutter. As the mainbeam is pointed off the velocity vector in azimuth, the range rate of the mainbeam return will decrease. When the mainbeam is pointing at 90° (either side of the aircraft), the mainbeam return is around zero range rate. For the beam pointing to the rear (180°), the zero-degree situation is reversed.

As shown in Figure 7-8, incoming targets have a positive range rate, while outgoing targets have a negative range rate (see Section 5.3). Having mapped the clutter at all radar mainbeam positions, we can be confident that a target in the mainbeam is out of the clutter, in the clutter-free zone, only if it is moving at range rates different from the clutter range rates. We know what those range rates are. Thus, to be out of the clutter, a target in the mainbeam of the radar must be moving toward or away from the radar at a speed greater than the radar aircraft over the ground. A target in front of the radar is out of the clutter if it is closing with the radar aircraft at a few meters/second or moving away from the radar aircraft at more than twice its range rate.

Most targets cannot be detected if they are within the Doppler filters containing the mainbeam clutter or the altitude return sidelobe clutter. However, targets with range rates similar to those associated with the sidelobe clutter may be detectable in the relatively small amplitude of the sidelobe clutter. Their detection requires high-resolution waveforms (in both Doppler and range) and very low antenna sidelobes. The antenna of the airborne pulse-Doppler systems have has some of the lowest sidelobes ever achieved—including ground-based antennas.

7.1.3 Moving Target Detection

MTD is a frequency domain signal processing technique used by ground-based radar systems to pass moving target signals while rejecting clutter signals

[Bouwman, 2009, Chapter 14; Schleher, 1991, Section 1.4; Skolnik, 2001, Section 3.6]. MTD signal processing is a combination of MTI (see Section 7.1.1) and Doppler processing (see Section 7.1.2), as shown in Figure 7-9. It provides improved clutter rejection compared with MTI or Doppler processing alone, improved target detection performance compared with MTI due to coherent integration of many pulses, and target Doppler "estimates" from the Doppler filter bank. MTD processing usually uses much fewer and wider Doppler filters than Doppler processing for range rate measurement (see Section 5.3).

7.1.4 Clutter Maps

A radar system can collect, process, and store a map of the large fixed-clutter sources not rejected by MTI or MTD processing. A clutter map is typically used by ground-based scanning radar systems, such as those for airport surveillance. The clutter map consists of the residual clutter signal in each range–azimuth resolution cell (see Section 5.5.3) in the search area of the radar system. The clutter map is incrementally developed over numerous (10–20) scans to ensure that the clutter has reached a steady-state value and that moving targets are eliminated (they do not stay in the same cell over time). The clutter map is incrementally updated to account for changes in the clutter signals over time. Bouwman [2009, Chapter 14], Richards et al. [2010, Section 17.6.1], Schleher [1991, Section 1.4.1], and Skolnik [2001, pp. 143–144] all discuss clutter maps in detail.

The clutter map is used to blank clutter areas and remove clutter detections and, unfortunately, coincident target detections as well. The blanking prevents "false alarms" from clutter signals and also tracks from being initiated on a clutter signal alone or a combined clutter and target signal. These tracks will eventually be identified as false and dropped. In the meantime, however, time and computer capacity are "wasted," especially when there are a large number of them.

Another use of a clutter map is to provide target detection in range–azimuth cells where the target signal sufficiently exceeds the clutter signal. Often called "super-clutter visibility," this can allow low-range rate (target Doppler shift within the clutter PSD) strong-target signals to be detected over weak clutter signals. As previously discussed, low range rate target signals are often rejected by traditional MTI and MTD processing. The clutter signal stored in each range-azimuth cell of the map is used to establish a detection threshold for the cell. Thus, the detection threshold is relative to the clutter signal instead of the receiver thermal noise (see Section 3.2.4) and is a form of constant false alarm rate (CFAR) processing (see Section 3.4.4). This detection approach helps maintain detection of and track on a target as it moves past the radar system. At this time the target velocity vector is orthogonal to the radar–target line of sight (LOS), resulting in a low or zero range rate.

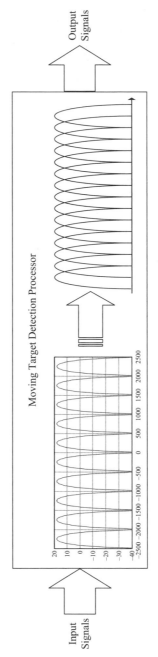

FIGURE 7-9 ■ Moving Target Detection Signal Processing

7.1.5 Clutter Rejection Metrics

If the depth of the null of the coherent MTI system is considered alone, the clutter rejection of the MTI system is very large (e.g., >80 dB). However, the width of the clutter PSD limits the clutter rejection. In addition, the target signal can be rejected (blind speeds) along with the clutter. Therefore, clutter rejection metrics must consider radar, clutter power spectral density, and target response. Two of the most common clutter rejection metrics are MTI improvement factor and subclutter visibility.

Moving Target Indicator Improvement Factor

The MTI improvement factor accounts for the impact on both the target signal as well as the clutter signal [Richards et al., 2010, Section 17.4.3; Skolnik, 2001, pp. 117–120]. The MTI improvement factor is the signal-to-clutter ratio at the output of the MTI divided by the signal-to-clutter ratio at the input of the MTI averaged over all target Doppler shifts of interest (Equation 7-6). The MTI improvement factor for an N-delay line MTI is given in Equation (7-7). This equation applies when the ratio of the standard deviation of the clutter PSD and PRF is small.

$$I_{fMTI} = \frac{\left(S/C\right)_{out}}{\left(S/C\right)_{in}} \tag{7-6}$$

$$I_{fMTI} \approx \frac{2^N}{N!} \left(\frac{1}{2\pi \left(\sigma_{Cpsd}/PRF\right)}\right)^{2N} \tag{7-7}$$

where:

I_{fMTI} = Moving target indicator improvement factor, no units
$(S/C)_{out}$ = Target signal-to-clutter ratio out of the moving target indicator, no units
$(S/C)_{in}$ = Target signal-to-clutter ratio into the moving target indicator, no units
N = Number of delay lines, no units
σ_{Cpsd} = Standard deviation of the clutter power spectral density, hertz
PRF = Pulse repetition frequency, hertz

The standard deviation of the clutter PSD contains all the factors contributing to less than perfect MTI performance. The three most important contributors are radar instabilities, actual Doppler spread of the clutter source, and the Doppler spread introduced by the scanning modulation of the antenna. Other contributors such as radar carrier frequency, intermediate frequency, and PRF instabilities can be made very small with careful radar design. The range rate spread of the source of the clutter tends to be small, less than ±1 to ±2 m/sec for ground clutter and ±3 to ±4 m/sec for most sea clutter. However, for higher frequency radar systems, even small range rates can equate to moderate Doppler shifts (see Section 5.3) (Equation 7-8). The Doppler spread from the antenna mainbeam scanning past a target is a function of the target illumination time (see Section 3.3) (Equation 7-9). The resultant standard deviation of the clutter PSD is the root sum square (RSS) of these three contributors (Equation 7-10). The MTI improvement factor as a function of the ratio of the standard deviation of the clutter PSD to the PRF is shown in Figure 7-10.

$$\sigma_s = \frac{2\,\sigma_{Rdots}}{\lambda} \tag{7-8}$$

$$\sigma_a = \frac{0.265}{T_{ill}} = \frac{0.265\,\theta_{dot}}{\theta_{3dB}} = \frac{0.265\,PRF}{n_p} \tag{7-9}$$

$$\sigma_{Cpsd} = \sqrt{\sigma_i{}^2 + \sigma_s{}^2 + \sigma_a{}^2} \tag{7-10}$$

where:

σ_s = Standard deviation of the clutter source power spectral density, hertz
σ_{Rdots} = Standard deviation of the clutter source range rate, meters/second
λ = Radar wavelength, meters
σ_a = Standard deviation of the antenna modulation power spectral density, hertz
T_{ill} = Target illumination time (see Section 3.3), seconds
θ_{dot} = Antenna scan rate, degrees/second or radians/second
θ_{3dB} = Antenna half-power (–3 dB) beamwidth (see Section 4.1.1), degrees or radians
PRF = Pulse repetition frequency, hertz
n_p = Number of pulses integrated (see Section 3.3), no units
σ_i = Standard deviation of the radar instabilities power spectral density, hertz

Subclutter Visibility

Subclutter visibility is the ratio by which the target signal may be weaker than the competing clutter signal and still be detected by the radar system [Richards et al., 2010, pg. 637; Schleher, 1991, p. 109]. For example, a subclutter visibility of 30 dB implies a moving target can be detected in the presence of clutter even though the clutter signal is 30 dB greater than the target signal. In addition to clutter characteristics, this complex metric also considers radar detection theory (see Chapter 3). However, now detection theory includes the clutter-to-noise ratio along with the usual factors such as signal-to-noise ratio (S/N); probability of detection (P_d); probability of false alarm (P_{fa}); multiple-pulse integration.

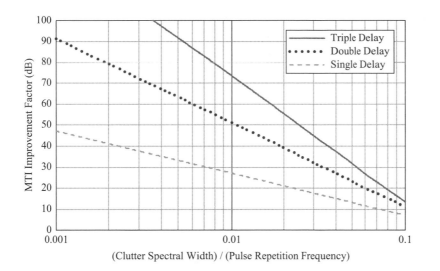

FIGURE 7-10 ■
Moving Target Indicator Improvement Factor as a Function of the Ratio of the Clutter Spectral Width to the Pulse Repetition Frequency

7.2 | SYNTHETIC APERTURE RADAR SYSTEMS

Among the most complex radar systems are synthetic aperture radars [Harger, 1970; Kovaly, 1976; Richards et al., 2010, Chapter 21; Stimson, 1998, Part VII "High-Resolution Ground Mapping and Imaging"; Sullivan, 2004, Part II "Imaging Radar"]. By measuring received signal strength in a mosaic of range and angle (cross-range) resolution cells, they produce detailed images of the region being illuminated by the radar system. The intensity, or "brightness," of each range–angle resolution cell is a function of the S/N in the cell. Numerous sample SAR images are available at the Sandia National Laboratories Synthetic Aperture Radar Homepage [2008]. Because array antenna theory (see Section 4.3) is the basis for SAR, the terms "synthetic aperture radar" and "synthetic array radar" are often used interchangeably.

Most of the families of SAR systems are used to make high-resolution ground images from a moving radar platform such as aircraft, unmanned aerial vehicles (UAVs), and satellites. However, because the target can also provide the necessary motion, the SAR concept can be used to image moving objects: satellites, asteroids, ships, etc., from the ground or air. This concept is sometimes called inverse SAR (ISAR).

Detailed images need fine resolution in both range and angle. Fine range resolution is straightforward using short pulse durations (see Section 5.2.1) or wideband pulse compression waveforms (see Section 5.2.2) (Equation 7-11). For example, a 1 μsec pulse width provides a range resolution of 150 meters, which may be acceptable for some natural-resource, remote-sensing applications. A 100 MHz pulse compression modulation bandwidth provides a very fine range resolution of 1.5 meters.

$$d_r = \frac{c\,\tau}{2} \qquad\qquad d_r = \frac{c}{2\,B_{pc}} \qquad\qquad (7\text{-}11)$$

where:

d_r = Range resolution, meters
c = Speed of light, 3×10^8 meters/second
τ = Transmitted pulse width, seconds
B_{pc} = Pulse compression modulation bandwidth, hertz

Fine angle resolution is a difficult matter. As discussed in Section 5.4.2, the cross-range (azimuth) resolution of a radar system is a function of the radar antenna half-power beamwidth (see Section 4.1.1) and the radar-to-target range (Equation 7-12). The radar system provides a "real beam image" as the cross-range resolution is provided by the beamwidth of the "real" radar antenna. Real beam images can be used when coarse cross-range resolution and short ranges are acceptable. For example, at a radar-to-target range of 20 km, a radar system with a 1° half-power beamwidth provides a cross-range resolution of about 349 meters.

$$d_a = R_{RT}\,\theta_{3dB} \qquad\qquad (7\text{-}12)$$

where:

d_a = Cross-range (azimuth) resolution, meters
R_{RT} = Radar-to-target slant range, meters
θ_{3dB} = Radar antenna half-power (–3 dB) beamwidth, radians

The approach to fine angle resolution for imaging radar systems can be viewed in a space-time frame of reference or a frequency domain frame of reference. In space-time, it can be viewed as using platform motion to provide the physically increased aperture required for fine angle resolution, essentially a synthetic aperture. In the frequency domain frame of reference, the platform motion may be viewed as providing a time history of Doppler returns from differing angles; processing the time history of Doppler shifts allows fine angle resolution.

Aside from their obvious complexity and enormous processing requirement (each range and angle cell of a rapidly moving radar antenna mainbeam must be addressed), SAR systems have several interesting characteristics. The finest angle resolution is obtained from the smallest radar antenna. The angle resolution obtainable (to first order) is independent of range. If there are any moving targets in the image, they show up at the wrong angle (cross-range). And no matter how low the grazing angle of the radar beam, the image reveals a plan view of the terrain.

Note that when discussing SAR systems it is common to use the two-way −4 dB antenna beamwidth to define the angle resolution [Stimson, 1998, p. 414]. This is the angular width of the antenna mainbeam where the gain is down 2 dB from the mainbeam. Therefore, the signal is down 4 dB overall: 2 dB on transmit and 2 dB on receive. The two-way −4 dB beamwidth for a uniformly illuminated antenna is given in Equation (7-13). Compared with other antenna beamwidth equations (see Table 4-1), using the two-way −4 dB beamwidth numerically simplifies many of the cross-range resolution equations, the first of which is given in Equation (7-14).

$$\theta_{4\,dB} = \frac{\lambda}{2\,D} \tag{7-13}$$

$$d_a = R_{RT}\,\frac{\lambda}{2\,D} \tag{7-14}$$

where:

$\theta_{4\,dB}$ = Antenna two-way −4 dB beamwidth, radians
d_a = Cross-range (azimuth) resolution, meters
λ = Wavelength, meters
D = Radar antenna dimension, meters

Synthetic Aperture Radar Fundamentals

To discover how SAR systems work, we will start with a simple unfocused synthetic aperture. As previously stated, the approach to providing fine cross-range resolution is to synthetically generate an "aperture" of the necessary length. Imagine an airborne radar system with a fixed antenna mainbeam pointed perpendicular to the velocity vector of the platform. Such a system is usually referred to as "side-looking" strip-map SAR.

A SAR takes advantage of the forward motion of the radar platform to produce a synthetic aperture (array) over time. Each "element" of the synthetic array is a discrete position of the radar antenna when a pulse is transmitted, as shown in Figure 7-11 along with the associated azimuth (horizontal) and elevation (vertical) plane geometry.

There are numerous range–cross-range (angle) resolutions cells within the swath being imaged. We will simplify our discussion of SAR by considering only one patch: a single range-cross-range resolution cell. The patch on the ground is illuminated by the radar antenna as the mainbeam moves past the patch.

In Section 4.3, we saw mainbeam formation for an array antenna depends on the relative phases of the signals from each element. We found the time when an

FIGURE 7-11 ■ A Synthetic Aperture Produced by a Moving Radar Platform and Associated Geometry

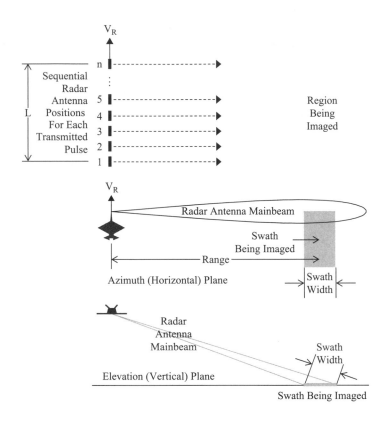

element radiated was not as important as its phase relationships with the other elements. We can therefore extend this line of thought to state that the far-field pattern of an array antenna would not be changed if only a single incremental element was placed at each element position in turn and was made to radiate with the proper phase—as long as there was a receiver in the far-field to store these radiations in their proper phase and add them all up at the end. One small, further step takes us to the situation where the receiver keeping tabs on the phases and adds them all up properly is at the same location as the incremental element, and the propagation path over which this is done is a round trip (i.e., to a target and back). This is the basis of synthetic aperture; we can give up time to obtain a large aperture. (It is not a violation of the First Law of Thermodynamics—getting something for nothing—because the time used to synthesize a large aperture is lost to surveillance and tracking that might have been done instead.)

Thus, we have synthetically generated an array of much longer length than the radar antenna dimension. The cross-range resolution provided by the synthetic array is a function of the synthetic array (aperture) length (Equation 7-15). Notice how we have just replaced the antenna dimension in Equation (7-14) with the synthetic aperture length.

$$d_a = R_p \frac{\lambda}{2 L} \qquad (7\text{-}15)$$

where:

d_a = Cross-range (azimuth) resolution, meters
R_p = Range to the patch on the ground, meters
L = Synthetic aperture (array) length, meters

A quick look at Equation (7-15) could lead us to think all we need for fine cross range resolution is a suitably long synthetic aperture length. However, there is a catch, the synthetic aperture length must be short relative to the range to the patch on the ground being illuminated. If it is comparatively long, then the ranges from the patch to each radar antenna (synthetic array element) position are essentially the same. Thus, the received signals from the patch at each radar antenna position have essentially the same phase (which is needed to form the synthetic array). However, as the synthetic array length increases, the range to the patch from each radar antenna position will diverge slightly. Since the radar wavelength is generally very short, very small range differences can result in considerable phase differences. The change in phase shift due to a range difference is given in Equation (7-16). These phase differences, also called phase errors, limit the ability to form the synthetic array (which as previously stated relies on the signals for each radar antenna position having essentially the same phase).

$$\Delta\phi = \frac{2\pi \Delta R}{\lambda} \equiv \frac{360 \Delta R}{\lambda} \qquad (7\text{-}16)$$

where:

$\Delta\phi$ = Difference in phase, radians (left) or degrees (right)
ΔR = Difference in range, meters

The question arises as to how much range difference, and thus associated phase error, is too much. Unfortunately, there are a few definitions of too much, however one of the most common is having a two-way (radar-to-patch and patch-to-radar) range difference of one-quarter wavelength ($\lambda/4$). Using the same approach as with finding the far-field distance (see Section 13.3) we can find the maximum effective synthetic array length associated with a $\lambda/4$ two-way range difference (Equation 7-17). The finest cross-range resolution provided by the maximum effective synthetic array (aperture) length is given in Equation (7-18). Most unfocused SAR systems operate within, not at the limits defined by, these equations.

$$L_e = \sqrt{\lambda R_p} \qquad (7\text{-}17)$$

$$d_{amin} = R_p \frac{\lambda}{2 L_e} = \frac{\sqrt{\lambda R_p}}{2} \qquad (7\text{-}18)$$

where:

L_e = Maximum effective synthetic aperture (array) length, meters

R_p = Range to the patch on the ground, meters

d_{amin} = Finest cross-range (azimuth) resolution for an unfocused SAR, meters

For an unfocused SAR system with a wavelength of 0.03 meters (10 GHz carrier frequency) and a range to the patch on the ground of 20 km, the maximum effective synthetic aperture length is 24.5 m, and the resultant cross-range resolution is 12.2 m (Equation 7-19). For longer ranges, the cross-range resolution is worse. The level of cross-range resolution provided by an unfocused SAR is acceptable for some imaging applications, but it is not for many others. For flexibility in most imaging applications, it would be nice if the cross-range resolution were independent of range. We will see how this is achieved by using a "focused" SAR and Doppler signal processing.

$$L_e = \sqrt{(0.03)(20 \times 10^3)} = 24.5 \text{ m}$$

$$d_{a\,min} = \frac{\sqrt{(0.03)(20 \times 10^3)}}{2} = 12.2 \text{ m} \qquad (7\text{-}19)$$

Focused Synthetic Aperture Radar

The limitation on maximum effective synthetic aperture length and associated cross-range resolution can be largely removed by "focusing" the synthetic aperture. In principle, to focus a synthetic aperture all we need to do is apply an appropriate phase correction to the return signals received over the duration of the synthetic aperture. The phase correction for any one radar antenna position is proportional to the geometry of the synthetic aperture and the range to the patch to the ground (similar to the far-field distance discussed in Section 13.3). With focusing, the synthetic aperture length can be greatly increased, and thus very fine cross-range resolution can be provided. Only focused SAR systems can provide the fine cross-range resolution required for many imaging radar applications.

However, the synthetic aperture concept works because the patch on the ground is illuminated each time we move the radar antenna (incremental element). Thus, the maximum attainable synthetic aperture has an upper limit based on the simple geometry of illuminating an individual patch on the ground. For a return signal to be received from the patch by all elements of the synthetic array, the mainbeam of the radar antenna must illuminate the patch over the entire length of the synthetic array, as shown in Figure 7-12.

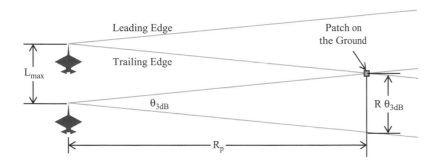

FIGURE 7-12 ▪
Synthetic Aperture
Radar Illuminating a
Patch on the Ground

Therefore, the maximum synthetic array (aperture) length is the same as the width of the radar antenna mainbeam at the range to the patch (Equation 7-20). The half-power (−3 dB) beamwidth of the radar antenna is approximately the wavelength divided by the radar antenna dimension (Equation 7-21). The resultant finest cross-range resolution for a focused SAR is given in Equation (7-22). Most focused SAR systems operate within, not at, the limits defined by these equations.

$$L_{max} = R_p\, \theta_{3dB} \tag{7-20}$$

$$\theta_{3dB} \approx \frac{\lambda}{D} \tag{7-21}$$

$$d_{amin} = R_p\, \frac{\lambda}{2\, L_{max}} = R_p\, \frac{\lambda}{2\left(R_p\, \lambda/D\right)} = \frac{D}{2} \tag{7-22}$$

where:

L_{max} = Maximum synthetic aperture (array) length, meters
R_p = Range to the patch on the ground, meters
θ_{3dB} = Antenna half-power (−3 dB) beamwidth, radians
λ = Wavelength, meters
D = Radar antenna dimension, meters
$d_{a\ min}$ = Finest cross-range (azimuth) resolution for a focused SAR, meters

Based on (Equation 7-22), for a side-looking strip-map–focused SAR, the finest cross-range resolution occurs by reducing the size of the radar antenna, thus providing a large beamwidth. However, what happens when we reduce the size of the radar antenna? As shown in Section 2.3, if the size of the radar antenna is reduced, the mainbeam gain of the antenna goes down, and the S/N of return signals goes down. When the S/N goes down, the intensity or "brightness" of the image goes down. Thus, the size of the radar antenna is constrained by both the cross-range resolution and image quality requirements. These constraints can be lessened in a "spotlight" SAR mode, where the mainbeam of the radar antenna is continuously pointed at the area on the ground being imaged.

Returning to our platform with its side-looking radar, we will address how frequency domain signal processing provides fine cross-range resolution. The majority of SAR systems obtain fine cross-range resolution by using frequency domain signal processing of the time history of Doppler returns from the region being illuminated by the radar. A Doppler filter bank (one filter for each cross-range resolution cell) at the output of each range resolution cell provides the mosaic of range-Doppler cells that produce the image. The Doppler shift and frequency domain signal processing in terms of a Doppler filter bank are discussed in Section 5.3.

The difference in the Doppler shift of two closely spaced patches on the ground is given in Equation (7-23). Note that this equation uses the small angle approximation for sine [$\sin\theta \approx \theta$ when θ is small and in radians]. Solving this equation for the angular separation gives Equation (7-24). Since the Doppler resolution is a function of the integration time (see Section 5.3.1) (Equation 7-25), we can rewrite the angular separation equation in terms of integration time (Equation 7-26). The integration time is the same as the time it takes the leading edge to the trailing edge of the radar antenna mainbeam (see Figure 7-12) to transit across the patch (Equation 7-27). The integration time is often called the time to form the synthetic aperture, or synthetic aperture time. Since the length of the synthetic aperture is the product of the platform speed and the integration time (Equation 7-28), we can rewrite the angular separation equation in terms of the synthetic aperture length (Equation 7-29).

$$\Delta f_d = \frac{2\,V_R\,\Delta\theta}{\lambda} \tag{7-23}$$

$$\Delta\theta = \frac{\lambda\,\Delta f_d}{2\,V_R} \tag{7-24}$$

$$\Delta f_d = \frac{1}{T_I} \tag{7-25}$$

$$\Delta\theta = \frac{\lambda}{2\,V_R\,T_I} \tag{7-26}$$

$$T_I = \frac{R_p\,\theta_{3dB}}{V_R} \tag{7-27}$$

$$L = V_R\,T_I \tag{7-28}$$

$$\Delta\theta = \frac{\lambda}{2\,L} \tag{7-29}$$

where:

Δf_d = Difference in Doppler shift between two patches on the ground, hertz

V_R = Radar speed, meters/second

$\Delta\theta$ = Angular separation between two patches on the ground, radians

λ = Wavelength, meters

T_I = Coherent integration time (time to form the synthetic aperture, time a patch on the ground is illuminated, and time it takes the leading edge to the trailing edge of the radar antenna mainbeam to transit across the patch), seconds

L = Synthetic aperture (array) length, meters

Using the definition of cross-range resolution—range times angle resolution (Equation 7-30)—we can develop the equations for cross-range resolution as a function of Doppler filter bandwidth, coherent integration time, and synthetic aperture length (Equation 7-31).

$$d_a = R_p \, \Delta\theta \tag{7-30}$$

$$d_a = R_p \, \frac{\lambda \, \Delta f_d}{2 \, V_R} = R_p \, \frac{\lambda}{2 \, V_R \, T_I} = R_p \, \frac{\lambda}{2 \, L} \tag{7-31}$$

where:

d_a = Cross-range (azimuth) resolution, meters

When a SAR system uses frequency domain signal processing, the cross-range resolution is often represented only in Doppler terms. The maximum Doppler spread as the radar antenna beamwidth moves across the patch on the ground is given in Equation (7-32). This equation uses the small angle approximation for sine [$\sin\theta \approx \theta$ when θ is small and in radians]. The number of individual Doppler filters (cross-range resolution cells) obtainable by this system is the maximum Doppler spread divided by the Doppler filter bandwidth (cross-range resolution in the frequency domain) (Equation 7-33). The finest attainable cross-range resolution is the beamwidth of the radar antenna divided by the number of Doppler filters (Equation 7-34). Most focused SAR systems operate within, not at, the limit defined by this equation.

$$f_{d\,max} = \frac{2 \, V_R \, \theta_{3dB}}{\lambda} \tag{7-32}$$

$$N_f = \frac{f_{d\,max}}{\Delta f_d} = \left(\frac{2 \, V_R \, \theta_{3dB}}{\lambda}\right)\left(\frac{R_p \, \theta_{3dB}}{V_R}\right) = \frac{2 \, R_p \, \theta_{3dB}^2}{\lambda} \tag{7-33}$$

$$d_{a\,min} = R_p \, \frac{\theta_{3dB}}{N_f} = \frac{\lambda}{2 \, \theta_{3dB}} \qquad\qquad \theta_{3dB} \approx \frac{\lambda}{D}$$

$$d_{a\ min} = \frac{D}{2} \qquad\qquad (7\text{-}34)$$

where:

f_{dmax} = Maximum Doppler spread across the mainbeam of the radar antenna, hertz

V_R = Radar speed, meters/second

θ_{3dB} = Antenna half-power (–3 dB) beamwidth, radians

λ = Wavelength, meters

N_f = Number of Doppler filters, no units

Δf_d = Doppler resolution (Doppler filter bandwidth), hertz

$d_{a\ min}$ = Finest cross-range (azimuth) resolution for a focused SAR, meters

D = Radar antenna dimension, meters

The cross-range resolution for a focused SAR is independent of range and speed, and small radar antennas provide the finest cross-range resolution. We saw this same relationship in Equation (7-22). The preceding discussion about small antennas, low antenna gain, and low S/N applies here as well.

The reciprocal properties of the SAR are interesting. The cross-range resolution obtainable far away in the image is equal to one-half the size of the radar antenna, while the lengths of the synthetic apertures created at the platform motion are equal to the mainbeam coverage of the radar antenna at any range strip on the image! These startling outcomes inspired the people at the University of Michigan in the 1950s to initiate their quest for high-resolution airborne SAR systems. SAR system research, development, implementation, and applications have continuously evolved ever since.

Synthetic Aperture Radar Limitations

The design of SAR waveforms must be accomplished within rather narrow ambiguity constraints. A PRF must be selected to avoid both range and Doppler ambiguities (see Sections 5.2.3 and 5.3.2, respectively). Numerically the PRF to avoid range ambiguities is larger than the PRF to avoid Doppler ambiguities, thereby resulting in a maximum and minimum PRF. The maximum PRF that will avoid range ambiguities either within the range extent of the image, often called the "swath width" (see Figure 7-11), or to the outer edge of the image, is given in Equation (7-35). Using the range extent of the image is common for SAR systems with a very low sidelobe antenna (see Section 4.1.2), because the return signals from outside the mainbeam will have very low amplitudes. Using the outer edge of the image to define range ambiguities is common in short-range SAR systems.

$$PRF_{max} = \frac{c}{2\ R_{sw}} \qquad\qquad PRF_{max} = \frac{c}{2\ R} \qquad\qquad (7\text{-}35)$$

where:

PRF_{max} = Maximum pulse repetition frequency that will avoid range ambiguities, hertz

c = Speed of light, 3×10^8 meters/second

R_{sw} = Range extent of the image or swath width, meters

R = Range to the outer edge of the image, meters

The minimum PRF that will avoid Doppler ambiguities across the Doppler extent of the image is defined by the null-to-null beamwidth (see Section 4.1.1) of the radar antenna (Equation 7-36). This equation uses the small angle approximation for sine [$\sin\theta \approx \theta$ when θ is small and in radians]. To avoid both range and Doppler ambiguities, the SAR system must have a PRF between the minimum PRF and maximum PRF (Equation 7-37). Notice how long ranges or wide swaths and high range rates tend to be incompatible, as do small radar antennas and long ranges.

$$PRF_{min} = \frac{2\, V_R\, \theta_{nn}}{\lambda} \approx \frac{4\, V_R}{D} \qquad \theta_{nn} \approx \frac{2\, \lambda}{D} \qquad (7\text{-}36)$$

$$PRF_{min} \leq PRF \leq PRF_{max} \qquad (7\text{-}37)$$

where:

PRF_{min} = Minimum pulse repetition frequency that will avoid Doppler ambiguities, hertz

V_R = Radar speed, meters/second

θ_{nn} = Radar antenna null-to-null beamwidth, radians

λ = Wavelength, meters

D = Radar antenna dimension, meters

PRF = Pulse repetition frequency to avoid both range and Doppler ambiguities, hertz

For a SAR to work properly, a reference frequency much finer than the Doppler filter bandwidth and a time reference much finer than the range resolution are necessary. Additionally, compensation for the three-dimensional motions of the radar antenna during flight is required. Radar antenna motion compensation is accomplished by one of three ways: (1) collecting antenna motion using a three-dimensional inertial measurement unit (IMU) on the radar antenna (periodically corrected by navigation sensors such as an altimeter and the global positioning system [GPS]); (2) corrected during image signal processing; or (3) combinations of these two. If the radar antenna mainbeam is fixed, corrections for only very limited platform motion are practical. If the radar antenna mainbeam is movable, either mechanically or electronically, real-time corrections can be made.

Because a SAR uses the Doppler generated by a moving platform surveying a fixed landscape, difficulties arise where there are moving objects on the landscape. In fact, such moving objects, depending on their range rates, may be displaced in angle (their ranges not being determined by Doppler will remain correct). Because the spectrum of Doppler frequencies available for SAR processing at microwave frequencies and platform speeds is not wide, moving ground objects can exceed the ambiguity limits of the radar waveform and therefore appear at any point on the image. Such wild excursions are unusual, however. A clever image interpreter can usually properly place moving vehicles or trains on their respective roads or railways. However, in a dense target environment where off-road vehicles may be encountered, the placement problem becomes complicated.

Some amelioration is available if the received signal is also processed in an MTI channel—often called ground moving target indicator (GMTI). In this way, moving objects can be identified and properly placed within the image. Initially, the accuracy with which moving objects can be placed was not good because the SAR beamwidth tends to be wide, giving poor instantaneous angle accuracy. Range will remain as good as with the SAR processing. Modern SAR/MTI systems and associated signal processing algorithms have greatly improved the moving object placement accuracy [Stimson, 1998, Chapter 24; Sullivan, 2004, Section 11.4.2].

All the previously mentioned requirements for data are added to the already prodigious processing associated with measuring received signal strength in every range and angle resolution cell of the radar over several square kilometers of area. Little wonder, then, how originally all SAR systems either had to record their data and process it after the flight or to dispatch it directly to the ground by wideband data link. Only at the end of the 1970s had the digital signal processing hardware advanced far enough to allow image processing and display on-board the aircraft in near real time. Real-time processing is currently commonplace.

Forming an image based on range and angle resolution cells provides SAR systems with a plan view of the region they are surveying. The image emerges as an accumulation of strips at constant range from the platform. Thus, because the grazing angle is not large, there is shadowing but no foreshortening, as there is in a photograph of the same region from the same angle. In fact, the two techniques are reciprocal. In the geometry where a photograph presents a plan view (straight overhead), the SAR would be foreshortened to the point of showing only height differences.

Images produced by SAR systems can never equal visible or infrared images for fine detail (ultimate resolution depends on λ^2), but they are superior in at least two ways: they furnish their own illumination; and they are all-weather.

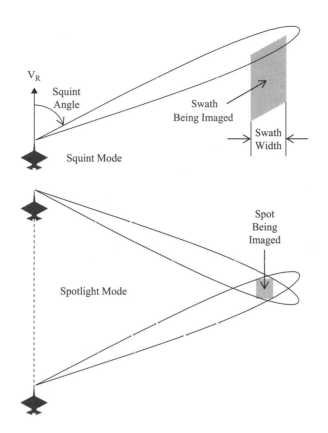

FIGURE 7-13 ▪
Synthetic Aperture
Radar: Squint and
Spotlight Modes

Synthetic Aperture Radar Modes

A SAR system is not limited to the creation of strip maps by side-looking perpendicular to the line of movement. Two other common SAR modes are squint and spotlight, as shown in Figure 7-13 [Richards et al., 2010, Chapter 21; Stimson, 1998, Chapters 32 and 33]. In squint mode the radar antenna can point at an acute angle off the platform velocity vector, often called a "squint angle," and still create a synthetic aperture. Popular uses of a squinted SAR are to take navigation fixes, search and rescue, and locating targets to attack.

If all other aspects of the SAR system are the same, the geometry associated with a squinted SAR will provide a shorter synthetic aperture length than a side-looking SAR, which results in a coarser cross-range resolution. When the SAR has a squint angle, the Doppler-related equations and their extensions as well as some geometry-related equations we developed for side-looking SAR will have to include the sine of the squint angle. We may as well also include the elevation angle to remove the assumption that it is small. An example of including the squint angle and elevation angle in an extension of a

Doppler-related equation, the cross-range resolution from Equation (7-31), is given in Equation (7-38).

$$d_a = R_p \frac{\lambda \Delta f_d}{2 V_R \sin\theta_p \cos\phi_p} = R_p \frac{\lambda}{2 V_R \sin\theta_p \cos\phi_p T_I} = R_p \frac{\lambda}{2 L} \quad (7\text{-}38)$$

where:

d_a = Cross-range (azimuth) resolution, meters

θ_p = Azimuth (horizontal plane) angle relative to the velocity vector to the patch on the ground (squint angle), degrees

ϕ_p = Elevation (vertical plane) angle relative to the velocity vector to the patch on the ground, degrees

The radar antenna can also be continuously pointed at a given area of interest, often called spotlight mode. With a spotlight mode the synthetic aperture length can be very long, as it is not limited by the radar antenna beamwidth, thus providing very fine cross-range resolution. Since the synthetic aperture length is not limited by the beamwidth of the radar antenna, a large antenna can be used. A large antenna provides a high mainbeam gain and thus high S/N or long range. High S/N provides better image quality, which provides a better capability for its end application.

Doppler Beam Sharpening

Doppler beam sharpening (DBS) gets its name because it uses frequency domain signal processing to separate the returns in the beamwidth of the radar antenna in Doppler—essentially providing an extremely narrow beam [Richards et al., 2010, Section 21.5.3; Stimson, 1998, p. 434]. DBS uses the radar antenna scan rate to provide the coherent integration time. Typically, the radar antenna continuously scans the ground about the aircraft flight path. The integration time is limited to the length of time a patch on the ground is illuminated by the scanning mainbeam of the antenna. Therefore, the cross-range resolution is coarser than can be achieved by a traditional SAR.

Additionally, the cross-range resolution is a function of range because the integration time is the same for all ranges. Since the region directly ahead of the radar, along the velocity vector, has very little Doppler spread within the beamwidth of the radar antenna, the cross-range resolution is very coarse in this region. Even with these cross-range resolution limitations, DBS is a very popular mode. It provides much finer cross-range resolution than a real beam image, other than for the regions along the velocity vector. Compared with SAR modes, DBS can provide a continuously updated image of a large area in front of the aircraft.

The cross-range resolution provided by DBS processing is given in Equation (7-39) [Schleher, 1991, Section 8.1]. Since DBS processing is not focused (see previous SAR discussion), the DBS cross-range resolution is limited to what is provided by an unfocused SAR (Equation 7-40).

$$\text{DBS}_a = \frac{\lambda \left(\frac{\theta_{dot}}{\theta_{3dB}}\right) R_p}{2 \, V_R \, \sin\theta_p \, \cos\phi_p} = \frac{\lambda \, \Delta f_d \, R_p}{2 \, V_R \, \sin\theta_p \, \cos\phi_p} \qquad (7\text{-}39)$$

$$\text{DBS}_a \geq \sqrt{\frac{\lambda \, R_p}{2}} \, \frac{1}{\sin\theta_p \, \cos\phi_p} \qquad (7\text{-}40)$$

where:

DBS_a = Doppler beam sharpening cross-range (azimuth) resolution, meters
λ = Wavelength, meters
θ_{dot} = Radar antenna scan rate, degrees/second
θ_{3dB} = Radar antenna half-power (−3 dB) beamwidth, degrees
R_p = Range to the patch on the ground, meters
V_R = Radar speed, meters/second
θ_p = Azimuth (horizontal plane) angle relative to the velocity vector to the patch on the ground (squint angle), degrees
ϕ_p = Elevation (vertical plane) angle relative to the velocity vector to the patch on the ground, degrees
Δf_d = Doppler resolution (Doppler filter bandwidth), hertz

As seen in these two equations the cross-range resolution is dependent on geometry, aircraft, and DBS characteristics. As such, while some conditions will result in a limitation in cross-range resolution, others will not. An exercise at the end of this chapter will numerically reinforce these concepts. A plot of DBS cross-range resolution as a function of azimuth angle to the patch is shown in Figure 7-14 for the conditions of this exercise.

Inverse Synthetic Aperture Radar

The first approach to inferring the characteristics of objects like satellites, ships, and vehicles from radar was to take Fourier transforms of the amplitude-time histories. This effort was mildly successful, but as finer range resolution and digital processing became available analyzing the phases of the radar returns became popular and successful.

The technique used to obtain angle resolution finer than the radar antenna beamwidth in the analysis of object signatures has been called by several names: cross-range analysis, differential Doppler processing, and inverse SAR [Sullivan, 2004, Chapter 6]. Conceptually, it is equivalent to SAR, where the Doppler spread occurring as an object changes in angle with respect to the radar line of sight is used to resolve the object into multiple Doppler resolution cells.

FIGURE 7-14 ■
Doppler Beam
Sharpening Cross-
Range Resolution
as a Function of
Azimuth Angle

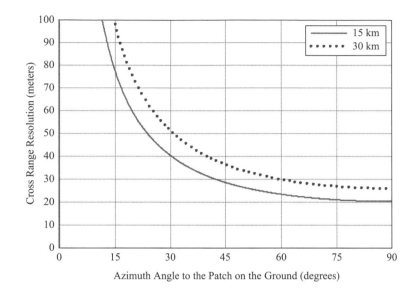

The received signal strength in each cell belongs at a particular distance from some center of rotation on the object. An assembly of the received signal strengths in their resolved locations plus the range resolution available from wideband pulse compression waveforms gives a three-dimensional image of the object. This image is seldom pretty because it arises from coherent micro-wave illumination rather than diffuse light and is at a wavelength five orders of magnitude longer than visible light.

The object must meet several conditions to be a good object for signature synthesis.

(1) It must have a number of scattering centers. That is, returns must emanate from more than one point on the body. This is an easy criterion to meet as even spheres may have enough irregularities to qualify.

(2) The body should be very much larger than a wavelength to give detail to the signature.

(3) The body must change its attitude with respect to the radar line of sight. The amount of angular rotation required depends on the dimensions of the body.

(4) The kinematics of the body must be correctly inferred.

The Satellite Object Identification (SOI) community has synthesized images of many satellites, the national defense applications of which are apparent. One interesting accomplishment was that, when the National Aeronautics and Space Administration (NASA) was uncertain of the orientation of its Skylab prior to committing it to reentry in 1979, the Massachusetts Institute of

Technology's Lincoln Laboratory SOI scientists were able to determine the attitude from their radar data.

7.3 | BISTATIC RADAR SYSTEMS

A bistatic radar system is one where the transmitter and receiver are separated by a significant distance [Cherniakov, 2007; Willis and Griffiths, 2007]. The separate transmitter, receiver, and target form a triangle, as shown in Figure 7-15. Bistatic radar systems offer economic and operational advantages over monostatic radar systems. Several receivers may share a single transmitter, and in being passive the receivers can be difficult to locate by electronic warfare (EW) receivers (see Chapter 8). Multistatic radar systems use multiple transmitters and/or receivers. Often bistatic radar systems are thought to be able to defeat reductions in monostatic RCS.

Traditional bistatic radar work has focused on using cooperative radar transmissions. Just like monostatic radar systems, bistatic radar systems can detect and measure targets in the range, Doppler, and angle domains. Depending on the signal amplitude and modulation used by the transmission source, one domain may be better (in terms of longer detection range, finer measurement resolution, and/or acceptable measurement ambiguities) than the others. A few experimental bistatic radar systems are currently demonstrating detection ranges of a few hundred kilometers (km), plenty adequate for most air surveillance applications. Developing accurate target measurements and tracks at comparable ranges from these detections is often the challenge.

Bistatic Radar Equation

We can develop the bistatic radar equation much the same way we developed the monostatic radar equation is Chapter 2. The transmitter directs the radar waveform at the target. The transmitted waveform propagates to the target and results in a power density at the target. The target intercepts this power density and reflects a portion of it to the receiver. The reflected power propagates to the receiver and results in a power density at the receive antenna. The receive

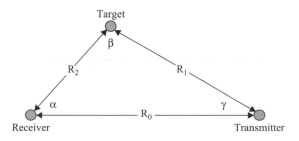

FIGURE 7-15 ■ Bistatic Radar Triangle

antenna effective area intercepts the received power density. The received target signal power is passed along to the receiver.

In the receiver, the received target signal power is processed. We can define the ratio of the processed target signal power to the receiver thermal noise power (see Section 2.3.1) (Equation 7-41). The integration of the received signal over a processing interval provides a signal processing gain. The signal processing gain is often defined by the time-bandwidth product, signal processing time × signal bandwidth (Equation 7-42) [Cherniakov, 2007, Section 2.7].

$$\left(\frac{S}{N}\right)_p = \frac{P_R \, G_{RtT} \, G_{RrT} \, \lambda^2 \, \sigma_b \, G_{sp}}{(4\pi)^3 \, R_1{}^2 \, R_2{}^2 \, F_R \, k \, T_0 \, B_R \, L_R} \tag{7-41}$$

$$G_{sp} = T_{sp} \, B_R \tag{7-42}$$

where:

$(S/N)_p =$ Processed target signal-to-noise ratio, no units

$P_R =$ Peak transmit power, watts

$G_{RtT} =$ Transmit antenna gain in the direction of the target, no units

$G_{RrT} =$ Receive antenna gain in the direction of the target, no units

$\sigma_b =$ Target bistatic radar cross section, square meters (m^2)

$G_{sp} =$ Signal processing gain, no units

$R_1 =$ Transmitter-to-target slant range, meters

$R_2 =$ Receiver-to-target slant range, meters

$F_R =$ Receiver noise figure, no units

$k =$ Boltzmann's constant, 1.38×10^{-23} watt-seconds/Kelvin

$T_0 =$ Receiver standard reference temperature, 290 Kelvin

$B_R =$ Receiver filter bandwidth, matched filter (see Section 5.1.1) bandwidth, hertz

$L_R =$ Bistatic radar system losses, no units

$T_{sp} =$ Signal processing interval, seconds

Bistatic radar systems do not have a detection range per se; rather, they have a detection range product. The bistatic radar equation can be solved for bistatic detection range product (Equation 7-43).

$$R_1 R_2 = \sqrt{\frac{P_R \, G_{RtT} \, G_{RrT} \, \lambda^2 \, \sigma_b \, G_{sp}}{(4\pi)^3 \, SNR_{dt} \, F_R \, k \, T_0 \, B_R \, L_R}} \tag{7-43}$$

where:

$R_1 R_2 =$ Bistatic detection range product, square meters (m^2)

$SNR_{dt} =$ Target signal-to-noise ratio required for detection (detection threshold), no units

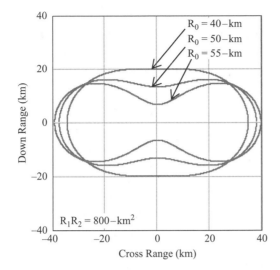

FIGURE 7-16 ▪
Bistatic Radar
Detection
Coverage—Ovals
of Cassini

Bistatic radar detection coverage is a function of not only the bistatic detection range product but also the transmitter–receiver separation. Bistatic radar detection coverage is often described as "ovals of Cassini." Figure 7-16 shows example bistatic radar detection coverage for different transmitter–receiver separations.

Bistatic Radar Cross Section

The bistatic RCS is a measure of the electromagnetic energy intercepted from the transmitter and reradiated to the receiver at the same wavelength by an object. Using the concepts defined in Chapter 6, bistatic RCS is forward scatter. There are many rules-of-thumb concerning bistatic RCS, many of them trying to relate the better understood monostatic RCS to bistatic RCS. When the bistatic angle (β in Figure 7-15) is small, the bistatic RCS is approximately equal to the monostatic RCS. Siegel et al. [1968] report that for simple smooth shapes the bistatic RCS is equal to the monostatic RCS at one-half the bistatic angle. Thus, if the bistatic angle is 50°, the bistatic RCS is equal to the monostatic RCS at 25°. When the bistatic angle is greater than 90° a commonly held belief is that the bistatic RCS becomes significantly larger than the monostatic RCS because for bistatic angles greater than 90° targets are likely to produce forward scatter. While this may be appropriate for smooth simple shapes, it does not necessary translate to complex targets. Cherniakov [2007, Chapters 14 and 15] provides a detailed discussion of bistatic RCS.

RCS predication codes addressing bistatic RCS can be used to determine which rule-of-thumb is applicable, if any. RCS is a classic example of conservation of energy; we cannot get something for nothing. While we may get a bistatic RCS greater than a monostatic RCS for some bistatic angles, some bistatic angles also result in the opposite. In addition, the receiver must be located in such a position as to receive the high reflected forward scatter from the target. Thus, not just bistatic RCS but also practical transmitter–target– receiver geometries are important factors. This is similar to seeing a fleeting specular reflection, a large forward scatter, off a car on a sunny day.

Passive Coherent Location Bistatic Radar Systems

Traditional bistatic radar work has focused on using cooperative radar transmissions. Recently the emphasis has shifted to commercial transmissions: analog frequency modulated (FM) radio, analog TV, cell phones, digital TV. Extensive digital signal processing is used to extract target returns produced by reflections from commercial coherent transmissions. Passive coherent location (PCL) is the name for such bistatic radar systems. The vast majority of PCL systems use Doppler processing extensively. A few experimental PCL systems are currently demonstrating detection ranges of a few hundred kilometers, plenty adequate for most air surveillance applications.

An example of a PCL bistatic radar system is the Manastash Ridge Radar (MRR), developed and used by University of Washington Radar Remote Sensing Laboratory for meteor detection [Lind, 1998]. The MRR uses the U.S. FM radio signal as the transmitter, and provides a range-Doppler map of the received signal. This map is capable of indicating the presence of targets up to 350 km away from the receiver. The MRR performs signal processing (combination of coherent Doppler processing and noncoherent averaging) to achieve a maximum Doppler resolution of ± 1.5 m/sec. The signal processing gain is due to coherent Doppler processing and thus is equal to the time-bandwidth product. The MRR performs Doppler processing of 0.25 seconds of data with a signal bandwidth of 20 kHz, which results in a coherent processing gain of approximately 37 dB. The MRR then noncoherently averages 10 seconds worth of these quarter second spectra, or 40 spectra. The resultant noncoherent processing gain is approximately the square root of 40, or about 8 dB. The net processing gain for the MRR is approximately 45 dB.

Another example of a PCL bistatic radar system is one built by Paul Howland [1997] using the European television signal as the transmitter. This PCL receiver system provides measurements of the Doppler shift and direction of arrival (DOA) of reflections from the TV video carrier. Signals are processed using the fast Fourier transform (FFT) to provide Doppler and DOA measurements. Each FFT requires 1 to 2 seconds of data. A constant false alarm rate (CFAR) detection scheme is used to identify target reflections and to reject noise and unwanted video carrier harmonics. The signal processing gain is accounted for by the narrow receiver noise bandwidth (Doppler processing using an FFT). The resultant range rate measurement resolution is 0.375 m/sec. This PCL bistatic radar detected airliner targets at ranges of several hundreds of kilometers.

7.4 | MODERN MULTI-FUNCTION, MODE, MISSION RADAR SYSTEMS

Modern multi-function, mode, mission radar systems share a few key features: active electronically steered array (AESA) for beam steering (see Section 4.3) and power management; waveform diversity and flexibility for enhanced detection (see Section 3.4) and measurement (see Chapter 5) performance; and adaptive mode control to support multiple missions. Modern radar systems use these key features to provide enhanced target detection, measurements, tracking, fire control, and weapon support performance; enhanced electronic protection (EP) (see Chapter 12) performance; and additional modes from the same radar system.

While electronic warfare (EW) systems (see Chapter 8) and radar systems have always chased each other, revolutionary changes are harder to catch up to than evolutionary changes. Collectively these few key radar features are revolutionary. Many of the newest radar systems are AESA, waveform diverse, with adaptive mode control. This started with airborne fire control radar systems, which always have had the challenge of having to do many functions with a single radar system, as shown in Figure 7-17. These multifunction radar systems provide near simultaneous air-to-air and air-to-ground modes such as search, track, imaging, and weapon support.

Multifunction ground-based target engagement radar systems for radar directed weapon systems followed, often combining traditional target acquisition and tracking into a single radar system. And recently, ground-based

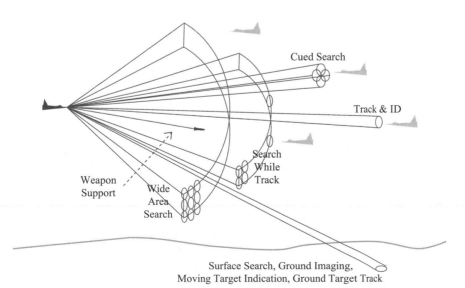

FIGURE 7-17 ◼
Airborne
Multifunction Radar
System

Cued Search

Track & ID

Search
While
Track

Weapon
Support

Wide
Area
Search

Surface Search, Ground Imaging,
Moving Target Indication, Ground Target Track

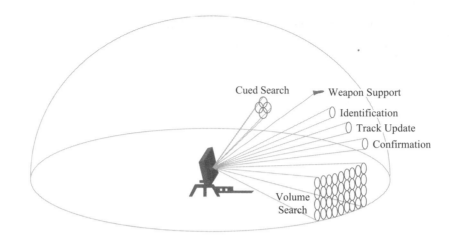

search (surveillance) radar systems incorporate these key features to be able to handle a large number and variety of air-breathing targets and theater ballistic missiles, as shown in Figure 7-18. These multifunction radar systems provide a combination of dynamic interleaved search, track, and weapon support modes.

The main challenge with multifunction radar systems is to provide the best use of the radar time budget to optimize the overall utility and performance of the radar system. A radar resource manager contains algorithms to provide optimal real-time mode and radar waveform sequencing and scheduling [Jeffery, 2009, Section 7.2.1]. As one could imagine, there are numerous radar resource management solutions, from simple to complex. However, the main objective is to optimize the radar time budget to share and allocate radar system resources for the numerous different modes of the radar system. The finite available time, both overall and for each task (mode), is the main driver for the resource manager.

The radar resource manager provides two main functions: task manager and dwell manager. The task manager determines the short-term sequence of tasks to be scheduled based on overall mode priorities and current target information. Different radar systems will have different overall mode priority orders, even within the same class of radar systems. An example mode priority order from highest to lowest is high priority search, track confirmation, high precision track, normal track, and low priority search. Task management is conducted as a "soft" time scale, on the order of a few seconds. The dwell manager translates tasks from the task manager into basic dwell requests and generates different candidate radar waveforms: pulse width, frequency, pulse repetition frequency, and duration. Dwell management is conducted as a "hard" time scale, on the order of a few milliseconds. Radar resource management functions are often adaptive to dynamic priority allocations, radar system workload, perceived target information (e.g., numbers, maneuvers, track quality, threat level), and environment (e.g., clutter, jamming) assessments. The radar resource management, task and dwell, is key to achieve and optimize radar system capability and performance.

Many multifunction radar systems use a combination of classic radar detection theory (see Chapter 3) and multiple event probability theory (see Section 3.4) to provide the overall required detection criteria: probability of detection (P_d); probability of false alarm (P_{fa}); and detection range within a required time period [Jeffery, 2009, Section 7.2.2]. The time requirement includes both the overall frame time and waveform dwell time. The correct use of multiple-event probability theory (i.e., multiple detection attempts) is provided by the beam agility and waveform diversity associated with the AESA. The beam agility provided by an AESA allows for the multiple detection attempts to be statistically correlated, a fundamental requirement for the correct use of multiple-event probability theory (see Section 3.4).

As discussed in Section 3.4, target search can be conducted using the alert–confirm (sequential) detection logic [Jeffery, 2009, pp. 149–150]. The confirmation waveform is transmitted only when the return from the alert waveform exceeds its associated detection threshold. When the confirmation waveform return successfully crosses the threshold, it results in a target detection. Thus, the alert–confirm detection logic saves time in the overall time budget by transmitting the confirmation waveform only when necessary.

Also as discussed in Section 3.4, binomial (M-out-of-N) probability and cumulative probability theory can also be used to provide the overall required detection criteria is less time than classic detection theory alone can provide. There are numerous ways to trade classic radar detection theory and multiple-event probability theory to achieve a desired result; a required detection criteria or minimize time (usually the primary goal), while providing the associated target measurements. Saving time allows the multifunction radar to perform additional tasks within a finite overall time budget.

Compared with a traditional high-power transmitter, the low-power transmit/receive (TR) modules (see Section 4.3.5) in an AESA can support a very wide range of radar waveforms: pulse width, frequency, modulation bandwidth (see Section 5.2.2), PRF, etc. It is much easier to generate a wide range of waveforms at low power than at high power. Additionally, direct digital synthesis (DDS) provides a radar waveform customized to each task: search, track, etc. For example, target search can use a pulse compression waveform (see Section 5.2.2) to enhance clutter rejection (see Section 7.1) performance and thus to simplify track initiation; the finer the range resolution, the easier it is to associate two plots together and thus to initiate a track [Jeffery, 2009, Section 5.5].

The radar resource manager can use current target measurements to develop subsequent radar waveforms—initial search waveform, track initiation waveform, track maintenance waveform—to support overall performance.

They are all designed dynamically to sequentially work together. Additionally, track maintenance can be performed using a waveform that is unambiguous (see Section 5.2.3) around the expected target range but is ambiguous at other ranges. This approach saves time by using a higher PRF (which results in range ambiguities), but shorter integration time required to provide the number of pulses required to detect the target (see Section 3.3). An exercise at the end of this chapter will numerically demonstrate this concept. Saving time allows the multifunction radar to perform additional tasks within a finite overall time budget.

The beam agility and waveform diversity of multifunction radar systems can provide an EP capability (see Chapter 12) by making it difficult for an EW receiver (see Chapter 9) to detect and identify the radar system [Stimson, 1998, Chapter 42]. The multifunction radar system transmits a sequence of diverse waveforms at many different beam positions, only a portion of which is in the direction of the EW receiver. Thus, the EW receiver senses only a portion of this sequence of diverse waveforms, which can appear somewhat "random" in the EW receiver's signal processor. Additionally, resource management can use power management based on the mode and current target tracks. For example, full power when searching for long-range targets and low power for track maintenance of close-in range targets. The combination of beam agility, waveform diversity, and power management provides a low probability of intercept (LPI) EP capability (see Section 12.1) for an advanced multifunction radar system.

In summary, advanced multifunction radar systems are driven by beam agility from an AESA, waveform diversity and flexibility, and adaptive mode control. Radar resource management and its associated algorithms are the "secret sauce" of advanced multifunction radar systems. Multiple waveform detection logic provides a wide range of options balancing radar characteristics, capability, and performance, which are constrained only by physics, detection theory, and probability theory. Waveforms can be adaptively designed for specific modes and performance. And dynamic beam agility and waveform diversity provide an inherent EP capability. Optimal use of time, both overall to complete all tasks and for each individual task, is the primary driver for multifunction radar systems.

7.5 | OVER-THE-HORIZON RADAR SYSTEMS

The over-the-horizon (OTH) radar systems discussed here bounce signals off the ionosphere, scatter them from targets at or near the ground, and detect these targets by observing the return path via the ionosphere [Fenster, 1977]. Technically, these are over-the-horizon backscatter (OTH-B) radar systems. Their operational concept is shown in Figure 7-19.

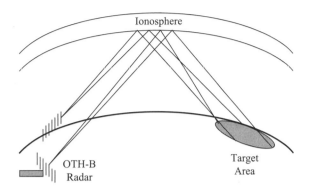

FIGURE 7-19 ■
Concept of OTH-B
Radar

High-frequency (HF) communications have made use of ionospheric refraction to obtain spectacularly long ranges for most of the twentieth century. Ham radio operators talk to the world with their relatively low-power rigs. Even though the very first radar systems were in the HF spectrum, this was viewed as a disadvantage. The idea of actually radiating via the ionosphere and detecting the backscatter returning by the same path did not seem credible until coherent processing became practical in the early 1960s. The next two decades saw slow but steady progress, as digital signal processing has permitted unlimited extension of coherent integration times. OTH-B radar systems exist, and the United States has deployed systems to provide warning of an air attack while the attackers are still several hundreds of kilometers away. OTH-B radar systems are also used to detect drug-smuggling aircraft.

What should interest us, however, is how the principles developed for radar systems at microwave frequencies apply at frequencies two to three orders of magnitude lower. I shall examine OTH-B antennas, waveforms, and hardware.

Antennas

Because the HF region available for OTH-B radar systems is about 6 to 30 MHz (the HF region being from 3 to 30 MHz), a 1 degree beam at the high end of the spectrum would require a 510 meter aperture. At 6 MHz, the antenna dimension would be 2550 meters. These antennas are phased arrays. Narrow beams in azimuth are desirable and wide elevation beams are acceptable, which is fortunate for the designer. A 25.5 meter antenna height at 30 MHz gives a 20 degree elevation beam. The resulting antenna is a phased array (see Section 4.3). In fact, it resembles the antenna farms of major U.S. Navy or Voice of America communications complexes.

OTH-B radar beams are required to propagate out at very low angles to the horizon. This is not a problem at microwave frequencies but is severe at HF where expensive conducting ground planes must be provided in front of the antenna. The electron density of the ionosphere and the available operating frequencies determine at what ranges the radar will operate. Skip distances can vary from several hundred kilometers outward to perhaps 2900 km, bounded by the take-off angle of the antenna beam, the height of the refracting ionosphere, and the curvature of the earth. The radar beam does not radiate into 4π steradians, even conceptually. It is as if the radar system were operating into a waveguide of infinite, gently curving, parallel planes. Where the one-way signal appears, it may be attenuated by less than $1/R^2$; where it does not exist, of course, attenuation is infinite. Yet the radar equation (see Chapters 2 and 3) does apply in its classic form to OTH-B radar systems.

Waveforms

Because the antenna, despite prodigious size, has fairly low gain and its efficiency is not good (its response must cover several octaves of bandwidth), the process of refracting or reflecting from the ionosphere is lossy, and the desired ranges are very long, large amounts of average power (see Section 5.1.3) are required for successful operation. This implies high duty cycle systems and, in fact, encourages 100% duty cycles, with some isolation provided by waveform modulation and the rest by having the receiver at a different location from the transmitter.

To provide adequate surveillance and warning, a good OTH-B system needs good range and Doppler information. Long pulses are no problem because the minimum range is several hundred kilometers. These factors make a long duration pulse compression waveform (see Section 5.2.2) the best choice. Bandwidths are limited to a few percent by physics: the ionosphere reflects differing frequencies to different locations. Although some coding schemes potentially provide equivalent performance, an obvious waveform solution is frequency modulated continuous wave (FM/CW), conceptually identical to the FM chirp waveform discussed in Section 5.2.2. An FM/CW waveform is shown in Figure 7-20.

The phenomenology associated with the OTH-B radar guarantees that it will look down on the earth's clutter. The clutter is a strain on the overall system design, but it becomes a mitigating factor in receiver design. Receiver noise figures and receiver antenna gain become parameters of less concern.

FIGURE 7-20 ■
Frequency
Modulated
Continuous Wave
Waveform

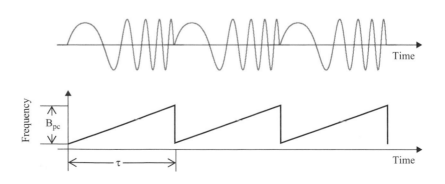

The OTH-B radar system's ability to reject clutter by range and angle resolution is poor by microwave radar standards. The OTH-B radar can exploit very long coherent processing times to get fine Doppler resolution. The radar system is not the limit here; the coherent processing times of the ionosphere and of the target are. Small wonder the OTH-B designers exploit modern digital signal processing.

The radar cross section is well behaved at HF for targets on the order of a few meters in dimension, in which no lobing, scintillation, or other major fluctuations are present to make the detection statistics difficult. Big transports (e.g., Boeing 747s) may be in the resonance region, but their RCS is sufficiently large to make them detectable anyway. However, small private planes or small missiles are in the Rayleigh region where λ^4 dependency is a quick forecloser. Note too, that at night, when refraction off the less dense ionosphere necessitates the use of lower frequencies, more targets will sink into the Rayleigh region and become more difficult to detect.

The relatively narrow absolute bandwidths and wide beams make OTH-B inherently poor in anti-jam performance despite extremely narrow band processing. However, given that the jammer (see Chapter 8) is not in physical proximity to the radar, it must find the right spot with respect to the skip distance or it is ineffective. In addition, the OTH-B, by a judicious selection of frequency, may quickly skip elsewhere.

Hardware

Many years of operating world-girdling communications at HF have kept the state-of-the-art moving in antennas, in very high-power, high-duty-cycle, tunable transmitters, and in adaptive receivers. This technology is directly applicable to OTH-B. Transmitters of several hundred thousand watts of average power at HF are available. Also on the market are accompanying ionospheric sounders for keeping track of the ionosphere in real time so the transmitting frequencies can be changed accordingly. FM/CW waveform generators and digital signal-processing technology are available from the microwave radar field, as are sophisticated system control and management equipment.

7.6 | RADAR ALTIMETERS

Radar altimeters were some of the first commercial applications of radar principles. They transmit a waveform downward from an aircraft, measure the time until a return is received, convert the time delay to altitude, and display or record it. They differ from big radar systems in several ways. Because they are relatively short range (≈ 30 km maximum), their effective radiated power (ERP) or power-aperture product can be small, their receivers noisy, and their signal processing less than optimum. Nevertheless, they must have the ability to report altitude to a fine resolution ($\approx<30$ meters) and follow faithfully (in a few milliseconds) rapid changes in altitude—say, the combination of an aircraft in a dive passing over a cliff. Such performance is not too difficult to attain, so the emphasis is on reliability and economy. Most important is minimum encroachment on the carrier platform, which means the radar system must be small, lightweight, and nonintrusive, requiring no changes in platform mass properties or aerodynamics.

Antennas

The radar altimeter needs an antenna beam sufficiently wide so that the altitude readings can be obtained over any specified flight attitude (perhaps a few tens of degrees for transport aircraft to close to $90°$ for highly maneuverable aircraft) or a steerable beam over those angles. The antenna needs to be flush mounted to provide smooth airflow. Horns, crossed dipoles, and slots potted in a dielectric are obvious approaches. Multiple-beam switched arrays are possible and could be made to meet all specifications, but they are generally too complex and expensive to be competitive except in special applications. Although radar altimeter antennas may have gain, they do not have to because the earth is a big, nearby target.

Radar Cross Section

The target for radar altimeters is the earth's surface: oceans, mountains, plains, and cities. Most of the time the radar altimeter beam is near perpendicular to the surface, making the material on clutter coefficients in Section 6.7 essentially inapplicable. At near-vertical angles of incidence, the clutter reflection coefficient may exceed 10 dBsm over smooth water; may be between 0 and 10 dBsm over smooth, open country; and may be near –10 dBsm over forests or dense undergrowth. Buildings and other man-made structures will provide strong specular returns, and there will be canceling and reinforcing interactions among objects of differing heights, shapes, and electrical characteristics. The resulting reflection coefficient should lie in the region below smooth water and above open country. The area illuminated by a wide beam can be very large; a $20°$ beamwidth at 10 km illuminates approximately 12.2 square kilometers. With a reflection coefficient of $1 \text{ m}^2/\text{m}^2$, this would be an RCS of $12.2 \times 10^6 \text{ m}^2$.

Radar Equation

The radar equation for an altimeter is essentially the same as for clutter, except for RCS. The RCS is not expressible in a single term in the altimeter radar equation. It is an integral or summation of all the contributing scatterers. Starting with the range equation (Equation 2-24) and modifying the clutter RCS (Equation 6-25), accordingly we get Equations (7-44) and (7-45). Combining these two equations results in the altimeter radar equation (Equation 7-46).

$$\frac{C}{N} = \frac{P_R \, G^2 \, \lambda^2 \, A_C \sum \sigma_i}{(4\pi)^3 \, R^4 \, F_R \, k \, T_0 \, B_R \, L_R} \tag{7-44}$$

$$A_C = (R \, \theta_{3dB})^2 \tag{7-45}$$

$$\frac{C}{N} = \frac{P_R \, G^2 \, \lambda^2 \, (R \, \theta_{3dB})^2 \sum \sigma_{0i}}{(4\pi)^3 \, R^4 \, F_R \, k \, T_0 \, B_R \, L_R} = \frac{P_R \, G^2 \, \lambda^2 \, (\theta_{3dB})^2 \sum \sigma_{0i}}{(4\pi)^3 \, R^2 \, F_R \, k \, T_0 \, B_R \, L_R} \tag{7-46}$$

where:

C/N = Single-pulse clutter signal-to-noise ratio, no units

$\sum \sigma_{0i}$ = Sum of the clutter reflection coefficients from all contributing scatterers, square meters/square meters (m^2/m^2)

This is a convenient way to view the radar altimeter and can be applicable to narrow beam, wide bandwidth systems. However, it does not stand up to detailed design of generalized systems. For example, with a wide beam and a relatively large bandwidth (for fine range resolution), the only RCS of interest is that of first return which may be quite small. As shown in Figure 7-21, the remaining illumination, spreading out in a set of concentric annular rings from beam center (widths determined by the effective pulse width of the altimeter), is of little interest and does not contribute to system performance.

Waveforms

Not surprisingly, a long-time popular waveform for radar altimeters is FM/CW, as shown in Figure 7-20. The reference ramp can be just the transmitted waveform fed into the receiver. FM/CW is simpler and cheaper to implement

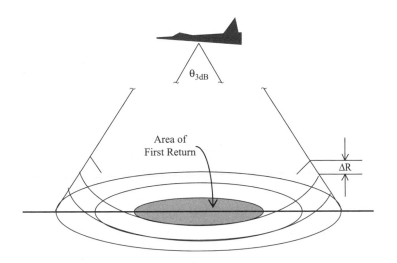

FIGURE 7-21 ■
Radar Altimeter
Ground Return

θ_{3dB}

Area of
First Return

ΔR

than pulsed waveforms and places less stringent requirements on amplifiers, switches, power supplies, and cooling.

Several unusual considerations are associated with the design of a suitable FM/CW waveform. One is time sidelobes (see Section 5.5.4). The FM modulation has sidelobes in the time domain whose power amplitude is of the $(\sin x/x)^2$ form. This means if the detection of the first return is to be made on relatively small RCS (low received signal power), the larger received signal powers (due to larger RCS) that will occur in the succeeding few resolution cells may obscure the first return. Time sidelobes can be suppressed, but at increased complexity. The practical effect becomes one of placing a lower bound on the first time sidelobe return instead of on the radar system sensitivity.

A second factor is unambiguous range (see Section 5.2.3). Ordinarily, the range of maximum altitude would be sufficient to describe this parameter. However, the large RCS of the smooth earth presents the possibilities for ambiguous (second-time-around) returns, which may be sufficient to interfere with the first return. Thus, it is prudent to select an unambiguous range greater than the maximum altitude, preferably over the horizon (more on the horizon in Section 13.1).

A third issue is one of resolution and associated accuracy. In the geometry of the system I have described, the ground return becomes an extended target. The measure of range accuracy derived in Section 5.5.1 tends to locate the "center of mass" of an extended target. This can be a disaster in an altimeter. One solution is to increase the FM bandwidth until the resolution by itself is sufficient to accommodate the accuracy specification.

Last, a range error is caused if the platform has a vertical velocity component. I discussed this in Chapter 5 and found such errors to be substantive in some cases. The design fix is an up–down FM ramp.

Hardware

Currently in operation are excellent FM/CW radar altimeters using less than 10 watts of power and having very modest antenna gains and noise figures. At additional expense and complexity, better-performing radar altimeters could be designed and produced. There is little general demand for such systems, especially now with GPS capabilities so readily available. Special needs, such as for space vehicles and cruise missiles, can be met on a case-by-case basis.

■■■ 7.7 | IONOSPHERIC RADAR SYSTEMS

In the late 1950s and early 1960s, the rapid increases in radar systems' ERP, or power-aperture product, created a need to understand the propagating medium better, particularly the ionosphere. As a result, ionospheric radar systems began to appear. Several located in the arctic regions had the specific objective of understanding the auroral ionosphere, but two located near the equator (one in

Peru and the other in Puerto Rico) had more generalized missions of iono-spheric physics and radio astronomy. We will discuss the one in Puerto Rico, known as the Arecibo Ionospheric Observatory [Air Force Office of Scientific Research, 1963].

Mission

The Arecibo radar has all the planets of our solar system out to Jupiter within range and can gather information about surface properties (RCS), rotational rates (Doppler), and orbits (range and range rate tracking). In fact, it has made a synthetic aperture radar map of the planet Venus [Rogers and Ingalls, 1970]. However, the radar's primary purpose is to study the earth's ionosphere. The method is to illuminate the ionosphere (from 50 km to several thousand kilometers altitude) and to examine the backscatter. By applying Maxwell's equations to the data in the radar returns, the electron density, the electron temperature, and the strength of the earth's magnetic field can be inferred.

Parameters

Arecibo's antenna is a 305 meter aperture spherical reflector located in a natural bowl. Because it is not a paraboloid, it needs a method of correcting for the spherical aberration. It does this by feeding the antenna with a line feed phased so the wavelets that are leaving it add up in the far field of the Arecibo antenna as a plane wave. Reciprocity applies, so we note (Figure 7-22) that when a plane wave is incident on a spherical cap then all ray tracings cross a line that passes through the sphere's center and is parallel to the plane wave. By placing a line feed in this position, making it an array, and phasing the array to provide the correct phase to each ray, a coherent antenna results. This is how the Arecibo feed works. It has the further attribute of being movable along a feed arm, so the antenna beam can be swung 20° off vertical in any azimuth with little loss in performance. Though not perfect, the antenna is excellent, especially considering the massiveness of its construction. The theoretical diffraction-limited beamwidth (1.22 λ/D) of the system is 2.8 milliradians. The Arecibo antenna achieves 2.9 milliradians at zenith.

Because the line feed at Arecibo has high ohmic losses and narrow band-width, a new dual reflector feed was developed at Cornell University with the help of Norwegians [Kildal et al, 1988].

Arecibo has a 2.5 MW peak power, 150 kW average power, klystron-driven transmitter. Operating frequency is 430 MHz. Pulse widths from 2 to 10 μsec and PRFs from 1 to 1000 Hz are available. The receiver is a low-noise parametric amplifier. Assuming a 50% efficient antenna, a system noise temperature of 300 Kelvin, and system losses of 10 dB, we can quickly calculate (with the tools we learned to use in Chapter 2) a matched filter receiver for a 10 μsec pulse from the Arecibo radar that will provide a 10 dB single-pulse S/N on a 10 m^2 target at over 19,000 km.

FIGURE 7-22 ■
Arecibo Spherical
Antenna

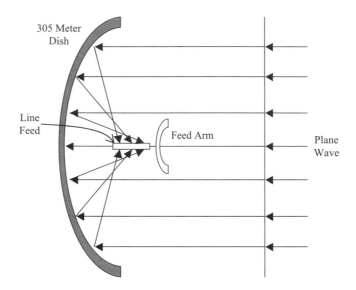

Although it has a very special mission, Arecibo is exemplary of the massive and high-powered radar systems built in the late 1950s and early 1960s. Others of this type include the other ionospheric radar in Peru, the extensive ballistic missile early warning system (BMEWS) now being replaced with phased arrays, the frequency diversity air defense radar systems, and several intelligence-gathering radar systems.

7.8 | LASER RADAR SYSTEMS

Laser is an acronym for "light amplification by stimulated emission of radiation." The first laser was built in 1960 using a ruby rod [Encyclopedia Britannica, 1984]. However, since then many materials have been caused to lase. As has been done with microwave devices for many years, the laser generates and radiates coherent electromagnetic energy. The principle of stimulated emission was recognized by Albert Einstein and is implicit in the photoelectric equation. When electrons, which have absorbed enough energy to jump to higher orbits, return to lower orbits, they emit photons of a particular frequency, depending on the energy difference between the orbits. Photons introduced into a medium (e.g., a tube containing a gas whose electrons have been excited) will stimulate a shower of photons as the electrons return to lower orbits. If a fully reflecting mirror is placed at one end of the tube at an integer number of wavelengths from a partially reflecting mirror at the other end of the tube, the system will radiate an intense coherent beam.

When a laser is augmented with a receiver and a clock, it becomes a laser radar (ladar) system or light detection and ranging (lidar). Lidars are extensively used for making precise measurements of distance. For example, the distance from the earth to the moon has been measured to an accuracy of 1 foot by bouncing a laser signal off a reflector placed on the moon by an astronaut. Other lidar applications include aircraft altimeters, measurers of atmospheric densities and currents, and vehicle speed measurement by law enforcement. The lidar can render high-resolution, three-dimensional images of objects using very short pulses and the narrow laser beam. A challenging potential application is for strategic defense in the form of a space-based lidar to provide long-range imaging of attacking missiles in midcourse.

The lidar has many advantages over conventional microwave radar as well as some shortcomings. Because wavelengths at optical frequencies are on the order of 100,000 times shorter than microwaves, all of the lidar features depending on wavelength—particularly aperture gain, measurement resolutions, waveform bandwidth, and target Doppler—are greatly enhanced. As you can calculate using knowledge gained from this book, a 44 dB gain aperture, which required a 15 meter diameter antenna at L band ($\lambda = 0.6$ meters), is only 0.024 millimeters across at the frequency of visible light ($\lambda = 0.5$ micrometers). Whereas bandwidths of 500 MHz are hard to come by at microwave frequencies, in the visible range even a 1% bandwidth gives 600 GHz. Because Doppler frequencies generated by even slowly moving targets are large, lidars do not need complex pulse-Doppler waveforms to do Doppler processing. Single short pulses are usually more than adequate.

Disadvantages are not trivial. Lidars have limited ability to penetrate rain, smoke, haze, and dust and are severely attenuated in clouds and fog. Their extremely narrow beams make them poor in surveillance applications (a 0.1 meter aperture lidar at a wavelength of 10 micrometers has a half-power beamwidth of 5.1×10^{-3} degree). In fact, even when a radar system provides fine angle and range information, it is challenging to acquire and track (particularly multiple) targets.

Lidar Radar Equation

Because light is electromagnetic radiation, the radar equation also applies to lidars. Many of the terms are identical, although translations may be required because lidars are in the domain of physicists specializing in optics whereas radar systems are in the province of electrical engineers. Lasers are not spoken of in terms of power-aperture product or effective radiated power (power times antenna gain) but instead in terms of "brightness," as defined in Equation (7-47) [Weiner, 1984]. The power intensity at any range can be calculated using (Equation 7-48). These expressions are practical because the narrow laser beam may not subtend the entire target.

$$\frac{P}{\theta^2} \quad \text{or} \quad \frac{P\,A_e}{\lambda^2} \tag{7-47}$$

$$\frac{P}{\theta^2\,(4\pi)\,R^2} \quad \text{or} \quad \frac{P\,A_e}{\lambda^2\,(4\pi)\,R^2} \tag{7-48}$$

where:

P = Power, watts
θ^2 = Effective aperture beam area, square radians
A_e = Effective aperture area, square meters (m^2)
λ = Wavelength, meters
R = Lidar-to-target slant range, meters

The receive antenna of a lidar will usually not be shared with the laser transmitter. A receive antenna has much less stringent tolerances, needing to focus the incoming radiation in a spot only the size of the photosensitive detectors—a square millimeter or so.

Lidar Signal-to-Noise Ratio

The two terms where the lidar radar equation deviates substantially from radar is in detection and radar cross section. At lidar frequencies, electromagnetic energy is more easily treated as quanta (photons) rather than waves. In enunciating the quantum theory, Max Planck noted that the energy in a photon was equal to Planck's constant times the frequency of the light. A lidar receiver such as a photomultiplier tube (PMT) simply counts photons during the lidar pulse width (as a matched filter responds to the incoming radar signal). The average number of signal photons arriving at the output of an optical receiver is given in Equation (7-49).

$$\bar{n}_s = \frac{\eta\,P_r}{h\,f} \tag{7-49}$$

where:

$\bar{n}_s$ = Average number of signal photons arriving at the output of an optical receiver, no units
η = Quantum efficiency (the percentage of incident photons converted into electrons), no units
P_r = Received energy, watts-seconds or Joules
h = Plank's constant, 6.6×10^{-34} Joule-seconds
f = Frequency of the laser light, hertz

Finding the noise term is more complicated than it is in radar. Thermal noise is swamped by photon noise beginning at frequencies around 10,000 GHz ($\lambda = 30$ micrometers). How much photon noise is present? Just as with radar, the total noise at the output of an optical receiver is increased by sources outside the receiver including the stars and planets, other stray light, and front-end losses. Nevertheless, the principal contributors are in the receiver, and we will use them to develop an expression for S/N to compare with radar [Dishington et al., 1963].

Unlike a radar receiver, the noise generated in an optical receiver is increased by the presence of a signal. Because photons arrive randomly, they obey Poisson statistics (a variation of the binomial distribution in which the chances of success in any given trial are very low). The power S/N for the optical receiver is given in Equation (7-50). Substituting in the received target signal power and assuming that the signal power is much larger than noise power in cases of interest gives Equation (7-51). Making the radar and lidar pulse durations equal (and thereby fixing the bandwidth), the expression on the right can be substituted in the radar equation (Equation 7-52).

$$\frac{S}{N} = \left(\frac{\bar{n}\,\eta}{\sqrt{\bar{n}_s}\,\eta}\right)^2 \tag{7-50}$$

$$\frac{S}{N} = \left(\frac{\dfrac{\eta\,P_r}{h\,f}}{\sqrt{\dfrac{\eta\,P_r}{h\,f}}}\right)^2 = \frac{\eta\,P_r}{h\,f} \tag{7-51}$$

$$\frac{S}{N} = \frac{P_r}{k\,T_s} \tag{7-52}$$

where:

S/N = Signal-to-noise ratio, no units
$\bar{n}$ = Variance of fluctuations in the Poisson distribution, no units
k = Boltzmann's constant, 1.38×10^{-23} Joule/Kelvin
T_s = Receiver system noise temperature, Kelvin

For a comparison of radar versus lidar, take the quotient (noise term) and insert some values. A microwave radar receiver with a system noise temperature of 300 K and a blue-green laser ($\lambda = 0.5$ micrometers), the radar would have approximately 96 times less noise than a lidar receiver (Equation 7-53).

$$k\,T_s = (1.38 \times 10^{-23})(300) = 4.14 \times 10^{-21} \text{ Joules}$$

$$\tag{7-53}$$

$$h\,f = (6.6 \times 10^{-34})\left(\frac{3 \times 10^8}{0.5 \times 10^{-6}}\right) = 396 \times 10^{-21} \text{ Joules}$$

Lidar Radar Cross Section

The differences between lidar and radar in radar cross section calculations arise from the wavelength dependence of RCS. The microscopic granularity of the reflecting surfaces is a powerful factor. In addition, objects in the resonant (or Mie) region and the Rayleigh region are molecular in size.

Because the beamwidth of a lidar is so narrow, real beam images with good cross-range resolution are feasible. However, because the size of surfaces that give specular reflections is so small, large amounts of high cross section returns (called "speckle") occur, often suffusing the image. Special limiting circuitry is used to suppress speckle.

The general equation for lidar RCS is the same as used in radar (Equation 7-54). However, in optics a term for the reflectivity of the target is inserted [Jelalian, 1977]. Adding in the expression for gain gives Equation (7-55).

$$\sigma = G\,A \tag{7-54}$$

$$\sigma = \rho \frac{4\pi\,A^2}{\lambda^2} \tag{7-55}$$

where:

$\sigma =$ Radar cross section, square meters (m^2)
$G =$ Gain back in the direction of the lidar, no units
$A =$ Effective intercept area of the target, square meters (m^2)
$\rho =$ Reflectivity of the target to the lidar wavelength, no units

For an isotropic scattering target, such as a perfectly conducting sphere, the radar engineer will set the gain of the target to unity; the optical physicist will substitute the gain of a diffuse optical (Lambertian) scatterer, which is four. Thus, the RCS at laser frequencies of a diffuse target is given in Equation (7-56).

$$\sigma = 4\,\rho\,A \tag{7-56}$$

The equations for corner reflectors at optical frequencies, given a reflectivity of one, are the same as for radar corner reflectors, but they are physically small for very high RCS (see exercise at the end of this chapter).

For the case where a lidar's beam is so narrow it does not illuminate the entire target, the RCS is for an "extended target," which is easily calculated because the intercept area is the solid angle of the transmit beam multiplied by the square of the range to the target (Equation 7-57). This equation assumes far-field illumination and a circular beam. Thus, the RCS for a partially illuminated target is given in Equation (7-58). When this equation is put into the radar equation, the power arriving back at the receiver becomes only range-squared dependent, as with the radar altimeter.

$$A = \frac{\pi}{4}\,(R\,\theta)^2 \tag{7-57}$$

$$\sigma = \pi\,\rho\,(R\,\theta)^2 \tag{7-58}$$

▰ 7.9 | SUMMARY

Chapters 2 through 6 laid the groundwork we can use to understand and analyze virtually any radar system. In this chapter, I discussed two classes of advanced radar systems: (1) a few applications marking the thrust of modern radar technology—clutter rejection, SAR, bistatic radar, and multifunction radar systems; and (2) somewhat unconventional applications—OTH, altimeters, ionospheric radar, and laser radar systems. Although all these radar systems represent extremely complex and intricate applications, their sophistication is achieved by manipulating the fundamental properties already discussed in the preceding chapters and/or the EW properties covered in the following chapters.

▰ 7.10 | EXERCISES

7-1. A radar system has the following characteristics: half-power beam-width $\theta_{3dB} = 2$ degrees; transmitted pulse width $\tau = 1$ microsecond; transmitted carrier frequency $f_c = 3$ GHz; pulse repetition frequency PRF — 500 Hz; and number of pulses integrated $n_p = 40$. (a) What is the signal-to-clutter ratio for the radar system trying to see a 1 square meter RCS target at 10 kilometers range in clutter having a clutter reflectivity σ_0 of –30 dBsm and a 1 degree grazing angle? (b) What is the moving target indicator improvement factor, I_{fMTI} (no units and dB), for a two-delay $N = 2$ MTI when the standard deviation of the clutter source range rate $\sigma_{Rdots} = 1$ m/sec and standard deviation of the radar instabilities power spectral density $\sigma_i = 5$ Hz? (c) What is the output signal-to-clutter ratio after the MTI, $(S/C)_{out}$ (no units and dB)?

7-2. Derive the equation for signal-to-clutter ratio (S/C) (Equation 7-1).

7-3. For pulse-Doppler radar systems on fast-moving aircraft, the clutter problems are complex. Qualitatively evaluate clutter from a hovering helicopter with such a radar system.

7-4. A focused side-looking SAR system, traveling at velocity V_R with plenty of sensitivity, has a λ/D radian beamwidth. (a) Calculate the Doppler spread across the beam, f_{dmax}, at range R and at range 2R. (b) What is time it takes for the beam to transit a target location T_I at those two ranges? (c) How many cross-range resolution cells (Doppler filters) at R and at 2R? (d) What is the cross-range resolution, d_a, at R, at 2R?

7-5. A focused side-looking SAR with a wavelength $\lambda = 0.03$ m is on an aircraft traveling at $V_R = 300$ m/sec. (a) What is the Doppler filter bandwidth needed to resolve two points separated by $\Delta\theta = 0.005$ degrees, Δf_d (hertz)? (b) What is the required integration time, T_I (seconds)? (c) What is the synthetic array length, L (meters)? (d) What is the cross-range resolution at a range $R_p = 50$ km, d_a (meters)? Repeat parts (c) and (d) if the radar antenna mainbeam is squinted at $\theta_p = 45$ degrees and the elevation angle to the patch on the ground is $\phi_p = 10$ degrees.

7-6.   A focused side-looking SAR system maps a 16 km wide by 16 km long swath in $T_I = 2$ minutes. With a pulse bandwidth $B_{pc} = 100$ MHz and a real beamwidth of $\theta_{3dB} = 2°$ at $f_c = 3000$ MHz, what would be the number of resolution cells processed per second?

7-7.   A side-looking focused SAR with an antenna $D = 1$ meter in diameter is imaging a range extent from 10 km to 100 km. (a) What is the maximum PRF required to avoid range ambiguities, PRF_{max} (hertz)? (b) At this PRF, what is the maximum platform speed to avoid Doppler ambiguities (assuming null-to-null beamwidth), V_R (m/sec)? Repeat part (b) if the radar antenna mainbeam is squinted at $\theta_p = 60$ degrees and the elevation angle to the patch on the ground is $\phi_p = 5$ degrees.

7-8.   A spotlight SAR operates at a carrier frequency $f_c = 5.4$ GHz on a platform with a speed of $V_R = 400$ m/sec. (a) What is the finest possible cross-range resolution for $T_I = 10$ sec of spotlight time at range $R_p = 185$ km, d_a (meters)? (b) What is the required minimum PRF to avoid Doppler ambiguities for a radar antenna null-to-null beamwidth $\theta_{nn} = 1.5$ degrees, PRF_{min} (hertz)? (c) What is the maximum slant range extent of the image, swath width, at the minimum PRF before range ambiguities occur, R_{sw} (meters)?

7-9.   A radar system with a Doppler beam sharpening (DBS) mode has the following characteristics: carrier frequency $f_c = 10$ GHz; and Doppler filter bandwidth $\Delta f_d = 15$ Hz. The radar speed is $V_R = 280$ m/sec. (a) What is the DBS cross-range resolution, DBS_a (meters), for the following geometries to the patch on the ground: range $R_p = 15$ km; elevation angle $\phi_p = 42$ degrees; at two azimuth angles $\theta_p = 60$ degrees and 30 degrees? (b) Repeat part (a) for the following geometry to the patch on the ground: range $R_p = 30$ km; and elevation angle $\phi_p = 30$ degrees? Compare your results with those in Figure 7-14.

7-10. A bistatic radar system uses an FM radio station for a transmitter: peak power $P_R = 50$ kW; transmit antenna gain in the direction of the target $G_{RtT} = 0$ dB; and frequency $f_c = 100$ MHz. The receiver has the following characteristics: receive antenna gain in the direction of the target $G_{RrT} = 10$ dB; signal processing gain $G_p = 15$; noise figure $F_R = 6$ dB; bandwidth $B_R = 200$ kHz; and losses $L_R = 7$ dB. If the detection threshold $SNR_{dt} = 13$ dB, what is the bistatic range product for a target bistatic radar cross section $\sigma_b - 5$ m^2?

7-11. A multifunction radar system uses a search waveform providing an unambiguous range $R_u = 150$ km and integrates $n_p = 32$ pulses. The resource manager needs to develop a waveform to update an existing track of a target at range $R_{RT} = 60$ km. (a) If the search waveform is used to provide the track update, what is the integration time, T_I (seconds)? (b) A customized update waveform can be developed based on the estimated target range. Since the target range is only an estimate, the customized update waveform is based on providing an unambiguous range 5% greater than the target range. What is the pulse repetition frequency, PRF (hertz) and integration time, T_I (seconds) for the customized update waveform? (c) How much time was saved by using the customized update waveform, and how many additional search waveforms (beam positions) could be transmitted in that time?

7-12. An over-the-horizon backscatter radar system has its receiver facility located 40 km from the transmitter facility and at right angles to the transmit antenna's boresight. The transmit and receive antenna beams scan ±30° simultaneously. Both are unweighted apertures of 20 dB gain in azimuth and 6 dB in elevation at 30 MHz. (a) At this frequency, what is the approximate isolation afforded by the geographic separation of the two facilities? (Hint: Consider a flat earth. Assume the sidelobe antenna gain is 1/100th [–20 dB] the mainbeam antenna gain.) (b) The system employs an unweighted FM/CW waveform using an FM ramp of 10 millisecond duration with a bandwidth of $B_{pc} = 1$ MHz. The minimum range is 740 km. What additional isolation is afforded by this waveform? (Hint: Assume waveform time sidelobes are $[\sin(x)/x]^2$.) (c) If the effective radiated power of the OTH-B system were 80 dBW, what signal level direct from the transmitter would enter the receiver front end? (d) Would it appear in the first range bin (740 km) at the output of the FM/CW processor? (e) A 10 m^2 target at 740 km range would enter the receiver at –133 dBW, assuming $1/R^4$ attenuation, and would appear at the processor output at about 10 dB below the transmitter power. What are your comments about the need for additional suppression of the direct transmitter signal?

7-13. What is the projected diameter of an optical reflector (cat's eye) with a laser radar cross section of 1 million square meters? (Assume the reflectivity $\rho = 1$ and use a wavelength of 0.5 micrometers.)

7.11 | REFERENCES

Air Force Office of Scientific Research, 1963, *Arecibo Ionospheric Observatory*, Arecibo, Puerto Rico: Air Force Office of Scientific Research.

Barton, D. K., 1988, *Modern Radar System Analysis*, Norwood, MA: Artech House.

Bouwman, Ronald, 2009, *Fundamentals of Ground Radar for Air Traffic Control Engineers and Technicians,* Raleigh, NC: SciTech Publishing.

Cherniakov, Mikhail (editor), 2007, *Bistatic Radar Principles and Practice*, West Sussex, UK: John Wiley & Sons.

Dishington, R., Hook, W., and Brooks, R., 1963, "The Performance of RF and Laser Radar Systems," Report 9200.2-137, Los Angeles: TRW Space Technology Laboratories.

Encyclopedia Britannica, 1984, "Lasers and Masers," Volume 10, 15th ed., pp. 686–689, Chicago: Benton.

Fenster, W., 1977, "The Application, Design, and Performance of Over-the-Horizon Radar in the HF Band," *International Conference, Radar-77*, Publication #155, pp. 36–40, London: IEE Conference.

Harger, R. O., 1970, *Synthetic Aperture Radar Systems Theory and Design*, New York: Academic Press.

Howland, Paul E., 1997, *Television Based Bistatic Radar*, Thesis, School of Electronic and Electrical Engineering, University of Birmingham, UK, September. Related information is also in: P. E. Howland, "A Passive Metric Radar Using a Transmitter of Opportunity," *International Conference on Radar 1994*, pp. 251–256, May 1994; H. D. Griffiths and N. R. W. Long, "Television Based Bistatic Radar," *IEE Proceedings* Part F, vol. 133, pp. 649–657, December 1986; H. D. Griffiths et al., "Bistatic Radar Using Satellite-Borne Illuminators of Opportunity," *IEE International Conference*, Radar 92, pp. 276–279, 1992; D. Poullin and M. Lesturgie, "Radar Multistatic a Emissions non Cooperatives," *International Conference on Radar 1994*, Paris, pp. 370–375, May 1994; and B. Carrara et al., "Le Radar MUET (Radar Multistatique Utilisant des Emetteurs de Television)," *International Conference of Radar 1994*, Paris, pp. 426–431, May 1994.

Jeffery, Thomas W., 2009, *Phased-Array Radar Design Application of Radar Fundamentals*, Raleigh, NC: SciTech Publishing.

Jelalian, A., 1977, "Laser Radar Theory and Technology," in E. Brookner (ed.), *Radar Technology*, Norwood, MA: Artech House.

Kildal, P., Petterson, T., Lier, E., and Aas, J., 1988, "Reflectors and Feeds in Norway," *IEEE Antennas & Propagation Society Newsletter*, Volume 30, no. 2, April.

Kovaly, J. J., 1976, *Synthetic Aperture Radar*, Dedham, MA: Artech House. Collection of thirty-three reports and papers providing in-depth, highly technical treatment of virtually all aspects of synthetic aperture radar.

Lind, Frank D., 1998, "Manastash Ridge Radar," University of Washington, Available at: http://rrsl.ee.washington.edu/Projects/Manastash.

Richards, M. A., Sheer, J. A., and Holm, W. A. (editors), 2010, *Principles of Modern Radar*, Volume 1: Basic Principles, Raleigh, NC: SciTech Publishing.

Rogers, E. E., and Ingalls, R. P., 1970, "Radar Mapping of Venus with Interferometric Resolution of the Range-Doppler Ambiguity," *Radio Science*, no. 2, pp. 425–433, February.

Sandia National Laboratories, 2008, "Synthetic Aperture Radar," Document Number: SAND99-0018. Available at: http://www.sandia.gov/RADAR/sar.html. Sandia National Laboratories Synthetic Aperture Radar Homepage with SAR applications, sample images, etc.

Schleher, D. Curtis, 1991, *MTI and Pulsed Doppler Radar*, Norwood, MA: Artech House.

Siegel, K. M., Crispin, J. W., and Newman, R. J., 1968, "RCS Calculation of Simple Shapes—Bistatic," in J. W. Crispin and K. M. Siegel (editors), *Methods of Radar Cross Section Analysis*, New York: Academic Press.

Skolnik, Merrill I., 2001, *Introduction to Radar Systems*, 3rd Edition, New York: McGraw-Hill.

Stimson, George W., 1998, *Introduction to Airborne Radar*, 2nd Edition, Raleigh, NC: SciTech Publishing. (Ground Moving Target Detection, Chapter 24; Part VII High Resolution Ground Mapping and Imaging, Chapters 31–33; Low Probability of Intercept (LPI) Chapter 42.)

Sullivan, Roger J., 2004, *Radar Foundations for Imaging and Advanced Concepts*, Raleigh, NC: SciTech Publishing. (Part II Imaging Radar, Chapters 6–9; Part III Chapter 11 Observation of Moving Targets by an Airborne Radar.)

Weiner, S., 1984, in Carter, A. B., and D. N. Schwartz (editors), *Ballistic Missile Defense*, Washington, DC: Brookings Institution, p. 94.

Willis, Nicholas J., and Griffiths, Hugh D., 2007, *Advances in Bistatic Radar*, Raleigh, NC: SciTech Publishing.

Electronic Warfare Overview

HIGHLIGHTS

- Overview of Electronic Warfare
- Electronic Support – Electronic Support Measures
- Electronic Attack – Electronic Countermeasures
- Electronic Protection – Electronic Counter-Countermeasures

Because radar systems find and make measurements of noncooperative targets, it motivates various countering actions. Motorists might want to deprive the highway patrolman of information about their speed to avoid a ticket. Certainly, nations want to deny other nations all kinds of information, particularly in times of tension or war.

Electronic warfare (EW) seeks to deny, degrade, and/or deceive an adversary's radar systems to ensure successful completion of the friendly force's mission [see the numerous references at the end of this chapter]. EW is divided into three components: electronic support (ES), electronic attack (EA), and electronic protection (EP) (Table 8-1). The purpose of ES is to intercept, identify, and locate radar systems. The purpose of EA is to "attack" radar systems to negatively affect their performance and/or capability. The purpose of EP is to protect radar systems from ES or EA. Traditionally, ES has been known as electronic support measures (ESM), EA has been known as electronic countermeasures (ECM), and EP has been known as electronic counter-countermeasures (ECCM). An international professional EW society called the Association of Old Crows (AOC) advances strategy, policy, and programs for EW and electromagnetic spectrum operations.

Since the fate of nations may depend on the effectiveness of ES, EA, and EP, details related to specific EW concepts and systems are highly classified. However, many useful general principles of ES, EA, and EP can be discussed and quantified.

TABLE 8-1 ▪ Electronic Warfare Components

Current Terms	Traditional Terms	Applications
Electronic Support	Electronic Support Measures	Signals Intelligence • Communications Intelligence • Electronic Intelligence Situational Awareness • Radar Warning Receiver • Jammer Receiver/Processor
Electronic Attack	Electronic Countermeasures	Self-Protection Jamming Support Jamming
Electronic Protection	Electronic Counter-Countermeasures	Passive Active

8.1 | ELECTRONIC SUPPORT: ELECTRONIC SUPPORT MEASURES

Two ES applications exist in signals intelligence collection and situational awareness (SA). Intelligence collection involves monitoring and analyzing the signals such as radio, data link, and radar that are emitted from an adversary's electronic systems. Technical information and intelligence are derived from the collected signals. SA is the sensing of the electromagnetic environment to determine what the adversary is doing and to provide the information necessary to formulate an EA response. SA is performed by EW receivers, radar warning receivers (RWR), and jammer receiver processors. EW receivers use a wide range of architectures and technologies: superheterodyne, scanning superheterodyne, channelized, instantaneous frequency measurement (IFM), digital, etc. Adamy [2001, Chapter 4] and Neri [2006, Chapter 4] discuss these different receiver architectures in detail. We will discuss EW receivers in detail in Chapter 9.

8.2 | ELECTRONIC ATTACK: ELECTRONIC COUNTERMEASURES

Two EA applications exist: self-protection jamming and support jamming. Often a coordinated jamming strategy with both self-protection and support jamming is used to maximize target protection. As shown in Figure 8-1, for self-protection jamming all the jamming resources are on the target being protected: internal systems, pods, passive and active decoys, etc. The primary goals of self-protection jamming are to reduce (preferably eliminate) weapon shot opportunities and to survive the shots that occur. Self-protection jamming attempts to

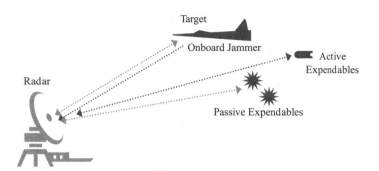

FIGURE 8-1 ■ Self-Protection Jamming

achieve these goals in three steps: (1) reduce the detection capabilities of radar systems; (2) prevent the weapon system from reaching a launch solution; and (3) protect the platform after weapon launch.

For vehicle and ship platforms, self-protection jammers are internal systems. For aircraft platforms, self-protection jammers are either internal systems or pods hung from pylons. An internal system includes all the jammer subsystems (e.g., antennas, receiver/processor, techniques generators, transmitters) installed in the platform. Often the jammer subsystems are distributed across the platform in multiple line replaceable units (LRUs). We will discuss self-protection jamming in detail in Chapter 10.

As shown in Figure 8-2, one, or a few, support jamming resources: aircraft, decoys, etc., protect many targets. Three types of support jamming

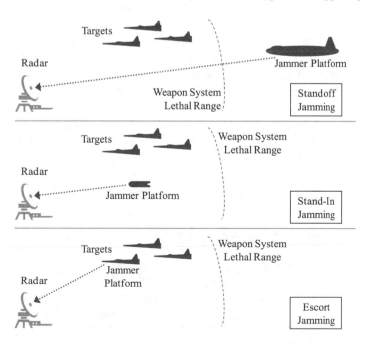

FIGURE 8-2 ■ Support Jamming

exist: (1) standoff, where the support jammer is standing out of range of the adversary's weapon systems; (2) stand in, where the support jammer is standing within range of the adversary's weapon systems; and (3) escort, where the support jammer accompanies the targets it is protecting. The primary goals of support jamming are to reduce detections, tracks, and weapon system shot opportunities against the targets being protected by the support jammer. Support jamming attempts to achieve these goals in three steps: (1) reduce the detection capabilities of radar systems; (2) reduce the efficiency in processing target detections into tracks; and (3) reduce (preferably eliminate) weapon system assignments against the protected targets.

Support jamming can be preemptive, reactive, or a combination of both. A preemptive jammer is "preprogrammed" with jamming techniques based on an estimate of the radar systems in the environment. A preemptive jammer is easy to implement, but it cannot respond to a changing radar environment. A reactive jammer uses receivers to sense the radar systems in the environment and then to respond with the necessary jamming techniques against the sensed radar systems. Reactive jamming is difficult to implement but of course can respond to a changing radar environment. We will discuss support jamming in detail in Chapter 11.

Since there are many similarities between self-protection and support jamming EA, we will describe the similarities now before going into the specifics in Chapters 10 and 11. Two main ways exist for defining and/or looking at EA: the type of jamming waveform; and the effect of the jamming on the radar system. There are two types of jamming waveforms: noise and false target (or pulsed). Noise and false target jamming have different effects on radar systems, as shown in Table 8-2. The effect of jamming on radar systems is categorized as either deny, degrade, or deceive. Deny is to control the information an adversary receives and prevent it from gaining accurate information about friendly forces. From a radar perspective, this is denying target detection and/or associated measurements. Degrade is to interfere with the adversary's efficient processing of information to limit attacks on friendly forces. From a radar perspective, this is degrading (e.g., slow down, miss) the effective processing of target detections, measurements, and tracks. Deceive is to confuse or mislead an adversary. From a radar perspective, this is inducing target measurement and/or track errors, thus deceiving the radar as to the actual target measurements and/or track.

TABLE 8-2 ■ Jamming Waveforms and Effect on Radar Systems

Jamming Waveform	Effect on Radar System
Noise	Deny/Degrade
False Target	Degrade/Deceive

Noise jamming is a straightforward denial/degrade EA technique. It seeks to deny detection or degrade (reduce) the detection range of the radar system (see Chapter 3). Noise jamming also seeks to deny measurements or degrade the radar system's ability to adequately measure target states: range, range rate, and/or angle (see Chapter 5). Additionally, noise jamming can also introduce noise power of sufficient magnitude into the radar receiver so that the noise jamming alone crosses the detection threshold, resulting in numerous false alarms that degrade the efficient processing of target detections.

A noise jammer transmits a wideband noise (random amplitude and phase) waveform at the radar system, with the desire to increase the noise in the radar receiver [Goj, 1993, Section 2.2.2.3; Neri, 2006, Sections 5.2.2.1, 5.2.3; Schleher, 1999, Section 3.3.1]. World War II saw the first use of this brute force EA technique against radar systems. Little else but the radar's carrier frequency and receiver bandwidth need to be known. The effect of noise jamming can be diminished by radar signal processing techniques: multiple-pulse integration, pulse compression, etc., just like receiver thermal noise (see Chapter 3). Noise jamming is often underestimated because of its simplicity. However, as we will see in Chapters 10 and 11 it can be very effective.

A noise jamming waveform has simple characteristics: power, carrier frequency, and bandwidth. A continuous wave (CW) jammer noise waveform is shown in Figure 8-3. A range gated noise jamming waveform transmits a wideband noise burst timed to coincide with the target return. Range gated noise is also called cover pulse or smart noise. The received jammer noise combines with the radar receiver thermal noise, thus increasing the noise in receiver. The increased noise due to the jammer reduces the detection range and affects the measurement capability of the radar system.

Noise jamming is often classified based on the ratio of the jamming noise bandwidth (B_J) to the radar receiver bandwidth (B_R). A large ratio, B_J is greater than approximately 10 times B_R, is called barrage jamming. A small ratio, B_J is less than approximately 10 times B_R, is called spot jamming. Barrage noise jamming attempts to cover a large portion of the spectrum to jam multiple radar systems, a frequency agile radar system, etc. Barrage jamming sacrifices high-power spectral density (power per unit frequency) for continuous wide frequency

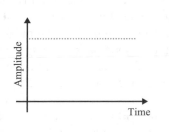 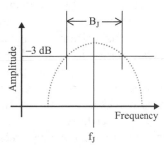

FIGURE 8-3 ■
Continuous Wave Jammer Noise Waveform

coverage. This simplifies the jammer receiver/processor, techniques generator, and transmitter compared with spot noise. Spot noise jamming concentrates the jammer power over a narrow bandwidth to maximize the power spectral density. This complicates the jammer compared with a barrage jammer, as a spot jammer usually must determine the frequency of the radar systems and tune the jammer to the radar frequency.

False target jamming is a complicated degrade/deceive EA technique [Goj, 1993, pp. 46–47; Neri, 2006, Sections 5.2.2.2, 5.2.3; Schleher, 1999, Section 3.3.2]. Against search radar systems, false target jamming seeks to degrade the ability of the radar system to effectively process true target signals using numerous false targets mixed in with the true targets. Against tracking radar systems false target jamming seeks to deceive the radar system to make measurements of and track a false target jamming signal, thereby generating measurement and/or track errors relative to the true target.

A false target jammer transmits a waveform similar to the actual radar waveform at the radar system with the desire that the radar receiver will contain both true target signals and false target jamming signals. False target jamming has been used since WWII. False target jamming is a finesse technique. The radar waveform, detection, measurement, and tracking characteristics need to be known, or at least have a good estimate available, for most false target jamming concepts. Because the false target jamming waveform is similar to the radar waveform, false target jamming can benefit from the radar signal processing.

A false target jamming waveform has the same characteristics as the radar waveform: carrier frequency, pulse width, pulse repetition interval, pulse modulation, etc. The radar waveform was shown in Figure 2-2. Since false target jamming depends on transmitting a high-quality representation of the radar waveform, the radar waveform is usually "captured" by the jammer receiver/processor. To degrade and/or deceive the radar system, most false target EA techniques add amplitude and/or time and/or frequency domain modulations to the captured radar waveform. The false target jamming signals emulate true target returns and hence are treated like true target signals in the radar receiver.

False target jamming is produced by repeater or transponder jammer systems. There are a few, sometimes contradictory, definitions of repeater and transponder jammers. We will use the most common, but make sure you know what definition applies to your jammer. A repeater jammer (Figure 8-4) collects the received radar waveforms, adds the modulation associated with the EA techniques, and transmits the modulated waveform back at the radar system.

A range false target is achieved with time modulation (delay) with respect to the received radar pulse. Repeaters respond to each received radar pulse and thus provide a delay with respect to each received radar pulse. The false target jamming pulse arrives later in time than the reflected pulse from the true target. Therefore, the repeater false target jamming will appear at a longer range than the true target (Figure 8-5). A range rate false target is achieved with frequency

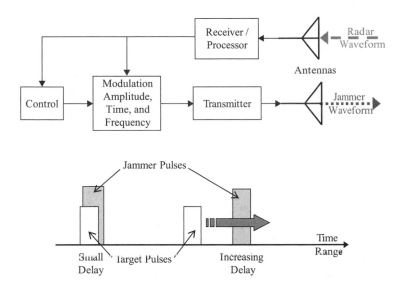

FIGURE 8-4 ■
Repeater False
Target Jammer

FIGURE 8-5 ■
Repeater
Response—Time/
Range Domain

modulation (shift) with respect to the received radar pulse. Amplitude modulation of the received radar pulse will generate angle errors in lobing (conical scan or sequential) angle trackers (see Chapter 5).

A transponder jammer (Figure 8-6) collects the received radar waveforms, stores them in memory, extracts them from memory, adds the modulation associated with the EA techniques, and transmits the modulated waveform back to the radar system. Analog jammer receiver/processors use recirculating delay lines to capture and store the radar waveform. Digital jammer receiver/processors use digital radio frequency memory (DRFM) to capture and store the radar waveform.

A range false target is achieved with time modulation (delay) with respect to the received radar pulse. Transponders store each received radar pulse and thus can determine the pulse repetition interval (PRI) of the radar waveform. With a time delay longer than the PRI, a false target jamming pulse can be made to arrive earlier in time than the reflected pulse from the true target. Therefore, the transponder false target jamming will appear at a shorter range

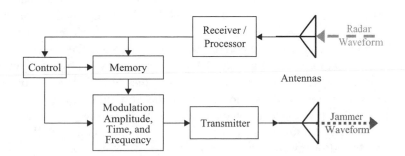

FIGURE 8-6 ■
Transponder False
Target Jammer

FIGURE 8-7 ▪
Transponder
Response—Time/
Range Domain

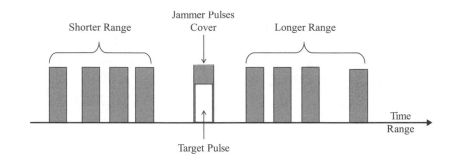

than the true target (Figure 8-7). This effect uses the concept of range measurement ambiguities (see Chapter 5). Multiple false targets can be generated from each stored radar pulse. A range rate false target is achieved with frequency modulation (shift) with respect to the received radar pulse. Amplitude modulation of the received radar pulse will generate angle errors in lobing (conical scan or sequential) angle trackers (see Chapter 5).

Expendables provide additional EA options beyond those provided by on-board systems. The purpose of self-protection expendables is to give the radar system something better than the target to detect and track. Self-protection expendables are discussed in detail in Chapter 10. Expendable support jammers can be used in missions that are too dangerous for manned platforms. Expendable support jammers are discussed in detail in Chapter 11.

Self-protection expendables include chaff, passive and active free-flight decoys, and passive and active towed decoys [Adamy, 2001, Chapter 10; Neri, 2006, Chapter 5; Schleher, 1999, Chapter 3; Stimson, 1998, Chapter 34]. Chaff is composed of small, thin pieces of aluminum foil or metalized fibers dispersed from the protected platform (the target). Chaff was developed in WWII. To the radar system, chaff will appear as a cluster of many secondary "targets" or as a "cloud" completely obscuring large volumes around the protected platform. As discussed in Chapter 6, the size and number of the chaff determine its RCS.

Free-flight decoys are designed to appear as targets to radar systems and can be powered or unpowered. Powered decoys can be small unmanned aircraft, missiles, or rockets. Unpowered decoys are launched from the protected platform and then either follow a ballistic free fall trajectory or use a parachute to slow their descent. The RCS of the passive free-flight decoy determines the decoy signal received by the radar system. The RCS of the decoy can be physically augmented, by adding a corner reflector, to increase the received decoy signal power. Active free-flight decoys contain a battery-powered jammer. Passive or active decoys can be towed by the platform they are protecting. Active towed decoys receive power from the tow platform. The receive antenna, receiver/processor, EA technique generator can be part of the decoy itself or on-board the towing platform. The "pros" and "cons" of chaff, free-flight, and towed decoys are shown in Table 8-3.

TABLE 8-3 ■ Expendable Electronic Attack Pros and Cons

Expendable	Pro	Con
Chaff	• Inexpensive • Can cover a large volume	• Limited resource • RCS is time dependent, it takes time to reach its peak value, and as the cloud dissipates the RCS decreases • Have to dispense chaff numerous times to provide continuous protection • Must be dispersed and timed to provide protection • The "cloud" is dependent on prevailing winds • Does not separate from slow-moving platforms • Fast-moving platforms quickly separate from the chaff
Free-Flight Decoy	• Consistent RCS • Accurate control of the decoy position relative to the protected platform	• Limited resource • Limited time close to the protected platform to deceive the radar system to the decoy; thus, need to dispense numerous decoys to provide continuous protection
Towed Decoy	• Consistent RCS • Always relatively close to the protected platform to deceive the radar system to the decoy • No propulsion or power needed • Can use the towing platform's receiver/processor	• Limited resource, towed decoys are usually not recoverable

████ 8.3 ████ | ELECTRONIC PROTECTION: ELECTRONIC COUNTER-COUNTERMEASURES

EP involves steps taken in the design and/or operation of a radar system to counter the effects of ES and EA. Many steps taken to improve radar system capability and/or performance also have EP benefits. For example, moving target indication (MTI) and pulse-Doppler processing are used to reduce clutter signal power (see Section 7.1); however, they are also effective at reducing the signal power of slow-moving chaff. While most EP techniques are passive, lethal EP concepts exist. For example, a second radar system can locate the jammer by trilateration, and a home-on-jam missile can be launched against the jammer. In this chapter we will introduce a few EP concepts: operator in the loop, waveform diversity, antenna-based signal processing techniques, and sophisticated target trackers.

Well-trained, experienced operators are hard to beat with EA. The role of an operator varies significantly from radar system to radar system. Radar systems have manual, semiautomatic, and automatic operation. In manual operation the operators do everything: control the radar modes and waveforms, declare target detection, initiate and maintain track, etc. Manual operation is most common in radar systems meant to handle only a few targets. In semiautomatic operation, the operators perform some functions, which vary from radar to radar, while the radar system automatically performs all others. Semiautomatic operation is most common in radar systems meant to handle only several targets. In automatic operation, the operators perform only high-level operation—including controlling modes and initiating use of the radar—of the radar system, and it automatically performs all functions. Automatic operation is most common in autonomous radar systems (e.g., active missile seekers and airborne interceptor) and in radar systems meant to handle a large number of targets (e.g., search, weather). An experienced operator can minimize the effects of EA. However, operators are "only human," and thus everyone is different, even across a group of people with the same training and experience. Accurately quantifying the EP benefit of operators requires careful inclusion of both radar-EW and human factors considerations.

Radar waveform diversity is one of the most effective EP techniques because it makes it very difficult for an EW system to detect, identify, and/or keep current with the radar waveform. All aspects of the radar waveform (see Section 2.2) can be diverse: carrier frequency, pulse width, modulation, pulse repetition interval and pulse repetition frequency, polarization. The overwhelming majority of EA techniques require measuring at least some of the radar waveform characteristics to be effective. Thus, without good, current radar waveform characteristics the full effect of the EA technique on the radar system may not be obtained. It is important to note many radar systems change waveform characteristics as part of implementing different modes such as ranging waveforms, Doppler waveforms, and MTI waveforms not just as an EP concept. Additionally, by employing low probability of intercept (LPI) techniques, a radar system can greatly reduce the range at which it can be detected by an EW receiver (see Chapter 9).

Antenna-based signal processing techniques attempt to reject support jamming signals entering the radar system through its antenna sidelobes (see Chapter 4). A sidelobe canceler (SLC) attempts to cancel noise jamming. The three main SLC types are closed loop (analog), open loop (digital), and adaptive nulling (electronically steered arrays). A sidelobe blanker (SLB) attempts to blank false target jamming signals. The SLB turns off the radar receiver output (range gate) when an unwanted signal appears in the radar antenna sidelobes.

With advancements in computer performance, sophisticated target trackers have become common in radar systems. A sophisticated target tracker, such as correlation and state trackers, can minimize the effectiveness of some EA techniques. Correlated measurement trackers consider multiple measurements, such as range and range rate, together when computing the overall target track. Uncorrelated measurements, often due to false target jamming, are "ignored" by the tracker. Target state trackers provide estimates of target state vectors such as position, velocity, and acceleration from individual range, range rate, and angle measurements. Examples of target state trackers are α-β and Kalman Filters. Target state trackers use current target measurements combined with previous target information and a target state model to compute a target state estimate. This helps the tracker "ignore" unrealistic or unexpected measurements, from both actual targets and EA signals.

There are a wide range of radar EP concepts, and we have only touched the surface. We will discuss radar EP concepts in detail in Chapter 12, after we have reviewed EW receivers, self-protection, and support jammers in Chapters 9, 10, and 11, respectively. As long as there is ES and EA, there will be EP and vice versa.

8.4 | SUMMARY

Electronic warfare seeks to deny, degrade, or deceive an adversary's radar system to ensure successful completion of the friendly force's mission. EW is divided into three components: electronic support (ES), electronic attack (EA), and electronic protect (EP). The purpose of ES is to intercept, identify, and locate radar systems. We will discuss EW receivers providing ES in detail in Chapter 9. The purpose of EA is to "attack" radar systems. We will discuss self-protection EA and support EA in detail in Chapters 10 and 11, respectively. The purpose of EP is to protect radar systems from ES and/or EA. We will discuss EP in detail in Chapter 12. The references in the next section provide a very good EW overview.

8.5 | REFERENCES

Adamy, David, 2000, *EW 101: A First Course in Electronic Warfare*, Norwood, MA: Artech House.
    Just what the title says, based on Adamy's long-running column in the *AOC Journal of Electronic Defense* magazine.

Chrzanowski, Edward, 1990, *Active Radar Electronic Countermeasures*, Norwood MA: Artech House.

Chapter 1 contains a very good EA-ECM overview.

Curry, G. Richard, 2001, *Radar System Performance Modeling*, Norwood MA: Artech House.

Chapter 10 contains an overview of EW with supporting math and associated Excel spreadsheets.

Farina, Alfonso, 1992, *Antenna-Based Signal Processing Techniques for Radar Systems*, Norwood MA: Artech House.

Chapter 1 contains a very good overview of EA-ECM and EP-ECCM. The rest of the book discusses the concepts of antenna-based EP-ECCM.

Neri, Filippo, 2006, *Introduction to Electronic Defense Systems*, 2nd Edition, Raleigh, NC: SciTech Publishing.

Chapter 1 contains a very good overview of electronic defense.

Schleher, D. Curtis, 1999, *Electronic Warfare in the Information Age*, Norwood, MA: Artech House.

Chapter 1 contains a very good EW overview; includes numerous associated MATLAB files.

Stimson, George W., 1998, *Introduction to Airborne Radar*, 2nd Edition, Raleigh, NC: SciTech Publishing.

This book contains a very good overview of EW: Chapter 34 ECM-EA; Chapter 35 ECCM-EP; and Chapter 36 EW Intelligence (ES-ESM).

Van Brunt, L. B., 1978, *Applied ECM*, Volume I and II, Dunn Luring, VA: EW Engineering, Inc.

Chapter 1 provides a good overview of EA-ECM and EP-ECCM, setting the stage for descriptions of myriad EA-ECM and EP-ECCM tactics and techniques.

Wiley, Richard, 2006, *ELINT: The Interception and Analysis of Radar Signals*, Dedham, MA: Artech House.

A comprehensive book on EW receivers from intercept to parameter extraction to identification to emitter location and everything in between.

Electronic Warfare Receivers

HIGHLIGHTS

- Define and discuss electronic warfare receivers: radar warning receivers and jammer receiver processors
- Develop the electronic warfare receiver equation, the starting point for our mathematical representations
- Describe detection of the received radar signal
- Describe electronic warfare receiver metrics to quantify its performance
- Describe signal parameter extraction, deinterleaving, identification, and output

9.1 | ELECTRONIC RECEIVER PRINCIPLES

Electronic warfare (EW) receivers—radar warning receivers and jammer receiver processors—are an element of Electronic Support (ES). EW receivers provide situational awareness (SA) about radar systems and other emitters (e.g., radios, data links) in the radio frequency (RF) environment [Neri, 2006, Chapter 4; Schleher, 1999, Chapter 6]. An EW receiver uses sensitive receivers to collect radar waveforms from the RF environment. It then detects radar signals in the presence of EW receiver thermal noise. Signal processors extract parameters from the detected radar signals and use them to identify specific radar systems and determine their location. The challenge for an EW receiver is to do all this quickly in a dense RF environments, millions of radar pulses per second and background (e.g., TV, radio, cell phone) emitters, at operationally useful ranges. The fundamental goal of all EW receivers is to "see them (radar systems) well before they see you (the platform with the EW receiver)."

■ 9.2 | ELECTRONIC WARFARE RECEIVER EQUATION

The EW receiver equation relates EW receiver, radar, and geometry/environment characteristics together to allow us to determine the radar signal power received by the EW receiver. It is the first step in determining EW receiver performance. We will use a radar warning receiver (RWR) as our example EW receiver; however, the equations apply to a jammer receiver processor as well. As shown in Figure 9-1, the received radar signal power is computed using incremental steps:

Step 1: radar effective radiated power (ERP)

Step 2: radar-to-target/RWR propagation

Step 3: radar power out of the RWR antenna

The radar system has an ERP (see Chapter 2) in the direction of target/RWR (Equation 9-1). The radar ERP propagates to the RWR, resulting in a radar power density (see Chapter 2) at the RWR (Equation 9-2). The incident radar power density is collected by the effective area of the RWR receive antenna in the direction of the radar, resulting in the radar signal power at the output of the RWR antenna (Equation 9-3). This equation starts with the effective area and finishes with the RWR antenna gain in the direction of the radar (see Chapter 2). Total RWR-related losses (see Chapter 2) include factors not explicitly taken into consideration in determining the received radar signal power: radar transmit loss; atmospheric attenuation (see Chapter 13); radar waveform polarization mismatch with the RWR antenna (see Table 9-1); RWR receive loss, etc. Different radar–RWR–environment combinations have different losses so we collect them up as we need them into a single term (Equation 9-4). When all RWR-related losses are considered, the resultant received single-pulse radar signal power is given in Equations (9-5), algebraic, and (9-6), logarithmic. The logarithmic form of this equation can be simplified by using the carrier frequency instead of wavelength and combining the constants (Equation 9-7). Since the EW community often uses the logarithmic forms of equations, we will present them along the traditional algebraic forms.

FIGURE 9-1 ■
Incremental
Build-Up of the
Received Radar
Signal Power

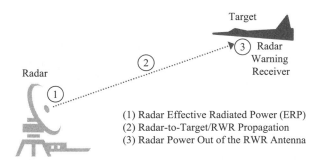

(1) Radar Effective Radiated Power (ERP)
(2) Radar-to-Target/RWR Propagation
(3) Radar Power Out of the RWR Antenna

TABLE 9-1 ▪ Polarization Mismatch Loss

Transmit Polarization	Receive Polarization	Polarization Mismatch Loss		
		Theoretical	Practical Horn	Practical Spiral
Vertical	Vertical	0 dB		N/A
Vertical	Slant (45° or 135°)	3 dB		N/A
Vertical	Horizontal	∞ dB	20–30 dB	N/A
Vertical	Right- or Left-Hand Circular	3 dB		
Horizontal	Vertical	∞ dB	20–30 dB	N/A
Horizontal	Slant (45° or 135°)	3 dB		
Horizontal	Horizontal	0 dB		N/A
Horizontal	Right- or Left-Hand Circular	3 dB		
Right-Hand Circular	Right-Hand Circular	0 dB		
Right-Hand Circular	Left-Hand Circular	∞ dB	20–30 dB	10–20 dB
Left-Hand Circular	Right-Hand Circular	∞ dB	20–30 dB	10–20 dB
Left-Hand Circular	Left-Hand Circular	0 dB		
Right- or Left-Hand Circular	Vertical, Horizontal, or Slant (45° or 135°)	3 dB		

$$ERP_R = \frac{P_R\,G_{RT}}{L_{Rt}} \tag{9-1}$$

$$\frac{P_R\,G_{RT}}{L_{Rt}}\left(\frac{1}{(4\pi)\,R_{RT}^2}\right) = \frac{P_R\,G_{RT}}{(4\pi)\,R_{RT}^2\,L_{Rt}} \tag{9-2}$$

$$\frac{P_R\,G_{RT}}{(4\pi)\,R_{RT}^2\,L_{Rt}}(A_e) = \frac{P_R\,G_{RT}}{(4\pi)\,R_{RT}^2\,L_{Rt}}\left(\frac{G_{RWR}\,\lambda^2}{(4\pi)}\right) = \frac{P_R\,G_{RT}\,G_{RWR}\,\lambda^2}{(4\pi)^2\,R_{RT}^2\,L_{Rt}} \tag{9-3}$$

$$L_{RWR} = L_{Rt}\,L_{RTa}\,L_{Rpol}\,L_{RWRr}\,L_{Rsp}\,\cdots \tag{9-4}$$

$$S_{RWR} = \frac{P_R\,G_{RT}\,G_{RWR}\,\lambda^2}{(4\pi)^2\,R_{RT}^2\,L_{RWR}} \tag{9-5}$$

$$\begin{aligned}
10\log(S_{RWR}) = {}&10\log(P_R) + 10\log(G_{RT}) + 10\log(G_{RWR}) \\
&+ 20\log(\lambda) - 20\log(4\pi) - 20\log(R_{RT}) \\
&- 10\log(L_{RWR})
\end{aligned} \tag{9-6}$$

$$\begin{aligned}
10\log(S_{RWR}) = {}&10\log(P_R) + 10\log(G_{RT}) + 10\log(G_{RWR}) \\
&- 20\log(f_c) - 20\log(R_{RT}) - 10\log(L_{RWR}) \\
&+ 147.5582
\end{aligned} \tag{9-7}$$

where:

$\mathrm{ERP_R}$ = Radar effective radiated power, watts

$\mathrm{P_R}$ = Radar peak transmit power, watts

$\mathrm{G_{RT}}$ = Radar transmit antenna gain in the direction of the target/RWR (anywhere in the radar antenna pattern, mainbeam or sidelobes), no units

$\mathrm{L_{Rt}}$ = Radar transmit loss, no units

$\mathrm{R_{RT}}$ = Radar-to-target/RWR slant range, meters

$\mathrm{A_e}$ = Radar warning receiver antenna effective area in the direction of the radar, square meters ($\mathrm{m^2}$)

$\mathrm{G_{RWR}}$ = Radar warning receiver antenna gain in the direction of the radar, no units

λ = Wavelength, meters

$\mathrm{L_{RWR}}$ = Total radar warning receiver-related losses, no units

$\mathrm{L_{RTa}}$ = Radar-to-target/RWR atmospheric attenuation loss, no units

$\mathrm{L_{Rpol}}$ = Radar-radar warning receiver polarization mismatch loss, no units

$\mathrm{L_{RWRr}}$ = Radar warning receiver receive loss, no units

$\mathrm{L_{RWRsp}}$ = Radar warning receiver signal processing loss, no units

$\mathrm{S_{RWR}}$ = Received single-pulse radar signal peak power, watts

$\mathrm{f_c}$ = Radar carrier frequency, hertz

9.2.1 Electronic Warfare Receiver Antennas

The purpose of the EW receiver antenna is the same as any other receive antenna: couple RF electromagnetic waves from the environment into the receiver. An EW receiver antenna has the same characteristics as radar antennas discussed in Chapter 4: mainbeam gain, beamwidth, sidelobes, antenna pattern. However, EW receiver antennas have different requirements than do radar antennas. EW receiver antennas must provide wide angular coverage and bandwidth, match the polarizations (see Chapter 2) associated with many radar systems, be small in size, and be low weight. These requirements usually lead to physically small antennas and therefore low antenna gain. EW receiver antennas can have very wide bandwidth, >10 GHz; however, the antenna gain falls off dramatically at the antenna frequency limits. Often, multiple EW receiver antennas are aligned to provide 360° azimuth coverage about the target, as shown in Figure 9-2.

As previously mentioned, there is a polarization mismatch loss when the transmit radar waveform and receive antenna polarizations are not the same. Table 9-1 contains the polarization mismatch loss for several radar waveform and receive antenna polarization pairs. To minimize polarization mismatch loss, EW receiver antennas typically have circular polarization: either right- or left-hand or both with dual antennas mounted next to each other. Two common

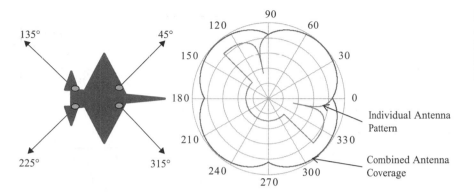

FIGURE 9-2 ▪
Angular Coverage
from Individual and
Multiple EW
Receiver Antennas

EW receiver antennas are circular and conical spirals, which also provide a wide bandwidth.

9.2.2 Electronic Warfare Receiver Thermal Noise

As discussed in detail in Chapter 2, all receivers contain thermal noise (random amplitude and phase with uniform power spectral density). The RWR receiver thermal noise is given in Equations (9-8), algebraic, and (9-9), logarithmic. The logarithmic form of this equation can be simplified by combining the constants (Equation 9-10).

$$N_{RWR} = F_{RWR}\, k\, T_0\, B_{RWR} \tag{9-8}$$

$$10\log(N_{RWR}) = 10\log(F_{RWR}) + 10\log(k) + 10\log(T_0) \\ + 10\log(B_{RWR}) \tag{9-9}$$

$$10\log(N_{RWR}) = 10\log(F_{RWR}) + 10\log(B_{RWR}) - 203.9772 \tag{9-10}$$

where:

N_{RWR} = Radar warning receiver thermal noise power, watts
F_{RWR} = Radar warning receiver noise figure (≥ 1), no units
  k = Boltzmann's constant, 1.38×10^{-23} watt-seconds/Kelvin
  T_0 = Receiver standard reference temperature, 290 Kelvin
B_{RWR} = Radar warning receiver bandwidth, hertz

The single-pulse radar signal power and receiver thermal noise as a function of radar-to-target/RWR range are shown in Figure 9-3. As seen in this figure and Equation (9-5), the received radar signal power varies as a function of $1/R^2$. The received radar signal power decreases by 6 dB (1/4) when the range doubles and increases by 6 dB (4 times) when the range halves. Note that the y-axis grid for this figure is in 10 dB steps, one order of magnitude (power

FIGURE 9-3 ■
Received Single-
Pulse Radar Signal
Power (S_{RWR}) and
Receiver Thermal
Noise (N_{RWR}) versus
Radar-to-Target/
RWR Range

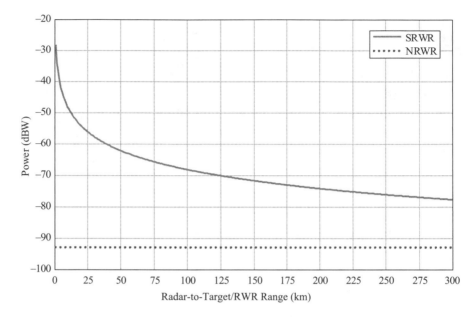

of 10), so the reader can easily interpret the change in power with range. Other figures in this chapter will use the same approach.

9.2.3 Radar Signal-to-Noise Ratio

The single-pulse radar signal-to-noise ratio (S/N) of the RWR, the ratio of the received signal power in one radar pulse to one sample of the RWR thermal noise is given in Equations (9-11), algebraic, and (9-12), logarithmic. The logarithmic form of this equation can be simplified by using the carrier frequency instead of wavelength and combining the constants (Equation 9-13).

$$\left(\frac{S}{N}\right)_{RWR} = \frac{P_R \, G_{RT} \, G_{RWR} \, \lambda^2}{(4\pi)^2 \, R_{RT}^2 \, F_{RWR} \, k \, T_0 \, B_{RWR} \, L_{RWR}} \tag{9-11}$$

$$\begin{aligned}
10\log\left[\left(\frac{S}{N}\right)_{RWR}\right] = {} & 10\log(P_R) + 10\log(G_{RT}) + 10\log(G_{RWR}) \\
& + 20\log(\lambda) - 20\log(4\pi) - 20\log(R_{RT}) \\
& - 10\log(F_{RWR}) - 10\log(k) - 10\log(T_0) \\
& - 10\log(B_{RWR}) - 10\log(L_{RWR})
\end{aligned} \tag{9-12}$$

$$\begin{aligned}
10\log\left[\left(\frac{S}{N}\right)_{RWR}\right] = {} & 10\log(P_R) + 10\log(G_{RT}) + 10\log(G_{RWR}) \\
& - 20\log(f_c) - 20\log(R_{RT}) \\
& - 10\log(F_{RWR}) - 10\log(B_{RWR}) \\
& - 10\log(L_{RWR}) + 351.5355
\end{aligned} \tag{9-13}$$

where:

$(S/N)_{RWR}$ = Single-pulse radar signal-to-noise ratio of the radar warning receiver, no units

The single-pulse radar S/N as a function of radar-to-target/RWR range is shown in Figure 9-4. As seen in this figure and Equation (9-11), the single-pulse radar S/N varies as a function of $1/R^2$. The single-pulse radar S/N decreases by 6 dB (1/4) when the range doubles and increases by 6 dB (4 times) when the range halves.

9.2.4 Detection of the Received Radar Signal

We have developed the equation for the received radar S/N but have not addressed how the RWR detects the received radar signal. Originally audio and visual cues were provided to the operator when a strong enough radar signal was received in a specific frequency band, much like a police radar detector. Audio tones, the demodulated radar pulse repetition frequency (PRF), can also be played in the operator's headphones. Basic spectrum analyzers were added to the display allowing the operator to detect, and possibly identify, the radar signal based on its spectral characteristics.

Even for well-trained and experienced operators, consistent, accurate, and timely detection of radar signals is best for a small number of diverse (e.g., time, frequency, angle, PRF) radar signals. To accurately, consistently, and quickly handle a large number and/or spatially, spectrally, or temporally diverse radar signals requires an automated detection process.

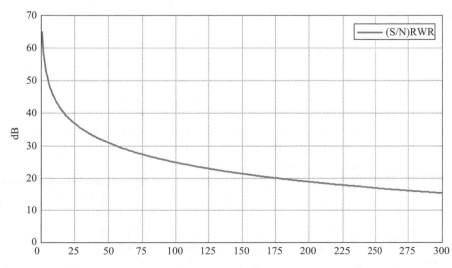

FIGURE 9-4 ■ Single-Pulse Radar Signal-to-Noise Ratio versus Radar-to-Target/RWR Range

FIGURE 9-5 ■
Threshold Detection
of Radar Signals

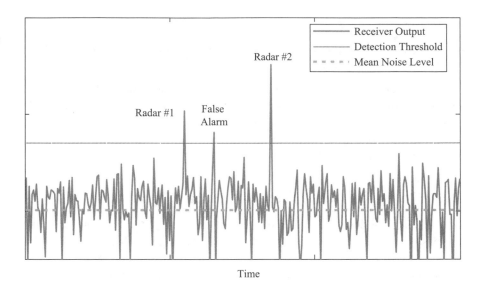

The RWR declares a detection when the output of the receiver, the radar signal plus thermal noise, exceeds the detection threshold, as shown in Figure 9-5. This is the same threshold detection process used by radar systems (see Chapter 3). When the threshold crossing is due to the radar signal, it is called a radar detection. When the threshold crossing is due to receiver thermal noise, it is called a false alarm. The detection threshold is set to maximize radar detections and minimize false alarms. The random nature of the receiver thermal noise, and possibly the radar signal, complicates setting the detection threshold. Thus, the detection threshold for EW receivers is set based on probability theory, just as it was for radar systems.

The basic descriptions of probability theory, probability density functions (PDF), and threshold detection given in Chapter 3 for radar systems apply here as well. A quick summary of threshold detection and associated probability theory is provided here from the RWR perspective. The probability of false alarm (P_{fa}) is the conditional probability that if no radar signal is present the noise alone exceeds the detection threshold. The probability of detection (P_d) is the conditional probability that if a radar signal is present the radar signal plus noise exceeds the detection threshold. Both valid and invalid threshold crossings are considered detections.

The probability integrals of the radar signal plus noise PDF and noise PDF can be solved for P_d and P_{fa}, respectively. The PDFs are quite complicated, especially for the radar signal plus noise. After several pages of calculus a relationship can be found between S/N, P_d, and P_{fa}. Using this relationship, or associated figures and tables, the RWR detection threshold, SNR_{dtRWR}, can be determined based on a required P_d and P_{fa}. The target detection relationships, figures, and tables discussed in Chapter 3 also apply to some wideband EW receivers.

RWR designers select a detection threshold to achieve a required P_d and P_{fa}. The radar signal is detected if the S/N exceeds the detection threshold. Some RWR systems require a very high P_d and extremely low P_{fa} to be operationally effective (e.g., detect and identify radar systems at the required range) and thus need a high detection threshold. Often an extremely low P_{fa} is required to minimize the workload of the RWR signal processor by limiting the number of false alarms entering the signal processor.

Low ERP radar systems present challenges for some wideband EW receivers to provide the necessary detection range. Advanced EW receiver architectures were developed to overcome these challenges: e.g., scanning superheterodyne, channelized receiver, etc. Neri [2006] and Tsui [1989, 2001, & 2005] both describe these advanced EW receiver architectures in detail. Detection theory for advanced EW receivers is a function of S/N, P_d, P_{fa}, and the ratio of the receiver's overall RF bandwidth (B_{RF}) and video processing bandwidth (B_V). The relationship between the RF bandwidth and video bandwidth is shown in Figure 9-6.

Increased signal processing performance has allowed signal processing algorithms to remove false alarms. Thus, a higher P_{fa} can be used resulting in a lower detection threshold for the same P_d. All other things being equal, a lower detection threshold results in a longer detection range. The detection threshold can also be dynamically set based on the perceived RF environment. This is similar to when you push scan on your car radio. The first time through the frequency band the threshold is high to pick up the local stations. While, the second time through the frequency band the threshold is low to pick up distant stations.

With all the different EW receiver architectures and associated detection concepts it is important to know how detection is performed by your specific EW receiver system. Tsui [1989, 2005, 2010] contains a detailed discussion of detection theory for different EW receiver architectures.

9.2.5 Receiver Sensitivity

The receiver community has long used receiver sensitivity as a measure of receiver and signal detection performance. In fact, the previous receiver

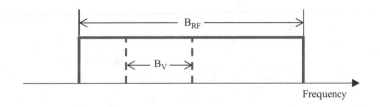

FIGURE 9-6 ■ Relationship between Radio Frequency and Video Bandwidths

thermal noise and detection threshold discussion is actually better aligned to with the radar community. There are two main receiver sensitivity definitions—minimum detectable signal (MDS) and minimum discernible signal (MDS)—unfortunately both with the same acronym. The minimum detectable signal sensitivity is a function of the RWR thermal noise and the detection threshold (Equation 9-14). The RWR will declare a detection when the output of the RWR exceeds the minimum detectable signal sensitivity. The minimum discernible signal sensitivity is the RWR thermal noise (Equation 9-15). An experienced operator can discern the presence of a radar signal due to a slight but perceivable difference in the thermal noise.

$$S_{\text{min dt}} = N_{\text{RWR}} \, SNR_{\text{dtRWR}} \tag{9-14}$$

$$S_{\text{min dis}} = N_{\text{RWR}} \tag{9-15}$$

where:

$S_{\text{min dt}}$ = Minimum detectable signal sensitivity, watts
SNR_{dtRWR} = Radar warning receiver detection threshold, no units
$S_{\text{min dis}}$ = Minimum discernible signal sensitivity, watts

The minimum detectable signal and minimum discernible signal sensitivities are shown in Figure 9-7. Since both of these sensitivities unfortunately have the same acronyms, one must be very careful in knowing which definition is being used and how detection theory is applied.

▮▮▮ 9.3 ▮ ELECTRONIC WARFARE RECEIVER METRICS

9.3.1 Electronic Warfare Receiver Detection Range

The single-pulse RWR S/N equation can be solved for the range at which the RWR will detect the presence of a radar pulse with a given S/N, which is the range at which the radar signal exceeds the receiver thermal noise by the detection threshold. The RWR detection range is given in Equations (9-16),

FIGURE 9-7 ▪
Minimum
Detectable Signal
and Minimum
Discernible Signal
Sensitivities

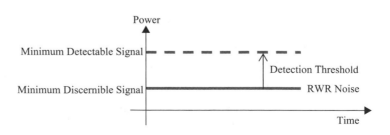

algebraic, and (9-17), logarithmic. The logarithmic form of this equation can be simplified by using the carrier frequency instead of wavelength and combining the constants (Equation 9-18).

$$R_{dtRWR} = \sqrt{\frac{P_R\, G_{RT}\, G_{RWR}\, \lambda^2}{(4\pi)^2\, SNR_{dtRWR}\, F_{RWR}\, k\, T_0\, B_{RWR}\, L_{RWR}}} \qquad (9\text{-}16)$$

$$\begin{aligned}
20\log(R_{dtRWR}) = {} & 10\log(P_R) + 10\log(G_{RT}) + 10\log(G_{RWR}) \\
& + 20\log(\lambda) - 20\log(4\pi) \\
& - 10\log(SNR_{dtRWR}) - 10\log(F_{RWR}) \\
& - 10\log(k) - 10\log(T_0) - 10\log(B_{RWR}) \\
& - 10\log(L_{RWR})
\end{aligned} \qquad (9\text{-}17)$$

$$\begin{aligned}
20\log(R_{dtRWR}) = {} & 10\log(P_R) + 10\log(G_{RT}) + 10\log(C_{RWR}) \\
& - 20\log(f_c) - 10\log(SNR_{dtRWR}) \\
& - 10\log(F_{RWR}) - 10\log(B_{RWR}) \\
& - 10\log(L_{RWR}) + 351.5355
\end{aligned} \qquad (9\text{-}18)$$

where:

R_{dtRWR} = Radar warning receiver detection range, meters
SNR_{dtRWR} = Radar warning receiver detection threshold, no units

The RWR detection range equation is often given in terms of receiver sensitivity. When using minimum detectable signal sensitivity, the RWR detection range is given in Equations (9-19), algebraic, and (9-20), logarithmic. The logarithmic form of this equation can be simplified by using the carrier frequency instead of wavelength and combining the constants (Equation 9-21).

$$R_{dtRWR} = \sqrt{\frac{P_R\, G_{RT}\, G_{RWR}\, \lambda^2}{(4\pi)^2\, S_{min\,dt}\, L_{RWR}}} \qquad (9\text{-}19)$$

$$\begin{aligned}
20\log(R_{dtRWR}) = {} & 10\log(P_R) + 10\log(G_{RT}) \\
& + 10\log(G_{RWR}) + 20\log(\lambda) - 20\log(4\pi) \\
& - 10\log(S_{min\,dt}) - 10\log(L_{RWR})
\end{aligned} \qquad (9\text{-}20)$$

$$\begin{aligned}
20\log(R_{dtRWR}) = {} & 10\log(P_R) + 10\log(G_{RT}) + 10\log(G_{RWR}) \\
& - 20\log(f_c) - 10\log(S_{min\,dt}) \\
& - 10\log(L_{RWR}) + 147.5582
\end{aligned} \qquad (9\text{-}21)$$

When using minimum discernible signal sensitivity, the RWR detection range is given in Equations (9-22), algebraic, and (9-23), logarithmic. The logarithmic form of this equation can be simplified by using the carrier frequency instead of wavelength and combining the constants (Equation 9-24).

$$R_{dtRWR} = \sqrt{\frac{P_R\, G_{RT}\, G_{RWR}\, \lambda^2}{(4\pi)^2\, SNR_{dtRWR}\, S_{min\,dis}\, L_{RWR}}} \qquad (9\text{-}22)$$

$$\begin{aligned}
20\log(R_{dtRWR}) = &\, 10\log(P_R) + 10\log(G_{RT}) + 10\log(G_{RWR}) \\
&+ 20\log(\lambda) - 20\log(4\pi) \\
&- 10\log(SNR_{dtRWR}) - 10\log(S_{min\,dis}) \\
&- 10\log(L_{RWR})
\end{aligned} \qquad (9\text{-}23)$$

$$\begin{aligned}
20\log(R_{dtRWR}) = &\, 10\log(P_R) + 10\log(G_{RT}) + 10\log(G_{RWR}) \\
&- 20\log(f_c) - 10\log(SNR_{dtRWR}) \\
&- 10\log(S_{min\,dis}) - 10\log(L_{RWR}) \\
&+ 147.5582
\end{aligned} \qquad (9\text{-}24)$$

All these detection range equations are consistent with the definitions and detection concepts in this chapter. However, as previously mentioned there are many different definitions and detection concepts across the EW receiver community. The reader needs to ensure that the equations are consistent with their application.

RWR systems can generally detect high ERP radar systems at very long ranges, especially when the mainbeam of the radar antenna is pointed at the RWR. The detection range when the sidelobes of the radar antenna are pointed at the RWR is significantly less. An exercise at the end of this chapter will numerically show the significant difference in mainbeam and sidelobe detection ranges. RWR detection range is often limited by the radar-to-target/RWR line of sight (LOS). Radar LOS is discussed in detail in Chapter 13.

Advanced RWR systems can generally detect low ERP radar systems at operationally sufficient ranges through the use of advanced receiver architectures and associated detection theory. RWR systems often have the highest sensitivity practical to facilitate other RWR functions. Efficient and accurate signal parameter (e.g., frequency, pulse width, angle of arrival) extraction, deinterleaving (the process of sorting radar signal parameters into specific emitters), and radar identification all require more S/N than detection of radar signals alone. Thus, the reader needs to be careful as to what S/N to use in these detection range equations.

9.3.2 Detection Range Ratio

As previously discussed, the fundamental goal of all EW receivers is to "see them (radar systems) well before they see you (the platform with the EW receiver)." Thus, the ratio of EW receiver detection range (Equation 9-16) to the radar detection range (Equation 3-43) provides a metric for this fundamental goal. The ratio of the detection ranges is given in Equation (9-25).

$$\frac{R_{dtRWR}}{R_{dt}} = \frac{\sqrt{\dfrac{P_R \, G_{RT} \, G_{RWR} \, \lambda^2}{(4\pi)^2 \, SNR_{dtRWR} \, F_{RWR} \, k \, T_0 \, B_{RWR} \, L_{RWR}}}}{\sqrt[4]{\dfrac{P_R \, G_{RT}^2 \, \lambda^2 \, \sigma \, G_{sp} \, G_I}{(4\pi)^3 \, SNR_{dt} \, F_R \, k \, T_0 \, B_R \, L_R}}} \tag{9-25}$$

where:

R_{dt} = Radar detection range, meters

SNR_{dt} = Radar detection threshold, no units

σ = Target radar cross section, square meters (m^2)

G_{sp} = Radar signal processing gain, no units

G_I = Radar integration gain, no units

F_R = Radar receiver noise figure ($\geq$1), no units

B_R = Radar receiver processing bandwidth, hertz

L_R = Radar-related losses, no units

▮ 9.4 | RADAR SIGNAL PARAMETER EXTRACTION, DEINTERLEAVING, IDENTIFICATION, AND OUTPUT

After the EW receiver detects a radar pulse, several other processes occur: signal parameter extraction, deinterleaving, identification, and output. Tsui [2010] and Wiley [2006] both have detailed descriptions of these processes. An EW receiver extracts parameters such as pulse width, carrier frequency, modulation of the carrier (e.g., pulse compression), received signal strength, time of arrival, and angle of arrival from the detected radar signal. The extracted parameters are stored in a pulse descriptor word (PDW). Parameter extraction measurement accuracy is a function of the radar pulse parameters, S/N, and EW receiver characteristics. The higher the S/N, the finer the measurement accuracy, which in turn improves deinterleaving and identification efficiency and performance. The EW receiver designer's choice of detection threshold is a trade between detection range and measurement accuracy. For radar pulses just exceeding the detection threshold, a lower detection threshold results in a longer detection range but coarser measurement accuracy, whereas a higher detection threshold results in a shorter detection range but finer measurement accuracy.

The EW receiver signal processor deinterleaves, or sorts, all the PDWs into specific emitters. The EW receiver will sort PDWs into cells based on PDW characteristics: carrier frequency, pulse width, angle of arrival, etc. The cell size is a function of the parameter extraction accuracy. The deinterleaving processing time depends on the number of PDWs within a resolution cell. The more PDWs in a cell, the longer the processing time—often nonlinearly. A time history of sorted PDWs is built by including time of arrival (TOA) information, which is used to determine the pulse repetition interval (PRI) and antenna scan rate and pattern of the emitter.

An EW receiver identifies specific radar systems by comparing the characteristics of specific sorted PDWs to a library (database) of known radar system characteristics, often called an emitter identification table. Radar identification performance depends on the quality of the information in the emitter identification table. In addition to identifying the radar system associated with the specific sorted PDWs, the EW receiver can also identify what mode (e.g., search, track) the radar system is in.

An EW receiver can also compute a rough estimate of the range to the identified radar system using a form of the EW receiver equation (Equation 9-26). Many of the terms in this equation are measured by the receiver or known by receiver. Additional radar characteristics are looked up in the emitter identification table. The emitter range is a rough estimate due to the parameter extraction accuracy, emitter identification table data inaccuracies, and EW receiver component uncertainties.

$$R_{RT} = \sqrt{\frac{P_R\ G_{RT}\ G_{RWR}\ \lambda^2}{(4\pi)^2\ S_{RWR}\ L_{RWR}}} \tag{9-26}$$

where:

R_{RT} = Radar-to-target/RWR slant range, meters
S_{RWR} = Received radar signal power, watts

The identified radar systems are then used to warn an operator on an RWR display or the jammer system (see Chapters 10 and 11). The RWR prioritizes the identified radar systems and displays the high-priority radar systems to the operator. The jammer receiver processor prioritizes the identified radar systems and provides the high-priority radar systems to a jammer so it can respond to them.

In summary, not all radar signals—only those exceeding the detection threshold—are detected by an EW receiver. Not all detected radar signals—only those whose parameters can be extracted and matched to the database of known radar characteristics—are identified. Not all identified radar signals—only those whose priority is high enough—are displayed or provide a warning to a jammer.

9.5 | SUMMARY

EW receivers—radar warning receivers and jammer receiver processors—are an element of ES, providing SA about emitters in the environment.

We developed the EW receiver equation, which allows us to determine EW receiver performance. We discussed the two main EW receiver metrics: (1) detection range, or how far away the EW receiver can detect a radar signal; and (2) detection range ratio, or the ratio of the EW receiver detection range to the radar detection range. We briefly discussed signal parameter extraction, signal processing, and display/output, the performance of which is extremely dependent on the contents of the emitter identification table. We have covered only the tip of the iceberg for EW receivers. The references at the end of this chapter all discuss EW receivers in detail.

9.6 | EXERCISES

9-1. A radar warning receiver (RWR) has the following characteristics: antenna gain in the direction of the radar $G_{RWR} = 3$ dBi and losses $L_{RWR} = 9$ dB. A radar system has the following characteristics: peak transmit power $P_R = 500$ kW; antenna gain in the direction of the target/RWR $G_{RT} = 35$ dBi; carrier frequency $f_c = 5$ GHz; and radar transmit loss $L_{Rt} = 2$ dB. The target/RWR is at a radar-to-target/RWR range $R_{RT} = 150$ km. Determine the following: (a) the transmitted radar effective radiated power ERP_R (watts and dBW); (b) the received radar power density at the RWR antenna; and (c) the received single-pulse radar signal power S_{RWR} (watts and dBW). Compare the received single-pulse radar signal power with the radar's transmitted effective radiated power.

9-2. The RWR from Exercise 9-1 has the following additional characteristics: minimum discernible signal sensitivity $S_{min\_dis} = -80$ dBW and receiver bandwidth $B_{RWR} = 6$ GHz. What is the receiver noise figure F_{RWR} (no units and dB)? What is the single-pulse radar signal-to-noise ratio of the RWR $(S/N)_{RWR}$ (no units and dB)? (Hint: use the results from Exercise 9-1.)

9-3. The radar warning receiver (RWR) from Exercises 9-1 and 9-2 has a signal-to-noise ratio required for detection (detection threshold) $SNR_{dtRWR} = 15$ dB. What is the RWR detection range, R_{dtRWR} (meters)? (Hint: use the results from Exercises 9-1 and 9-2.)

9-4. The first sidelobe of the radar antenna is pointing at the RWR from Exercises 9-1, 9-2, and 9-3. The radar antenna has a uniform illumination function (see Section 4.1 and Table 4-1). What is the RWR detection range R_{dtRWR} (meters)? (Hint: use the results from Exercises 9-1, 9-2, and 9-3.)

9-5. An RWR has the following characteristics: antenna gain in the direction of the radar $G_{RWR} = 4$ dBi and losses $L_{RWR} = 8$ dB. The RWR receives a single-pulse radar signal peak power $S_{RWR} = -60$ dBW. The RWR detects the radar pulse, extracts the radar carrier frequency $f_c = 10$ GHz, identifies the radar system, and looks up the following radar characteristics from the emitter identification table: peak transmit power $P_R = 100$ kW; and mainbeam antenna gain $G_{RT} = 38$ dBi. What is the estimated radar-to-target/RWR range R_{RT} (meters)?

9.7 | REFERENCES

Neri, Filippo, 2006, *Introduction to Electronic Defense Systems*, 2nd Edition, Raleigh, NC: SciTech Publishing. EW receivers are discussed in Chapter 4 with concepts, technologies, and supporting math.

Schleher, D. Curtis, 1999, *Electronic Warfare in the Information Age*, Norwood, MA: Artech House. EW receivers are discussed in Chapter 6 with concepts, technologies, and supporting math; includes MATLAB files for some of the concepts.

Tsui, James, 2005, *Microwave Receivers with Electronic Warfare Applications*, Raleigh, NC: SciTech Publishing. The first book to go to when you need detailed EW receiver concepts, technology, and supporting math.

Tsui, James, 1989, *Digital Microwave Receivers: Theory and Concept*, Norwood, MA: Artech House. The second book to go to when you need detailed EW receiver concepts, technology, and supporting math.

Tsui, James, 2001, *Digital Techniques for Wideband Receivers*, Norwood, MA: Artech House. The third book to go to when you need detailed EW receiver concepts, technology, and supporting math.

Tsui, James, 2010, *Special Design Topics in Digital Wideband Receivers*, Norwood, MA: Artech House. The book to go to when you need detailed design topics for digital receivers and encoders (radar waveform parameter extraction), with numerous examples and the supporting math.

Wiley, Richard, 2006, *ELINT The Interception and Analysis of Radar Signals*, Norwood, MA: Artech House. A comprehensive book on EW receivers, from intercept to parameter extraction to identification to emitter location and everything in between.

Self-Protection Jamming Electronic Attack

HIGHLIGHTS

- Defining and discussing self-protection jamming electronic attack
- Develop the self-protection jamming equation, the starting point for our mathematical representations
- Noise self-protection jamming, its effect on the radar, and metrics to describe its performance
- False target self-protection jamming, its effect on the radar, and metrics to describe its performance
- Expendable self-protection jamming, passive and active
- Other self-protection jamming techniques used against the target tracker

As discussed in Chapters 1 and 8, a self-protection jammer is collocated with the platform (the target) it is protecting. In this chapter we will discuss self-protection jamming principles along with the supporting math [Chrzanowski, 1990; Goj, 1993, Chapter 2; Neri, 2006, Chapter 5; Schleher, 1999; Chapter 4]. The self-protection jamming waveform most often enters the radar antenna through its mainbeam as the radar system is attempting to detect the target, as shown in Figure 10-1. The effect of the self-protection jamming is to obscure true target signals or introduce false targets within the radar's mainbeam. The electronic attack (EA) community often refers to the received target signal as the "skin return" as it is produced by the reflection off the skin of the target.

10.1 | SELF-PROTECTION JAMMING PRINCIPLES

Self-protection jamming principles will be described in the following sections: self-protection jamming equation, noise jamming, and false target jamming.

FIGURE 10-1 ■
Self Protection
Jamming

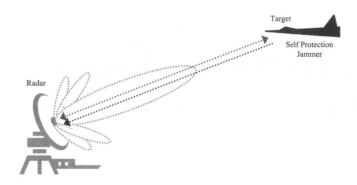

The self-protection jamming equation section addresses the jamming power equations for constant power and constant gain jammers. The noise jamming section addresses noise jamming in the radar receiver and noise jamming metrics—burnthrough range and jamming-to-signal ratio (J/S). The false target jamming section addresses false target jamming in the radar receiver and false target jamming metrics—J/S and detection range.

10.2 | SELF-PROTECTION JAMMING EQUATION

The self-protection jamming equation relates jammer, radar, and geometry/environment characteristics together to allow us to determine the jammer power received by the radar system. The self-protection jammer equation is the first step in determining radar performance (impact of the jammer) in the presence of the self-protection jammer. As shown in Figure 10-2, the received jammer power is computed using incremental steps:

Step 1: jammer effective radiated power (ERP)

Step 2: jammer-to-radar propagation

Step 3: jammer power out of the radar antenna

FIGURE 10-2 ■
Incremental Build-
Up of Received Self
Protection Jammer
Power

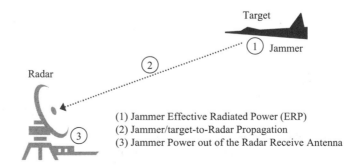

(1) Jammer Effective Radiated Power (ERP)
(2) Jammer/target-to-Radar Propagation
(3) Jammer Power out of the Radar Receive Antenna

These steps are similar to those used to develop the target signal power equation in Chapter 2. Thus, we will efficiently go through the steps here and readers can refresh themselves with the details in Chapter 2 as needed.

The self-protection jammer has an ERP in the direction of the radar (Equation 10-1). The self-protection jammer ERP propagates to the radar resulting in a jammer power density at the radar (Equation 10-2). The incident self-protection jammer power density is collected by the effective area of the radar receive antenna in the direction of the jammer, resulting in the jammer signal power at the output of the radar antenna (Equation 10-3). This equation uses radar antenna gain in the direction of the jammer instead of effective area. Total jammer-related losses include factors not explicitly taken into consideration in determining the received self-protection jammer power: jammer transmit loss; atmospheric attenuation (see Chapter 13); jammer waveform polarization mismatch with the radar antenna (see Table 9-1), radar receive loss, etc. Different radar–jammer–environment combinations have different losses, so we collect them up as we need them into a single term (Equation 10-4). When all jammer-related losses are considered, the resultant received self-protection jammer power is given in Equations (10-5), algebraic, and (10-6), logarithmic. The logarithmic form of this equation can be simplified by using the carrier frequency instead of wavelength and combining the constants (Equation 10-7). Since the EW community often uses the logarithmic forms of equations, we will present them along the traditional algebraic forms.

$$\mathrm{ERP_J} = \frac{\mathrm{P_J \, G_{JR}}}{\mathrm{L_{Jt}}} \tag{10-1}$$

$$\frac{\mathrm{P_J \, G_{JR}}}{\mathrm{L_{Jt}}} \left(\frac{1}{(4\pi) \, \mathrm{R_{RT}}^2} \right) = \frac{\mathrm{P_J \, G_{JR}}}{(4\pi) \, \mathrm{R_{RT}}^2 \, \mathrm{L_{Jt}}} \tag{10-2}$$

$$\frac{\mathrm{P_J \, G_{JR}}}{(4\pi) \, \mathrm{R_{RT}}^2 \, \mathrm{L_{Jt}}} (\mathrm{A_e}) = \frac{\mathrm{P_J \, G_{JR}}}{(4\pi) \, \mathrm{R_{RT}}^2 \, \mathrm{L_{Jt}}} \left(\frac{\mathrm{G_{RT} \, \lambda^2}}{(4\pi)} \right) = \frac{\mathrm{P_J \, G_{JR} \, G_{RT} \, \lambda^2}}{(4\pi)^2 \, \mathrm{R_{RT}}^2 \, \mathrm{L_{Jt}}} \tag{10-3}$$

$$\mathrm{L_J} = \mathrm{L_{Jt} \, L_{RTa} \, L_{Jpol} \, L_{Rr} \, L_{Rsp}} \cdots \tag{10-4}$$

$$\mathrm{J} = \frac{\mathrm{P_J \, G_{JR} \, G_{RT} \, \lambda^2}}{(4\pi)^2 \, \mathrm{R_{RT}}^2 \, \mathrm{L_J}} \tag{10-5}$$

$$\begin{aligned} 10 \log(\mathrm{J}) = {} & 10 \log(\mathrm{P_J}) + 10 \log(\mathrm{G_{JR}}) + 10 \log(\mathrm{G_{RT}}) \\ & + 20 \log(\lambda) - 20 \log(4\pi) - 20 \log(\mathrm{R_{RT}}) \\ & - 10 \log(\mathrm{L_J}) \end{aligned} \tag{10-6}$$

$$\begin{aligned} 10 \log(\mathrm{J}) = {} & 10 \log(\mathrm{P_J}) + 10 \log(\mathrm{G_{JR}}) + 10 \log(\mathrm{G_{RT}}) \\ & - 20 \log(\mathrm{f_c}) - 20 \log(\mathrm{R_{RT}}) - 10 \log(\mathrm{L_J}) \\ & + 147.5582 \end{aligned} \tag{10-7}$$

where:

ERP_J = Jammer effective radiated power, watts

P_J = Jammer peak transmit power, watts

G_{JR} = Jammer transmit antenna gain in the direction of the radar, no units

L_{Jt} = Jammer transmit loss, no units

R_{RT} = Radar-to-target/jammer slant range, meters

A_e = Radar receive antenna effective area in the direction of the target/jammer, square meters (m^2)

G_{RT} = Radar receive antenna gain in the direction of the target/jammer, no units

λ = Wavelength, meters

L_J = Total jammer-related losses, no units

L_{RTa} = Target/jammer-to-radar atmospheric attenuation loss, no units

L_{Jpol} = Jammer-radar polarization mismatch loss, no units

L_{Rr} = Radar receive loss, no units

L_{Rsp} = Radar signal processing loss, no units

J = Received self-protection jammer peak power, watts

f_c = Radar carrier frequency, hertz

Since we now have an equation for the received self-protection jammer power, we can compare it to the target signal power and radar receiver thermal noise power (both from Chapter 2). The single-pulse/sample received target signal power (S), radar receiver thermal noise (N), and received self-protection jammer power as a function of radar-to-target/jammer range are shown in Figure 10-3. As seen in this figure and Equation (10-5), the received self-protection jammer power varies as a function of $1/R^2$. The received self-protection jammer power decreases by 6 dB (1/4) when the range doubles and increases by 6 dB (4 times) when the range halves. In contrast, the received target signal

FIGURE 10-3 ■
Target Signal (S), Receiver Thermal Noise (N), and Jammer (J) Power versus Radar-to-Target/Jammer Range

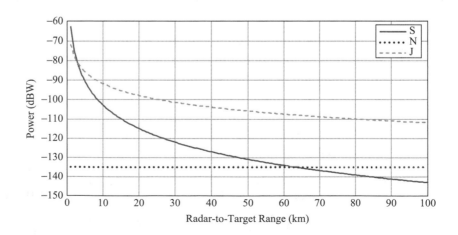

power at the radar varies as a function of $1/R^4$ (see Chapter 2). The received target signal power decreases by 12 dB (1/16) when the range doubles and increases by 12 dB (16 times) when the range halves. Note: the y-axis grid for this figure is in 10 dB steps, one order of magnitude (power of 10), so the reader can easily interpret the change in power with range. Other figures in this chapter will use the same approach.

10.2.1 Constant Gain Jammer

Most jammers have a constant power output, as we have described so far. Constant power jammers are simple to implement, and they are always transmitting the maximum available (saturated) power. Some jammers operate by amplifying the received radar signal power by a constant gain and transmitting it back to the radar. This type of jammer is called a constant gain or repeater jammer [Chrzanowski, 1990, Sections 2.10, 2.11]. A constant gain jammer is analogous to the combination of a microphone, fixed gain amplifier, and speaker. Constant power and constant gain jammers are shown in Figure 10-4. The transmitted output power for constant gain jammer is directly proportional to the power of the intercepted radar signal.

The self-protection jamming equation for a constant gain jammer relates jammer, radar, and geometry/environment characteristics together to allow us to determine the jammer power received by the radar system. As shown in Figure 10-5, the constant gain self-protection jammer ERP is computed using incremental steps:

Step 1: incident radar power density
Step 2: received radar power
Step 3: amplified radar power
Step 4: jammer ERP

The constant gain self-protection jammer ERP is given in Equation (10-8). For clarity in this and subsequent equations, the losses are initially excluded. Total jammer-related losses are included at the conclusion of the self-protection

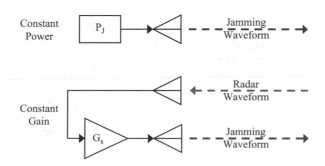

FIGURE 10-4 ■
Constant Power and Constant Gain Jammers

FIGURE 10-5 ■
Incremental Build-
Up of Constant Gain
Self Protection
Jammer Effective
Radiated Power

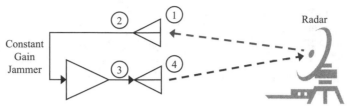

(1) Incident Radar Power Density   (2) Received Radar Power
(3) Amplified Radar Power   (4) Jammer Effective Radiated Power (ERP)

jammer equation development. A constant gain jammer will linearly amplify
the received radar signal until the jammer transmitter reaches its maximum
(saturated) power. Thus, the power from a constant gain jammer is limited to be
less than or equal to the saturation power of its transmitter (Equation 10-9).
When the amplifier is driven into saturation, the constant gain jammer becomes
a constant power jammer. The constant gain jammer ERP propagates to the
radar resulting in an incident power density (Equation 10-10). The incident
constant jammer power density is collected by the radar receive antenna
(Equation 10-11). The received constant gain self-protection jammer power
including jammer-related losses is given in Equation (10-12).

$$\text{ERP}_J = \frac{P_R\, G_{RT}\, G_{JRr}\, \lambda^2\, G_s}{(4\pi)^2\, R_{RT}{}^2} (G_{JRt}) \tag{10-8}$$

$$\underbrace{\frac{P_R\, G_{RT}\, G_{JRr}\, \lambda^2\, G_s}{(4\pi)^2\, R_{RT}{}^2} (G_{JRt})}_{\leq\, P_{Js}} \tag{10-9}$$

$$\underbrace{\frac{P_R\, G_{RT}\, G_{JRr}\, \lambda^2\, G_s}{(4\pi)^2\, R_{RT}{}^2} (G_{JRt})}_{\leq\, P_{Js}} \left[\frac{1}{(4\pi)\, R_{RT}{}^2}\right] \tag{10-10}$$

$$\underbrace{\frac{P_R\, G_{RT}\, G_{JRr}\, \lambda^2\, G_s}{(4\pi)^2\, R_{RT}{}^2}}_{\leq\, P_{Js}} \left[\frac{G_{JRt}}{(4\pi)\, R_{RT}{}^2}\right] \left[\frac{G_{RT}\, \lambda^2}{(4\pi)}\right] \tag{10-11}$$

$$J_{cg} = \underbrace{\frac{P_R\, G_{RT}\, G_{JRr}\, \lambda^2\, G_s}{(4\pi)^2\, R_{RT}{}^2}}_{\leq\, P_{Js}} \left[\frac{G_{JRt}\, G_{RT}\, \lambda^2}{(4\pi)^2\, R_{RT}{}^2\, L_J}\right] \tag{10-12}$$

where:

ERP_J = Jammer effective radiated power, watts

P_R = Radar transmitter peak power, watts

G_{RT} = Radar antenna gain in the direction of the target/jammer, no units

G_{JRr} = Jammer receive antenna gain in the direction of the radar, no units

λ = Wavelength, meters

R_{RT} = Radar-to-target/jammer slant range, meters

G_s = Jammer system gain, no units

G_{JRt} = Jammer transmit antenna gain in the direction of the radar, no units

P_{Js} = Jammer saturation peak power, watts

J_{cg} = Received constant gain jammer power, watts

L_J = Total jammer-related losses, no units

Normally the constant gain (linear) region of a constant gain jammer occurs at large distances from the radar. This is because at large distances the incident radar power density is small and thus less likely to drive the jammer amplifier into saturation. A constant gain jammer transitions to a constant power (saturated) jammer as the target/jammer approaches the radar, as shown in Figure 10-6.

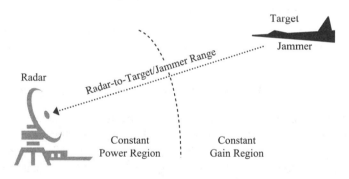

FIGURE 10-6 ■ Constant Power and Constant Gain Regions

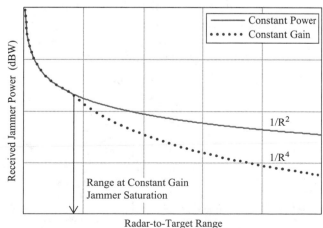

FIGURE 10-7 ■ Received Self Protection Jammer Power—Constant Power and Constant Gain

The received self-protection jammer power for constant power and constant gain jammers as a function of radar-to-target/jammer range is shown in Figure 10-7. As seen in this figure, the received self-protection jammer power for a constant gain jammer varies with range just like a target $(1/R^4)$. When the amplifier is driven into saturation, the constant gain jammer becomes a constant power jammer.

10.3 | SELF-PROTECTION NOISE JAMMING

For a noise self-protection jammer, the jammer transmits a noise (random amplitude and phase with a uniform power spectral density) waveform (see Chapter 8) at the radar system. The received jammer noise combines with the radar receiver thermal noise (see Chapter 2) to create an interference signal. Noise jamming is a denial/degrade form of EA. Noise jamming can deny detection of the target or can degrade (reduce) the detection range of the radar. Noise jamming can also deny target measurements or can degrade the radar's ability to measure target states. Noise jamming can also introduce noise power of sufficient magnitude into the radar receiver so that the noise alone exceeds the detection threshold, resulting in numerous false alarms that degrade the efficient processing of target detections.

10.3.1 Self-Protection Noise Jamming in the Radar Receiver

The jammer noise waveform is usually not perfectly matched to the radar waveform (and associated matched filter receiver; see Chapters 2 and 3), as shown in Figure 10-8. The mismatch can be in carrier frequency and/or bandwidth. Since the radar receiver accepts signals only within its bandwidth, only the portion of the received jammer noise power within the radar receiver bandwidth passes through the radar receiver.

To account for the mismatch between the radar receiver and jammer noise bandwidths, the received self-protection jammer power is reduced by

FIGURE 10-8 ■
Jammer Waveform
Mismatch to the
Radar Waveform
and Receiver
Bandwidth

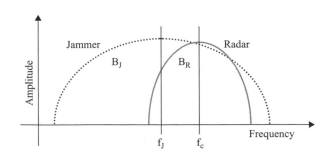

the ratio of the radar receiver bandwidth to the jammer noise bandwidth (Equation 10-13). This equation assumes the jammer noise bandwidth overlaps the radar receiver bandwidth; it is greater than the radar receiver bandwidth and the carrier frequency of the jammer noise is approximately equal to the radar carrier frequency. This assumption also ensures the jammer noise is wideband with respect to the receiver thermal noise, the receiver bandwidth, and target signal. Inside the radar receiver there is a complex (amplitude and phase) combination of the self-protection jammer noise and radar receiver thermal noise. The result of this combination is called the interference signal (Equation 10-14). The interference signal equation uses the concept of additive means from probability theory; here its mean radar receiver thermal noise plus mean jammer noise.

$$ J_N = \frac{P_J\, G_{JR}\, G_{RT}\, \lambda^2}{(4\pi)^2\, R_{RT}{}^2\, L_J} \left(\frac{B_R}{B_J} \right) \tag{10-13}$$

$$\text{Note}: \text{assumes } B_J \geq B_R \text{ and } f_J \approx f_c$$

$$ I = N + J_N \tag{10-14}$$

where:

J_N = Jammer noise power out of the radar receiver, watts
B_J = Jammer noise bandwidth, hertz
B_R = Radar receiver processing bandwidth, hertz
f_J = Jammer transmitted frequency, hertz
f_c = Radar carrier or transmitted frequency, hertz
I = Interference signal power, watts
N = Radar receiver thermal noise power, watts

The single-pulse/sample received target signal power, radar receiver thermal noise, self-protection jammer noise, and interference signal power as a function of radar-to-target/jammer range are shown in Figure 10-9. For the self-protection jammer to be effective its noise must be significantly greater than the radar receiver thermal noise ($J_N \gg N$), at the vast majority of ranges. Thus, the interference signal power equals, or is slight more than, the jammer noise for the vast majority of ranges. This is clearly the case for the example shown in this figure.

The single-pulse signal-to-interference ratio (S/I) is a measure of the strength of the received target signal power relative to the interference signal. The S/I is similar to the target signal-to-noise ratio (S/N) discussed in Chapters 2 and 3, except the radar receiver thermal noise is combined with the self-protection jammer noise in the denominator. The S/I is the ratio of the power from a single received target pulse to the power in one sample of the interference noise signal (Equation 10-15). Including the equations for the target

FIGURE 10-9 ▪
Target Signal (S),
Receiver Thermal
Noise (N), Self
Protection Jammer
Noise (J_N), and
Interference (I)
versus Radar-to-
Target/Jammer
Range

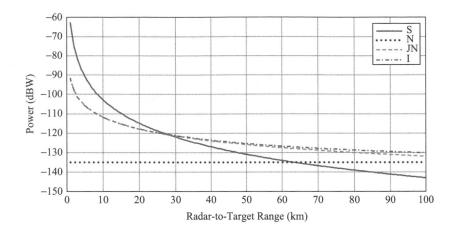

signal power (see Chapter 2), radar receiver thermal noise (see Chapter 2), and self-protection jammer noise results in Equation (10-16). As discussed in Chapter 3, radar systems improve their detection performance by integrating multiple pulses. The improvement in S/I provided by integration is quantified by the integration gain (see Chapter 3). The S/I after integration of multiple pulses is given in Equation (10-17).

$$\frac{S}{I} = \frac{S}{N + J_N} \tag{10-15}$$

$$\frac{S}{I} = \frac{P_R \, G_{RT}^2 \, \lambda^2 \, \sigma \, G_{sp}}{(4\pi)^3 \, R_{RT}^4 \, L_R \left[F_R \, k \, T_0 \, B_R + \dfrac{P_J \, G_{JR} \, G_{RT} \, \lambda^2}{(4\pi)^2 \, R_{RT}^2 \, L_J} \left(\dfrac{B_R}{B_J} \right) \right]} \tag{10-16}$$

$$\left(\frac{S}{I} \right)_n = \frac{P_R \, G_{RT}^2 \, \lambda^2 \, \sigma \, G_{sp} \, G_I}{(4\pi)^3 \, R_{RT}^4 \, L_R \left[F_R \, k \, T_0 \, B_R + \dfrac{P_J \, G_{JR} \, G_{RT} \, \lambda^2}{(4\pi)^2 \, R_{RT}^2 \, L_J} \left(\dfrac{B_R}{B_J} \right) \right]} \tag{10-17}$$

where:

S/I = Single-pulse target signal-to-interference ratio, no units
P_R = Radar peak transmit power, watts
σ = Target radar cross section, square meters (m^2)
G_{sp} = Radar signal processing gain, no units
L_R = Radar-related losses, no units
F_R = Radar receiver noise figure (≥ 1), no units
k = Boltzmann's constant, 1.38×10^{-23} watt-seconds/Kelvin

$T_0 =$ IEEE standard reference temperature for radio frequency receivers, 290 Kelvin

$(S/I)_n =$ Target signal-to-interference ratio after integration of multiple pulses, no units

$G_I =$ Radar integration gain, no units

Figure 10-10 shows the multiple-pulse signal-to-noise ratio $(S/N)_n$ (see Chapter 3), detection threshold (see Chapter 3), and $(S/I)_n$ as a function of radar-to-target/jammer range. Note that the change in $(S/I)_n$ is essentially (when $J_N > N$) $1/R^2$, whereas the change in $(S/N)_n$ is $1/R^4$. The EA designer often strives for $(S/I)_n < 0$ dB, while the radar engineer would note the radar is essentially ineffective when the $(S/I)_n$ is less than the detection threshold.

10.3.2 Self-Protection Noise Jamming Metrics

We have developed the $(S/I)_n$ and compared it with the $(S/N)_n$ and detection threshold but have not addressed the effectiveness of the self-protection noise jammer. The effectiveness of the self-protection noise jamming is addressed in terms of self-protection jamming metrics. There are two main self-protection noise jamming metrics: burnthrough range and jamming-to-signal ratio (J/S). Target detection achieved in the presence of a noise jammer is called "burn-through," and the radar-to-target/jammer range at which it occurs is called the "burnthrough range." Thus, burnthrough range is the detection range of the radar in the presence of the self-protection noise jammer. It is the primary self-protection noise jamming metric. A secondary self-protection noise jamming metric is J/S, the ratio of the jamming noise power to the target signal power, which tells us how strong or weak the jamming is relative to the received target signal.

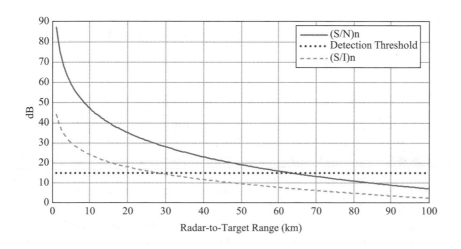

FIGURE 10-10 ■ Multiple-Pulse Signal-to-Noise Ratio $(S/N)_n$, Detection Threshold, and Signal-to-Interference Ratio $(S/I)_n$ versus Radar-to-Target/Jammer Range

Burnthrough Range

The $(S/I)_n$ equation can be solved for the range at which the radar will detect the target signal in the presence of the noise jammer: the burnthrough range. Directly solving the $(S/I)_n$ equation for range requires us to find the positive real root of a fourth-order equation. While this is not a problem to compute if we are using a computer, we can avoid this difficulty by assuming the self-protection jammer noise (J_N) is much greater than the receiver thermal noise (N), that is, $J_N \gg N$. After all, if J_N were not much greater than N, the jammer would have little impact on the radar. When we make this assumption we remove the receiver thermal noise from the $(S/I)_n$ equation (Equation 10-18). We now have a quadratic equation in range, which is much simpler to solve for the detection range in the presence of the noise jammer, the range at which the target signal exceeds the jammer noise by the detection threshold. The burnthrough range is given in Equations (10-19), algebraic, and (10-20), logarithmic.

$$\left(\frac{S}{J_N}\right)_n = \frac{P_R\, G_{RT}\, \sigma\, G_{sp}\, G_I}{(4\pi)\, R_{RT}{}^2\, L_R}\left(\frac{B_J}{B_R}\right)\frac{L_J}{P_J\, G_{JR}} \tag{10-18}$$

$$R_{bt} = \sqrt{\frac{P_R\, G_{RT}\, \sigma\, G_{sp}\, G_I}{(4\pi)\, SNR_{dt}\, L_R}\left(\frac{B_J}{B_R}\right)\frac{L_J}{P_J\, G_{JR}}} \tag{10-19}$$

$$\begin{aligned}
20\,\log(R_{bt}) = &\ 10\,\log(P_R) + 10\,\log(G_{RT}) + 10\,\log(\sigma)\\
&+ 10\,\log(G_{sp}) + 10\,\log(G_I) + 10\,\log(B_J)\\
&+ 10\,\log(L_J) - 10\,\log(4\pi) - 10\,\log(SNR_{dt})\\
&- 10\,\log(L_R) - 10\,\log(B_R) - 10\,\log(P_J)\\
&- 10\,\log(G_{JR})
\end{aligned} \tag{10-20}$$

where:

$(S/J_N)_n$ = Target signal-to-jamming ratio after integration of multiple pulses, no units

R_{bt} = Radar burnthrough range, meters

SNR_{dt} = Radar detection threshold, no units

Remember that these burnthrough range equations are approximations $(J_N \gg N)$, so we need to be careful using them. For the majority of radar–target–jammer–geometry/environment combinations, the difference between the approximate and actual burnthrough range is slight. Figure 10-11 shows the burnthrough range and detection range overlaid on plots of the $(S/N)_n$ and $(S/I)_n$.

The burnthrough range is a function of many target, jammer, and radar parameters. As the reader who does the problems at the end of the chapter will find, only a small mainbeam jammer ERP (relative to the radar ERP) is sufficient to reduce the detection range of a long-range radar system, particularly if

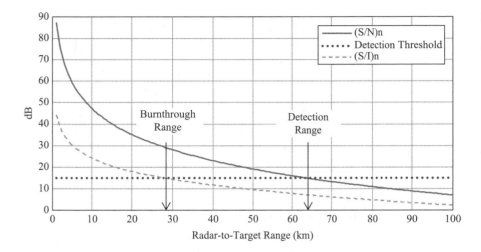

FIGURE 10-11 ■
Burnthrough Range
and Detection Range

the radar cross section (RCS) of the target is low. This is because the jammer signal has a one-way propagation, $1/R^2$, while the target signal has a two-way propagation, $1/R^4$.

We can look at the sensitivity of the burnthrough range to a few main target, jammer, and radar parameters and/or relationships. The sensitivity of the burnthrough range to target RCS is shown in Figure 10-12. In this figure, the burnthrough range is normalized to the value at an RCS of 0 dBsm (1 m²). The sensitivity of burnthrough range to the jammer-radar bandwidth mismatch (B_J/B_R) is shown in Figure 10-13. In this figure, the burnthrough range is normalized to the value at $B_J/B_R = 1$. The sensitivity of burnthrough range to jammer antenna gain in the direction of the radar is shown in Figure 10-14. In this figure, the burnthrough range is normalized to the value at the jammer antenna

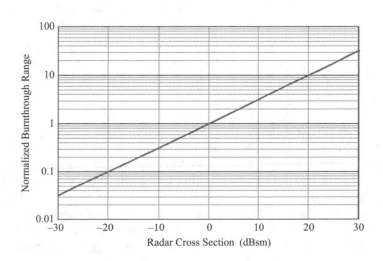

FIGURE 10-12 ■
Sensitivity of
Burnthrough Radar
to Target Radar
Cross Section

FIGURE 10-13 ■
Sensitivity of
Burnthrough Radar
to Jammer-Radar
Bandwidth
Mismatch

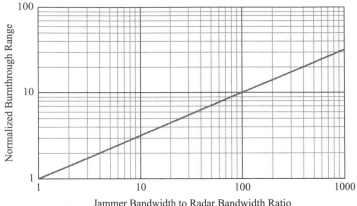

FIGURE 10-13 ■
Sensitivity of
Burnthrough Radar
to Jammer-Radar
Bandwidth
Mismatch

FIGURE 10-14 ■
Sensitivity of
Burnthrough Radar
to Jammer Antenna
Gain in the Direction
of the Radar

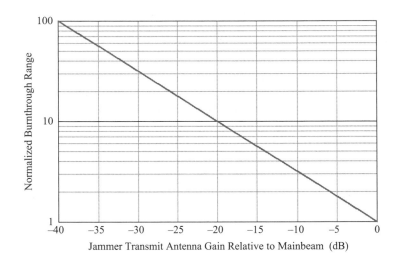

mainbeam gain and depicts the increase in burnthrough range as a function of reduced jammer antenna gain. The sensitivities in Figure 10-13 and Figure 10-14 assume that the self-protection noise jamming is much greater than the radar receiver thermal noise.

Jamming-to-Signal Ratio

J/S, the ratio of the jamming noise to the target signal power, is a long-standing electronic warfare (EW) metric. It quantifies the power relationship between the jamming noise and the target signal (see Chapter 2) after the integration of multiple pulses (see Chapter 3). The multiple-pulse J/S is given in Equations (10-21), algebraic, and (10-22), logarithmic. A J/S greater than one, or positive in dB, tells us the noise jammer power is higher than the target signal power. A J/S less than one, or negative in dB, tells us the noise jammer power is lower

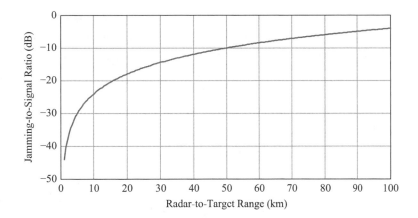

FIGURE 10-15 ▪
Jamming-to-Signal
Ratio as a Function
of Radar-to-Target/
Jammer Range

than the target signal power. Figure 10-15 shows the J/S (dB) as a function of radar-to-target/jammer range. In this figure the J/S is always negative (in dB), a very common situation with noise jammers.

$$\left(\frac{J_N}{S}\right)_n = \frac{P_J \, G_{JR}}{L_J} \left(\frac{B_R}{B_J}\right) \frac{(4\pi) \, R_{RT}^2 \, L_R}{P_R \, G_{RT} \, \sigma \, G_{sp} \, G_I} \tag{10-21}$$

$$\begin{aligned}
10 \log\left[\left(\frac{J_N}{S}\right)_n\right] = {} & 10 \log(P_J) + 10 \log(G_{JR}) + 10 \log(B_R) \\
& + 10 \log(4\pi) + 20 \log(R_{RT}) + 10 \log(L_R) \\
& - 10 \log(L_J) - 10 \log(B_J) - 10 \log(P_R) \\
& - 10 \log(G_{RT}) - 10 \log(\sigma) - 10 \log(G_{sp}) \\
& - 10 \log(G_I)
\end{aligned} \tag{10-22}$$

where:

$(J_N/S)_n$ = Noise jamming-to-target signal ratio after integration of multiple pulses, no units

Summary

Burnthrough range is the principal noise jamming metric. It shows the reduced detection range as a result of the noise jammer. J/S is a secondary noise jammer metric. Many people assume (and you know what is said to happen when we assume) the jamming must be greater than the target signal for the jammer to be effective. As shown in Figure 10-16, we do not necessarily need a positive (in dB) J/S (i.e., the jamming is greater than the target signal) for a noise jammer to be effective in reducing the detection range! This is because the effect of noise jamming is relative to the receiver thermal noise, not the target signal power. Thus, J/S can be a misleading metric for noise jammers. To avoid making an "assumer" out of you or others, use burnthrough range as the metric for noise jammers.

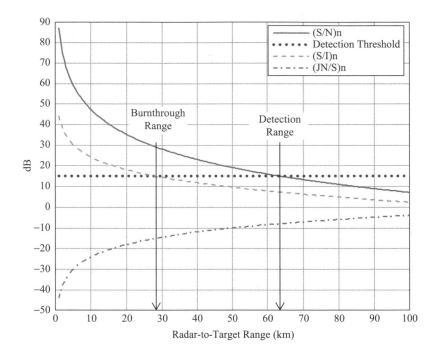

Burnthrough range and J/S are both dynamic quantities, dependent on the instantaneous conditions of the radar–target/jammer engagement. Specifically, there is a dynamic radar–target/jammer geometry due to the target/jammer movement and maneuvers. Radar mode, target RCS, radar antenna gain, jammer antenna gain, propagation path, and polarization mismatch can all change over the course of an engagement. Moreover, dynamic radar–jammer interactions (e.g., manual, semiautomatic, or automatic jammer) and radar operations occur over the course of the engagement. Hence, the jammer performance metrics discussed provide us with a "snapshot" of the interactive dynamics during the course of the overall engagement. Thorough radar or jammer performance predictions account for all of these various dynamics in the form of sophisticated computer models.

▐ 10.4 ▌ | SELF-PROTECTION FALSE TARGET JAMMING

For a self-protection false target jammer, the jammer transmits a false target waveform (see Chapter 8) similar to the actual radar waveform at the radar. The false target jamming waveform is received and processed just like a received target signal. False target jamming is a degrade/deceive type of EA: it degrades the radar's ability to effectively process true targets by introducing multiple false targets with the true target; and it deceives the radar to make measurements and tracks of false targets.

10.4.1 Self-Protection False Target Jamming in the Radar Receiver

The self-protection false target jamming signal arrives at the radar's receiver. Sometimes the false target jamming signal is more "attractive" (e.g., stronger, nonfluctuating) to the radar than is the true target signal. However, it is very difficult, if not impossible, for the false target jamming signal to perfectly match a true target signal. Most often, this mismatch occurs for radar systems employing pulse compression modulation (frequency or phase) waveforms (see Chapter 5). The false target jamming signal mismatch is often reflected in the radar signal processing gain (see Chapter 2) and/or the integration gain (see Chapter 3). In most cases, detailed hardware testing and evaluation (T&E) and/or high-fidelity modeling, simulation, and analysis (MS&A) are required to determine the radar signal processing gain and integration gain for a mismatched false target jamming signal.

In the radar receiver, a perfectly matched false target jamming signal appears the same as a true target signal. Thus, we will describe it in the same way as we do with a true target signal (see Chapter 2). However, instead of a target S/N we have a J/N, which is the ratio of the power from a single received false target jamming pulse to the power of one sample of radar receiver thermal noise (see Chapter 2) (Equation 10-23). This equation includes the effect of radar signal processing (see Chapter 2) on the J/N. As discussed in Chapter 3, radar systems improve their detection performance by integrating multiple pulses. The improvement in S/I provided by integration is quantified by the integration gain (see Chapter 3). The J/N after integration of multiple pulses is given in Equation (10-24).

$$\frac{J}{N} = \frac{P_J \, G_{JR} \, G_{RT} \, \lambda^2 \, G_{sp}}{(4\pi)^2 \, R_{RT}^2 \, F_R \, k \, T_0 \, B_R \, L_J} \tag{10-23}$$

$$\left(\frac{J}{N}\right)_n = \frac{P_J \, G_{JR} \, G_{RT} \, \lambda^2 \, G_{sp} \, G_I}{(4\pi)^2 \, R_{RT}^2 \, F_R \, k \, T_0 \, B_R \, L_J} \tag{10-24}$$

where:

J/N = Single pulse false target jamming-to-noise ratio, no units

G_{sp} = Radar signal processing gain, no units

$(J/N)_n$ = False target jamming-to-noise ratio after integration of multiple pulses, no units

G_I = Integration gain, no units

FIGURE 10-17 ■
Multiple-Pulse
Signal-to-Noise
Ratio (S/N)$_n$,
Detection Threshold,
and Jamming-to-
Noise Ratio (J/N)$_n$
versus Radar-to-
Target/Jammer
Range

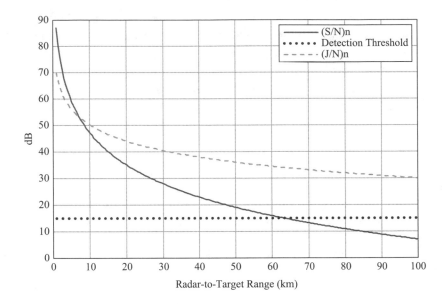

Figure 10-17 shows the multiple-pulse target signal-to-noise ratio $(S/N)_n$ (see Chapter 3), the radar detection threshold (see Chapter 3), and the $(J/N)_n$ as a function of radar-to-target/jammer range. Note that the change in $(J/N)_n$ is $1/R^2$, whereas the change in $(S/N)_n$ is $1/R^4$. The $(J/N)_n$ decreases by 6 dB (1/4) when the radar-to-target/jammer range doubles and increases by 6 dB (4 times) when the radar-to-target/jammer range halves. The $(S/N)_n$ decreases by 12 dB (1/16) when the radar-to-target range doubles and increases by 12 dB (16 times) when the radar-to-target range halves.

10.4.2 Self-Protection False Target Jamming Metrics

We have developed the $(J/N)_n$ and compared it with the $(S/N)_n$ and detection threshold but have not addressed the effectiveness of the self-protection false target jammer. Its impact on the radar system is examined in terms of two main self-protection false target jamming metrics: J/S and detection range. J/S is the ratio of the self-protection false target jamming signal power to the target signal power, and it quantifies the power relationship between the false target jamming and target signal. J/S is the primary self-protection false target jamming metric. Detection range is the radar-to-target/jammer range at which the radar will detect the false target jamming signal in the presence of the radar receiver thermal noise; how far away is the self-protection jammer when the false targets are detected. Detection range is a secondary self-protection false target jamming metric.

Jamming-to-Signal Ratio

J/S, the ratio of the self-protection false target jamming signal power and the target signal power (see Chapter 2) after integration of multiple pulses (see Chapter 3), is a long-standing EW metric. The $(J/S)_n$ is given in Equations (10-25), algebraic, and (10-26), logarithmic. Figure 10-18 shows an example of the J/S as a function of the radar-to-target/jammer range.

$$\left(\frac{J}{S}\right)_n = \frac{P_J \; G_{JR}}{L_J} \frac{(4\pi) \; R_{RT}^2 \; L_R}{P_R \; G_{RT} \; \sigma} \tag{10-25}$$

$$\begin{aligned} 10 \log\left[\left(\frac{J}{S}\right)_n\right] = {} & 10 \log(P_J) + 10 \log(G_{JR}) + 10 \log(4\pi) \\ & + 20 \log(R_{RT}) + 10 \log(L_R) - 10 \log(L_J) \\ & - 10 \log(P_R) - 10 \log(G_{RT}) - 10 \log(\sigma) \end{aligned} \tag{10-26}$$

where:

$(J/S)_n$ = False target jamming-to-target signal ratio after integration of multiple pulses, no units

Figure 10-19 shows the constant gain jammer J/S as a function of radar-to-target/jammer range. The J/S remains constant as a function of range until the radar-to-jammer range is reached where the jammer transmitter reaches its maximum output power (saturation). At this range, the constant gain jammer can put out only its maximum power and is a constant power jammer from this range on. The constant gain J/S then "varies" the same as the J/S of a constant power jammer, as shown in the figure.

The J/S is a function of many target, jammer, and radar parameters. As the reader who does the problems at the end of the chapter will find, a small

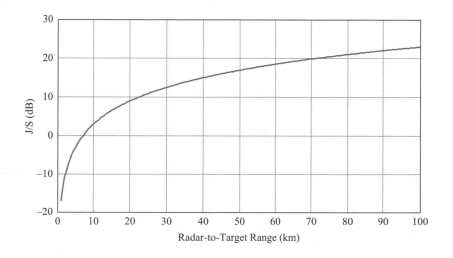

FIGURE 10-18 ■ False Target Jamming-to-Signal Ratio as a Function of Radar-to-Target/ Jammer Range

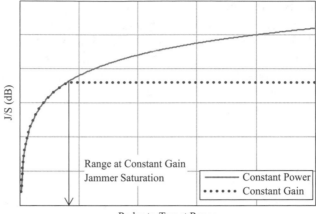

FIGURE 10-19 ■
Comparison of the
Jamming-to-Signal
Ratio from Constant
Power and Constant
Gain Jammers

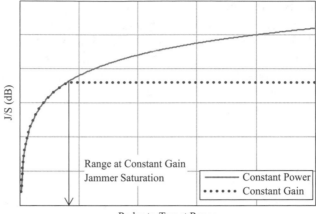

mainbeam jammer ERP (relative to the radar ERP) will produce a J/S > 1 (positive in dB) at long radar-to-target/jammer ranges, particularly if the radar cross section of the target is low. This is because the jammer signal has a one-way propagation, $1/R^2$, while the target signal has a two-way propagation, $1/R^4$.

We will look at the sensitivity of the J/S to two main target and jammer parameters. Its sensitivity to target RCS is shown in Figure 10-20, in which the J/S is normalized to the value at an RCS of 0 dBsm (1 m^2). The sensitivity of J/S to jammer transmit antenna gain in the direction of the radar is shown in Figure 10-21, in which the J/S is normalized to the value at the jammer antenna mainbeam gain and depicts the decrease in the J/S as a function of reduced jammer antenna gain.

J/S is often used as an indicator of false target jamming effectiveness. The false target jamming is "assumed effective" if the J/S exceeds a minimum

FIGURE 10-20 ■
Sensitivity of
Jamming-to-Signal
Ratio to Target
Radar Cross Section

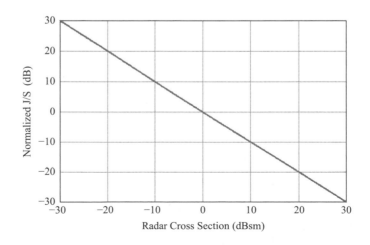

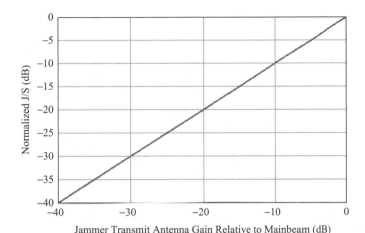

FIGURE 10-21 ■
Sensitivity of
Jamming-to-Signal
Ratio to Jammer
Antenna Gain in the
Direction of the
Radar

value. The minimum J/S required for effectiveness is a complicated function of the particular EA technique, the part of the radar being attacked, and, if present, the operator skill level. The minimum J/S required for effectiveness is determined based on detailed T&E and/or MS&A. If there is sufficient J/S to provide the necessary jamming "effectiveness," increasing the J/S will not necessarily increase the effectiveness and could in fact be harmful if it triggers the radar system's electronic protection (EP) capabilities or alerts the operator to the presence of jamming (see Chapters 8 and 12).

J/S is often misunderstood and misused. It is easy to calculate but difficult to measure! Because the individual equations are relatively straightforward, it's easy to show charts with the individual target, jammer, clutter, and thermal noise signals. However, the radar receiver contains the complex (amplitude and phase) combination of target, jammer, clutter, and thermal noise signals. It is very hard, if not impossible, to accurately and consistently separate the individual signals from the receiver output in the real world, such as we can with equations and plots. J/S is very difficult to measure in the controlled environment of a laboratory and almost impossible to measure in the dynamic environment of an open-air range. Care must be taken, therefore, in identifying and understanding assumptions made in expressions of the J/S metric.

Detection Range

The radar-to-target/jammer range at which the radar will detect the false target jamming signal is a secondary self-protection false target jamming metric. The $(J/N)_n$ equation is solved for the radar-to-target/jammer range at which the false target jamming signal exceeds the radar receiver thermal noise by the detection threshold. The false target detection range is given in Equations (10-27), algebraic, and (10-28), logarithmic. The logarithmic form of this equation can be simplified by using the carrier frequency instead of wavelength and combining the constants (Equation 10-29).

$$R_{dt} = \sqrt{\frac{P_J\, G_{JR}\, G_{RT}\, \lambda^2\, G_{sp}\, G_I}{(4\pi)^2\, SNR_{dt}\, F_R\, k\, T_0\, B_R\, L_J}} \qquad (10\text{-}27)$$

$$
\begin{aligned}
20\,\log(R_{dt}) = {}& 10\,\log(P_J) + 10\,\log(G_{JR}) + 10\,\log(G_{RT}) \\
& + 20\,\log(\lambda) + 10\,\log(G_{sp}) + 10\,\log(G_I) \\
& - 20\,\log(4\pi) - 10\,\log(SNR_{dt}) \\
& - 10\,\log(F_R) - 10\,\log(k) - 10\,\log(T_0) \\
& - 10\,\log(B_R) - 10\,\log(L_J)
\end{aligned}
\qquad (10\text{-}28)
$$

$$
\begin{aligned}
20\,\log(R_{dt}) = {}& 10\,\log(P_J) + 10\,\log(G_{JR}) + 10\,\log(G_{RT}) \\
& - 20\,\log(f_c) + 10\,\log(G_{sp}) + 10\,\log(G_I) \\
& - 10\,\log(SNR_{dt}) - 10\,\log(F_R) \\
& - 10\,\log(B_R) - 10\,\log(L_J) + 351.5355
\end{aligned}
\qquad (10\text{-}29)
$$

where:

R_{dt} = Radar detection range, meters
SNR_{dt} = Radar detection threshold, no units

Figure 10-22 shows an example comparison of the false target detection range to the true target detection range. As can be seen in this example, there is generally ample opportunity for the false target jamming signal to be detected by the radar system.

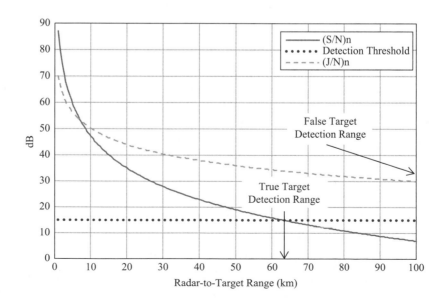

FIGURE 10-22 ■ True Target and False Target Jamming Detection Ranges

Summary

J/S is the principal self-protection false target jamming metric. Often, a specific J/S is needed for the false target jammer to be effective. Detection range is a secondary self-protection false target jamming metric. Detection range shows how far away the self-protection jammer is when the false target jammer signal is detected by the radar system. J/S and detection range are both dynamic quantities dependent on the instantaneous conditions of the radar-target-jammer engagement. There is a dynamic radar-target/jammer geometry due to the radar and/or target/jammer movement or maneuvers. Radar mode, target RCS, radar antenna gain, jammer antenna gain, propagation path, and polarization mismatch can all change over the course of an engagement. Moreover, dynamic radar–jammer interactions such as manual, semiautomatic, and automatic modes of operation occur over the course of engagement. Hence, the performance metrics discussed provide us with a "snapshot" of these dynamics during the course of the engagement. Thorough radar or jammer performance predictions account for all of these various dynamics using sophisticated computer models.

■ 10.5 | SELF-PROTECTION EXPENDABLES

As discussed in Chapter 8, self-protection expendables give the radar system something better than the target to detect, measure, and track. In this section we will discuss both passive (i.e., chaff, free-flight decoys, and towed decoys) and active (i.e., free-flight and towed decoys) self-protection expendables.

10.5.1 Passive Expendables

The principal forms of passive EA are chaff and decoys. Both Neri [2006] and Schleher [1999] describe chaff and decoys in detail. Use of either necessitates a variety of approaches. The radar characteristics of simple chaff (half-wave dipoles), including peak and average RCS of both a single dipole and a cloud of n dipoles, are treated in Chapter 6. Because dipoles need be only a half-wavelength long and thick enough to maintain their structure, a few pounds can create very large radar cross sections (as in the answer to Chapter 6, Exercise 5: 100,000 L-band dipoles produce an average RCS of 1,350 m^2). Such a large RCS would mask the presence of targets in its vicinity.

Chaff is popular because it is inexpensive, light, and simple to make, although it is temperamental to deploy. The dispensing of chaff is critical to its effectiveness. The "bloom time" is how long it takes for the chaff to reach its peak RCS. The chaff RCS decays as the chaff dipoles disperse in the atmosphere. Figure 10-23 shows the chaff RCS as a function of time. Because of this time dependency, multiple chaff drops (shown in Figure 10-24) may be required to sustain a specific RCS to protect the target.

FIGURE 10-23 ■
Chaff Radar Cross
Section as a
Function of Time

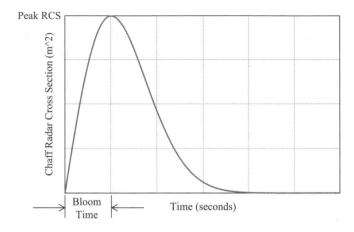

FIGURE 10-24 ■
Multiple Chaff Drops
over Time

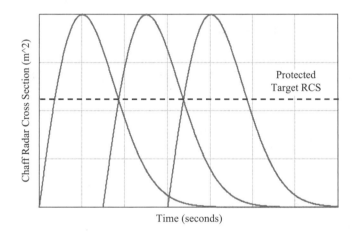

A chaff cloud may contain several lengths of chaff dipoles to be resonant in the frequency bands of several radar systems. It is used in ground and air operations and as part of the penetration systems of ballistic missiles. In the atmosphere, chaff's lightness causes it to stop quickly and hang in the air, so a moving target indicator (MTI) or Doppler radar can differentiate it from moving targets (see Chapter 7). In space, however, chaff moves along with the targets it is masking until it encounters the palpable atmosphere.

Chaff can be designed to be retro-directive to provide gain back toward the radar system illuminating it. The radar corner reflector discussed in Chapter 6 is retro-directive. Figure 10-25 shows two examples of retro-directive chaff: eight trihedral corner reflectors nested together; and a jack. The RCS of a corner reflector varies with angle but is substantial everywhere. The nested corner reflectors provide 4π steradians angular coverage. The RCS of a trihedral corner reflector was treated in Chapter 6. At X band ($\lambda = 0.03$ m), a corner

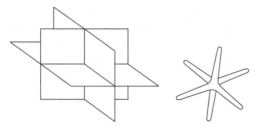

FIGURE 10-25 ■
Retro-Directive
Reflectors: A 4π
Corner Reflector and
a Jack

reflector with 0.3 meter sides gives an RCS of 339 m^2. A simpler chaff-like structure, also providing returns over multiple angles, is the jack. The jack is three-dimensional crossed dipoles. When dispensed in large numbers, retro-directive reflectors can wreak havoc on a radar system, flooding adjoining resolution cells in range, angle, and Doppler sidelobes (see Section 5.5.4) and potentially saturating the radar processors.

Whereas chaff is used principally for masking targets, decoys are used to simulate them. Decoys can be effective only if they can provide target simulation at weights, complexities, and costs that are much less than for real targets. To decoy a radar system requires simulating RCS and range rate, generally with a small vehicle with RCS enhancements. In simple decoys, these can take the form of embedded reflectors with gain in the directions of likely radar viewing aspect angles.

Decoys accompanying aircraft, helicopters, ships, or even ground vehicles on their operational missions are relatively expensive and are operationally complicated. If they accompany missile warheads, they may not have operational problems, but the strains on credibility as they sustain exponential decelerations in the atmosphere are severe. More practically, decoys are carried along, to be dispensed at crucial periods, reducing the time they need to be credible target simulators. Such decoys can be free flight (i.e., ballistic free fall or powered) or towed. Since free-flight decoys separate from the target, they are sequentially dispensed with the hope that one of them will capture the radar's tracker. Towed decoys solve the problems associated with having to sequentially dispense multiple free-flight decoys. And since they are reasonably close to the target, they continuously provide the opportunity to capture the radar's tracker.

10.5.2 Passive Expendable Equation

Since a passive expendable is an object with an RCS, we can use the radar equation from Chapter 2 for the passive expendable. The passive expendable signal power is given by Equation (10-30). Its signal-to-noise ratio is the ratio of the power from a single received passive expendable pulse to the power of one sample of radar receiver thermal noise (see Chapter 2). The single-pulse passive expendable signal-to-noise ratio (J_{PE}/N) is given in Equation (10-31). The J_{PE}/N equation often includes the effect of signal processing techniques used by radar systems (see Chapter 2). As discussed in Chapter 3, radar systems improve their detection performance by integrating multiple pulses. The improvement in J_{PE}/N provided by integration is quantified by the integration gain (see Chapter 3). The J_{PE}/N after integration of multiple pulses is given in Equation (10-32).

$$J_{PE} = \frac{P_R \, G_{RE}{}^2 \, \lambda^2 \, \sigma_{PE}}{(4\pi)^3 \, R_{RE}{}^4 \, L_R} \tag{10-30}$$

$$\frac{J_{PE}}{N} = \frac{P_R \, G_{RE}{}^2 \, \lambda^2 \, \sigma_{pe} \, G_{sp}}{(4\pi)^3 \, R_{RE}{}^4 \, F_R \, k \, T_0 \, B_R \, L_R} \tag{10-31}$$

$$\left(\frac{J_{PE}}{N}\right)_n = \frac{P_R \, G_{RE}{}^2 \, \lambda^2 \, \sigma_{pe} \, G_{sp} \, G_I}{(4\pi)^3 \, R_{RE}{}^4 \, F_R \, k \, T_0 \, B_R \, L_R} \tag{10-32}$$

where:

J_{PE} = Received passive expendable signal peak power, watts

P_R = Radar peak transmit power, watts

G_{RE} = Radar antenna gain in the direction of the passive expendable, no units

λ = Wavelength, meters

σ_{pe} = Passive expendable radar cross section, square meters (m^2)

R_{RE} = Radar-to-passive expendable slant range, meters

L_R = Total radar-related losses, no units

J_{PE}/N = Single pulse passive expendable signal-to-noise ratio, no units

N = Radar receiver thermal noise, watts

F_R = Radar receiver noise figure (≥ 1), no units

k = Boltzmann's constant, 1.38×10^{-23} watt-seconds/Kelvin

T_0 = Receiver standard reference temperature, 290 Kelvin

B_R = Radar receiver filter bandwidth, hertz

G_{sp} = Radar signal processing gain, no units

$(J_{PE}/N)_n$ = Passive expendable signal-to-noise ratio after integration of multiple pulses, no units

G_I = Radar integration gain, no units

10.5.3 Passive Expendable Metric

J/S is the passive expendable metric, and it quantifies the power relationship between the passive expendable and target signals. It is the ratio of the passive expendable signal power and the target signal power (see Chapter 2) after integration of multiple pulses (see Chapter 3). The $(J_{PE}/S)_n$ is given in Equations (10-33), algebraic, and (10-34), logarithmic.

$$\left(\frac{J_{PE}}{S}\right)_n = \frac{G_{RE}^2}{G_{RT}^2}\frac{\sigma_{pe}}{\sigma}\frac{R_{RT}^4}{R_{RE}^4} \tag{10-33}$$

$$10\log\left[\left(\frac{J_{PE}}{S}\right)_n\right] = 20\log(G_{RE}) + 10\log(\sigma_{pe}) + 40\log(R_{RT}) \tag{10-34}$$
$$- 20\log(G_{RT}) - 10\log(\sigma) - 40\log(R_{RE})$$

where:

$(J_{PE}/S)_n$ = Passive expendable jamming-to-target signal ratio after integration of multiple pulses, no units

R_{RT} = Radar-to-target slant range, meters

G_{RT} = Radar antenna gain in the direction of the target, no units

σ = Target radar cross section, square meters (m^2)

As previously discussed with the J/S metric, a jammer (here a passive expendable) is assumed to be effective when the J/S exceeds a specific value, normally greater than one or positive in dB. As seen in Equation (10-33), $(J_{PE}/S)_n$ is a function of passive expendable and target RCSs and geometries (ranges and angles to determine the antenna gains). Because a passive expendable starts at the target, for this initial geometry the $(J_{PE}/S)_n$ is the ratio of the passive expendable and target RCS values. Thus, there is a great emphasis on the RCS of the passive expendable to ensure that it is greater than the RCS of the target. If the passive expendable is unsuccessful in degrading or deceiving the radar system in this initial geometry, the later geometries do not really matter because the radar is tracking the target. Additional passive expendable J/S comments are summarized in Table 10-1.

TABLE 10-1 ■ Passive Expendable Jamming-to-Signal Ratio Comments

Passive Expendable	RCS over Time	Relative Geometry over Time (Range, Range Rate, and Angle)	
		Slow-Moving Targets	Fast-Moving Targets
Chaff	Changes	Fairly Constant	Changes
Free-Flight Decoy	Fairly Constant	Changes	Changes Significantly
Towed Decoy	Fairly Constant	Changes	

10.5.4 Active Expendables

Active expendables solve the main limitation of passive expendables; there rarely is enough passive expendable RCS to go around (e.g., aspect angles, radar frequencies). Just like their passive cousins, active expendables can be free flight or towed. Active expendables have a small false target jammer in the decoy that provides an "electronic RCS" over a wide range of aspect angles that are based on the antenna pattern of the jammer and radar frequencies that are based on a wide jammer bandwidth. Active free-flight decoys are powered by a battery, while active towed decoys are most often powered by the towing platform. The active decoy jammer can be a simple constant power or constant gain repeater or produce sophisticated false target waveforms.

10.5.5 Active Expendable Equation

The active expendable jamming waveform is generated by the active decoy jammer and arrives at the radar's receiver. Sometimes the active expendable jamming signal is more "attractive" (e.g., stronger, nonfluctuating) to the radar than the true target signal. However, it is very difficult, if not impossible, for the active expendable jamming signal to perfectly match a true target signal. Most often, this active expendable jamming mismatch occurs for radar systems employing pulse compression modulation (frequency or phase) waveforms (see Chapter 5). The active expendable jamming signal mismatch is often reflected in the radar signal processing gain (see Chapter 2) and/or the integration gain (see Chapter 3). In most cases, detailed hardware T&E and/or high-fidelity MS&A is required to determine the radar signal processing gain and integration gain for a mismatched active expendable jamming signal.

Since an active expendable is a false target jammer, we can start with the false target jammer equation from earlier in this chapter and add the geometry uniqueness associated with an active decoy. The received active expendable signal power is given by Equation (10-35). In the radar receiver, a perfectly matched active expendable signal appears the same as a target signal. Thus, we will describe it in the same way as we do with a true target signal: as a signal-to-noise ratio (see Chapter 2). The active expendable signal-to-noise ratio is the ratio of the power from a single received active expendable pulse to the power of one sample of radar receiver thermal noise. The single-pulse active expendable signal-to-noise ratio (J_{AE}/N) is given in Equation (10-36). The J_{AE}/N equation often includes the effect of signal processing techniques used by radar systems (see Chapter 2). As discussed in Chapter 3, radar systems improve their detection performance by integrating multiple pulses. The improvement in J_{AE}/N provided by integration is quantified by the integration gain (see Chapter 3). The J_{AE}/N after integration of multiple pulses is given in Equation (10-37).

$$J_{AE} = \frac{P_J \, G_{JR} \, G_{RE} \, \lambda^2}{(4\pi)^2 \, R_{RE}{}^2 \, L_J} \tag{10-35}$$

$$\frac{J_{AE}}{N} = \frac{P_J \, G_{JR} \, G_{RE} \, \lambda^2 \, G_{sp}}{(4\pi)^2 \, R_{RE}{}^2 \, F_R \, k \, T_0 \, B_R \, L_J} \tag{10-36}$$

$$\left(\frac{J_{AE}}{N}\right)_n = \frac{P_J \, G_{JR} \, G_{RE} \, \lambda^2 \, G_{sp} \, G_I}{(4\pi)^2 \, R_{RE}{}^2 \, F_R \, k \, T_0 \, B_R \, L_J} \tag{10-37}$$

where:

J_{AE} = Received active expendable signal peak power, watts
P_J = Jammer peak transmit power, watts
G_{JR} = Jammer transmit antenna gain in the direction of the radar, no units
G_{RE} = Radar receive antenna gain in the direction of the active expendable, no units
λ = Wavelength, meters
R_{RE} = Radar-to-active expendable slant range, meters
L_J = Total jammer-related losses, no units
J_{AE}/N = Single pulse active expendable jamming-to-noise ratio, no units
G_{sp} = Radar signal processing gain, no units
$(J_{AE}/N)_n$ = Active expendable jamming-to-noise ratio after integration of multiple pulses, no units
G_I = Integration gain, no units

10.5.6 Active Expendable Metric

J/S is the active expendable metric, and it quantifies the power relationship between the active expendable jamming and target signal. It is the ratio of the active expendable jamming signal power and the target signal power (see Chapter 2) after integration of multiple pulses (see Chapter 3). The $(J_{AE}/S)_n$ is given in Equations (10-38), algebraic, and (10-39), logarithmic.

$$\left(\frac{J_{AE}}{S}\right)_n = \frac{P_J \, G_{JR} \, G_{RE}}{R_{RE}{}^2 \, L_J} \, \frac{(4\pi) \, R_{RT}{}^4 \, L_R}{P_R \, G_{RT}{}^2 \, \sigma} \tag{10-38}$$

$$\begin{aligned}
10 \log\left[\left(\frac{J_{AE}}{S}\right)_n\right] = {} & 10 \log(P_J) + 10 \log(G_{JR}) + 10 \log(G_{RE}) \\
& + 10 \log(4\pi) + 40 \log(R_{RT}) + 10 \log(L_R) \\
& - 20 \log(R_{RE}) - 10 \log(L_J) - 10 \log(P_R) \\
& - 20 \log(G_{RT}) - 10 \log(\sigma)
\end{aligned} \tag{10-39}$$

where:

$(J_{AE}/S)_n$ = Active expendable jamming-to-target signal ratio after integration of multiple pulses, no units

As previously discussed with the J/S metric, a jammer (here an active expendable) is assumed to be effective when the J/S exceeds a specific value, normally greater than one or positive in dB. As seen in Equation (10-38), $(J_{AE}/S)_n$ is a function of received active expendable and target signal powers and geometries (ranges and angles to determine the antenna gains). Because an active expendable starts at the target, for this initial geometry the $(J_{AE}/S)_n$ is a function of only the signal powers. Thus, there is a great emphasis on the ERP of the active expendable to ensure that it is greater than the reflected target signal power due to the RCS of the target. If the active expendable is unsuccessful in degrading or deceiving the radar system in this initial geometry, the later geometries do not really matter because the radar is tracking the target. Additional active expendable J/S comments are summarized in Table 10-2.

TABLE 10-2 ■ Active Expendable Jamming-to-Signal Ratio Comments

Active Expendable	Jammer Power over Time	Relative Geometry over Time (Range, Range Rate, and Angle)	
		Slow-Moving Targets	Fast-Moving Targets
Free-Flight Decoy	Based on Battery Life	Changes	Changes Significantly
Towed Decoy	Constant	Changes	

■ 10.6 | OTHER SELF-PROTECTION JAMMING DECEPTION TECHNIQUES

So far, we have discussed only classic noise and false target jamming. Still to be considered are tactics and techniques generating false targets to deceive tracking radar systems, the primary emphasis of self-protection jamming [Chrzanowski, 1990, Chapters 5, 6; Neri, 2006, Section 5.3; Schleher, 1999, Chapters 3, 4; Van Brunt, 1978]. To deceive a tracking radar system (i.e., hardware, software, or operator as applicable), a self-protection jammer will generally transmit a small number of false targets.

As discussed in Chapter 8, repeater and transponder jammers capture the incident radar signal and return it to the radar. They may reradiate it immediately or delay the return. By delaying the return, a false target is generated in range. The jammer may store the radar pulse and repeat it several times or alter the delay over time to present changing ranges. The jammer may change the frequency to simulate different range rates or change the amplitude to attack a lobing angle tracker.

A radar system that performs target tracking by keeping the centroid of the signal in the center of a tracking gate (see Chapter 5) is vulnerable to a false target jammer that covers the target's return with a large amplitude signal whose characteristics are gradually changed, thus moving the tracking gate away from the true target. Thus, the jammer attempts to steal the range, range rate, or angle tracking gates either singularly or in combination. A gate-stealing jamming technique deceives the tracker using jamming waveforms that are contrived to capture the tracking gate. Of course, detailed radar tracker information or estimates are required to design effective deception jamming waveforms.

The tracker is "captured" when the tracking gate follows the false target jamming signal. Depending on the jamming waveform characteristics, the tracker can be dropped, held at a specific point, or steered into clutter. The tracker is dropped "far" from the true target so that it does not quickly reacquire the true target. Successful tracking gate stealing EA captures the tracker resulting either in a "break lock" or tracking error. A tracker breaks lock when it is captured and the jamming signal goes away or is "dropped." With only receiver thermal noise remaining in the tracking gate, the tracking loop is not closed on any signal (false target jamming or true target). Tracking error, relative to the true target, is generated when the tracking loop is closed on the false target jamming signal or a combination of true target and false target jamming signals. The amount of tracking error generated is a complicated function of J/S, the specific EA technique, and tracker characteristics such as hardware, software, and operator.

The range gate pull off (RGPO) or range gate stealing (RGS) EA technique covers the target return with a large amplitude signal whose time delay is gradually increased (using a repeater) or decreased (using a transponder), moving the range tracking gate away from the true target. If the jammer signal is sufficiently stronger than the true target signal, the tracker will follow the jammer signal. The rate at which the time delay can be changed (increased or decreased), the "pull rate" (Δsec/sec), is a function of the range tracker characteristics. If this rate exceeds the range tracker's maximum rate (bandwidth), the tracker will "low-pass filter" out the jamming signal. This is analogous to our inability to visually track a bird quickly crossing our field of view. Since there is no way for the jammer to know if the RGPO-RGS waveform has successfully captured the range tracking gate, it repeats itself over time, continually attempting to capture the tracking gate. The combination of the tracker bandwidth and jamming waveform repetition constrains how much range tracking error can be achieved.

Similarly, a range rate (Doppler) track can be attacked by a false target jammer generating a strong signal at the target Doppler and gradually changes its frequency, thus moving the tracking gate off the true target. This EA technique is called velocity gate pull off (VGPO) or velocity gate stealing (VGS). If the jammer signal is sufficiently stronger than the true target signal, the tracker will follow the jammer signal. The rate at which the frequency can be changed (increased or decreased), the "pull rate" (Hz/sec), is a function of the range rate tracker characteristics. If this rate exceeds the range rate tracker's maximum rate (bandwidth) the tracker will "low pass filter" out the jamming signal. Since there is no way for the jammer to know if the VGPO-VGS waveform has successfully captured the range rate tracking gate, the VGPO-VGS waveform repeats itself over time, continually attempting to capture the tracking gate. The combination of the tracker bandwidth and jamming waveform repetition constrains how much range rate tracking error can be achieved.

Angle deception is the most powerful jamming technique for three reasons: weapon guidance is very sensitive to angle; there are constraints on how much range and/or range rate error can be achieved; and when the radar antenna mainbeam is pulled off the target the S/N gets very small very fast. Angle deception against lobing (conical scan and sequential lobing) angle trackers includes amplitude modulation (AM) and inverse gain. AM is bursts of strong synchronized jamming pulses transmitted during the minimums of the angle tracking sequence. The inverse gain technique has jamming signal power modulated with the inverse of the received radar signal power. Lobing angle trackers are also vulnerable to an EA waveform which frequency sweeps upward or downward across its lobing frequency.

Since a monopulse angle tracker (see Chapter 5) forms an angular error estimate on each return pulse, it is insensitive to pulse-to-pulse amplitude modulations from jamming signals. A monopulse tracker can track jamming signals designed to attack lobing trackers; this is often referred to as "home-on-jam." Angle deception against monopulse trackers either exploits imperfections in the monopulse design and/or implementation or attacks fundamental monopulse concept weaknesses using multiple source techniques. In general, it is better to attack fundamental concept weaknesses than to rely on design and implementation weaknesses. Exploiting the latter implies a detailed understanding of the victim radar system, which is susceptible to modification of the design or implementation to correct those weaknesses. Example monopulse angle deception jamming techniques include cross-polarization (cross-pol), cross-eye, terrain bounce, and decoys.

Chrzanowski [1990], Neri [2006], Schleher [1999], and Van Brunt [1978] all discuss these deception self-protection jamming techniques in detail. As previously discussed, jammers, which attack the radar system—hardware, software, operator, and/or tracker characteristics—usually require detailed knowledge or estimates of the radar's mechanisms for range, range rate, and angle tracking. Thus, it is very challenging to generically describe the effectiveness of deception self-protection jamming techniques.

10.7 | SUMMARY

This chapter discussed self-protection jamming and developed the fundamental on-board self-protection jamming equation. We examined noise (wideband random amplitude and phase) self-protection jamming, false target (radar-like waveform) self-protection jamming, and expendable self-protection jamming (passive and active), their effects on the radar, and the metrics used to describe their performance. We concluded by looking at other self-protection jamming techniques used to deceive target tracking radar systems.

Of course, the radar community does not sit idly by while the EW community renders its radar system useless. Numerous EP techniques have been developed in an attempt to preserve the radar system's target detection, measurement, and tracking capability and performance in the presence of EA. EP techniques are discussed in detail in Chapter 12.

10.8 | EXERCISES

10-1. A self-protection jammer has the following characteristics: peak transmit power $P_J = 25$ watts; transmit antenna gain in the direction of the radar $G_{JR} = 3$ dBi; total jammer-related losses $L_J = 15$ dB; and jammer transmit loss $L_{Jt} = 2$ dB. The jammer is on the target and jamming the radar system from Exercises 2-5 and 3-15. Compute the following: (a) the transmitted jammer effective radiated power ERP_J (watts and dBW), (b) the received jammer power density at the radar receive antenna, and (c) the received jammer power J (watts and dBW). Compare the received jammer power with the jammer transmitted effective radiated power. (Hint: use the results from Exercises 2-5 and 3-15.)

10-2. The self-protection jammer from Exercise 10-1 uses a noise waveform with a bandwidth $B_J = 200$ MHz centered about the radar carrier frequency. Compute the following: (a) the jammer noise power in the radar receiver J_N (watts and dBW), (b) the interference signal power I (watts and dBW), and (c) the target signal-to-interference ratio after integration of multiple pulses $(S/I)_n$ (no units and dB). Compare the target signal-to-interference ratio with the target signal-to-noise ratio after integration of multiple pulses from Exercise 3-15. (Hint: use the results of Exercises 2-5 and 3-15.)

10-3. Compute the burnthrough range R_{bt} (meters) and jamming-to-signal ratio after integration of multiple pulses $(J_N/S)_n$ (no units and dB) for the self-protection noise jammer from Exercise 10-2. Compare the burnthrough range to the detection range from Exercise 3-15. (Hint: use the results from Exercises 2-5 and 3-15.)

10-4. The self-protection jammer from Exercise 10-1 uses a false target waveform perfectly matched to the radar waveform. Compute the false target jamming-to-noise ratio after integration of multiple pulses $(J/N)_n$ (no units and dB). Compare the false target signal-to-noise ratio with the target signal-to-noise ratio after integration of multiple pulses from Exercise 3-15. (Hint: use the results from Exercises 2-5 and 3-15.)

10-5. Compute the false target jamming-to-signal ratio after integration of multiple pulses $(J/S)_n$ (no units and dB) and the false target radar detection range R_{dt} (meters) for the self-protection false target jammer from Exercise 10-4. Compare the false target radar detection range with the radar detection range for the true target in Exercise 3-15. (Hint: use the results from Exercises 2-5 and 3-15.)

10-6. To determine if the noise jamming power will be above the thermal noise in the receiver, calculate the ratio of jamming noise to receiver thermal noise out of the radar receiver. The target/jammer is at a range $R_{RT} = 200$ km, power $P_J = 10$ watts, antenna gain in the direction of the radar $G_{JR} = 1$ dBi, noise bandwidth $B_J = 50$ MHz, and jammer-related losses $L_J = 3$ dB. The radar has an antenna gain in the direction of the target/jammer $G_{RT} = 40$ dBi, receiver noise figure $F_R = 7$ dB, wavelength $\lambda = 0.05$ meters, and the receiver filter bandwidth matched to a 1 μsec pulse.

10-7. The jammer in the preceding exercise (10-6) is aboard an aircraft flying toward the radar which has a peak power $P_R = 1$ MW, radar-related losses $L_R = 6$ dB, signal processing gain $G_{sp} = 1$, and integration gain $G_I = 15$ dB. What is the target signal-to-jamming ratio after integration of multiple pulses, $(S/J_N)_n$ (no units and dB), if the aircraft has a nose-on radar cross section $\sigma = 1$ m^2, of 10 m^2?

10-8. Assuming a signal-to-jamming ratio $S/J = 6$ dB is required for detection and using the radar and jammer characteristics from the previous two exercises (10-6 and 10-7), what is the burnthrough range of the radar for a target with a radar cross section $\sigma = 1$ m^2? With a radar cross section $\sigma = 10$ m^2?

10-9. A radar system attempts to track a target with a radar cross section $\sigma = 5$ m^2 at a radar-to-target range $R_{RT} = 15$ km. The target releases chaff passive expendable, and the radar system tracks the chaff as it separates from the target. The radar mainbeam antenna gain in the direction of the chaff passive expendable is $G_{RE} = 35$ dBi. The target is separated in angle from the chaff by one-half of the antenna half-power beamwidth. The chaff is 50 meters farther in range than the target. What is the passive expendable jamming-to-signal ratio after integration of multiple pulses $(J_{PE}/S)_n$ (no units and dB)? Use the average radar cross section of the chaff from Exercise 6-5.

10.9 | REFERENCES

Chrzanowski, Edward, 1990, *Active Radar Electronic Countermeasures*, Norwood MA: Artech House. Contains a very detailed description of self protection jamming concepts, technology, and supporting math.

Goj, Walter, 1993, *Synthetic-Aperture Radar and Electronic Warfare*, Norwood MA: Artech House. Contains a very clear description and supporting math for jamming pulse compression and pulse-Doppler radar systems.

Neri, Filippo, 2006, *Introduction to Electronic Defense Systems*, 2nd Edition, Raleigh, NC: SciTech Publishing. Contains a very detailed description of self protection jamming concepts, technology, and supporting math.

Schleher, D. C., 1999, *Electronic Warfare in the Information Age*, Norwood, MA: Artech House. Contains a very detailed description of self protection jamming concepts, technology, and supporting math, including numerous MATLAB files.

Van Brunt, L. B., 1978, *Applied ECM*, Volume I & II, Dunn Luring, VA: EW Engineering, Inc. Contains clear descriptions and details on myriad EA-ECM and EP-ECCM tactics and techniques.

Support Jamming
Electronic Attack

HIGHLIGHTS

- Definition and discussion of support jamming electronic attack
- Development of the support jamming equation, the starting point for our mathematical representations
- Noise support jamming, its effect on the radar, and metrics to describe its performance
- False target support jamming, its effect on the radar, and metrics to describe its performance
- Expendable support jamming

As discussed in Chapters 1 and 8, a support jammer is separated from the platforms (the targets) it is protecting. In this chapter we will discuss support jamming principles along with the supporting math [Chrzanowski, 1990; Neri, 2006, Chapter 5; Schleher, 1999; Chapter 4; Van Brunt, 1978]. The effect of the support jamming is to obscure true target signals or introduce false targets. A successful support jammer does this over a wide angular extent about the radar system. The support jamming waveform most often enters the radar antenna through a sidelobe as the radar system is attempting to detect the targets through its mainbeam, as shown in Figure 11-1. Support jamming requires sufficient incident jammer power density to overcome the low radar antenna sidelobe gain. Sufficient incident jammer power density is obtained with a high jammer effective radiated power (ERP) and/or a short radar-to-jammer range. High jammer ERP is obtained either with a high power transmitter or a high gain antenna on the jammer that then must be steered at the radar or a combination of both.

FIGURE 11-1 ■
Support Jamming

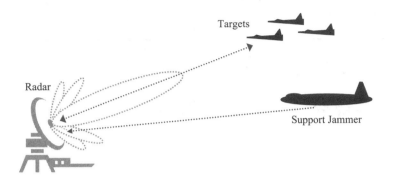

11.1 | SUPPORT JAMMING PRINCIPLES

The support jamming equation, noise jamming, and false target jamming will be described in the following sections. The support jamming equation section addresses the support jamming power equations for constant power and constant gain jammers. The noise jamming section addresses noise jamming in the radar receiver and noise jamming metrics: burnthrough range and jamming-to-signal ratio (J/S). The false target jamming section addresses false target jamming in the radar receiver and false target jamming metrics: detection range and J/S.

11.2 | SUPPORT JAMMING EQUATION

The support jamming equation relates jammer, radar, and geometry/environment characteristics together to allow us to determine the jammer power received by the radar system. The support jammer equation is the first step in determining radar performance (impact of the jammer) in the presence of the support jammer. As shown in Figure 11-2, the received jammer power is computed using incremental steps:

Step 1: jammer ERP

Step 2: jammer-to-radar propagation

Step 3: jammer power out of the radar antenna

These steps are similar to those used to develop the target signal power equation in Chapter 2. Thus, we will efficiently go through the steps here and the reader can refresh themselves with the details in Chapter 2 as needed.

The support jammer has an ERP in the direction of the radar (Equation 11-1). The support jammer ERP propagates to the radar resulting in a jammer power

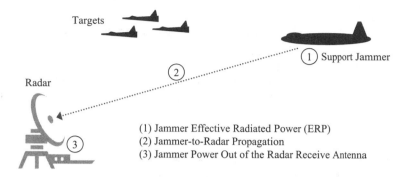

FIGURE 11-2 ■
Incremental
Build-Up of
Received Support
Jammer Power

(1) Jammer Effective Radiated Power (ERP)
(2) Jammer-to-Radar Propagation
(3) Jammer Power Out of the Radar Receive Antenna

density at the radar (Equation 11-2). The incident support jammer power density is collected by the effective area of the radar receive antenna in the direction of the support jammer, resulting in the jammer power at the output of the radar antenna (Equation 11-3). This equation uses antenna gain in the direction of the support jammer instead of effective area in the direction of the support jammer. Total jammer-related losses include factors not explicitly considered in determining the received support jammer power: jammer transmit loss; atmospheric attenuation (see Chapter 13); jammer waveform polarization mismatch with the radar antenna (see Table 9-1), radar receive loss, etc. Different radar–jammer–environment combinations have different losses, so we collect them up as we need them (Equation 11-4). When all jammer-related losses are considered, the resultant received support jammer power is given in Equations (11-5), algebraic, and (11-6), logarithmic. The logarithmic form of this equation can be simplified by using the carrier frequency instead of wavelength and combining the constants (Equation 11-7). Since the electronic warfare (EW) community often uses the logarithmic forms of equations, we will present them along the traditional algebraic forms.

$$\mathrm{ERP_J} = \frac{P_J\ G_{JR}}{L_{Jt}} \tag{11-1}$$

$$\frac{P_J\ G_{JR}}{L_{Jt}} \left(\frac{1}{(4\pi)\ R_{RJ}{}^2} \right) = \frac{P_J\ G_{JR}}{(4\pi)\ R_{RJ}{}^2\ L_{Jt}} \tag{11-2}$$

$$\frac{P_J\ G_{JR}}{(4\pi)\ R_{RJ}{}^2\ L_{Jt}} (A_e) = \frac{P_J\ G_{JR}}{(4\pi)\ R_{RJ}{}^2\ L_{Jt}} \left(\frac{G_{RT}\ \lambda^2}{(4\pi)} \right) = \frac{P_J\ G_{JR}\ G_{RT}\ \lambda^2}{(4\pi)^2\ R_{RJ}{}^2\ L_{Jt}} \tag{11-3}$$

$$L_J = L_{Jt}\ L_{JRa}\ L_{Jpol}\ L_{Rr}\ L_{Rsp} \dots \tag{11-4}$$

$$J = \frac{P_J\ G_{JR}\ G_{RJ}\ \lambda^2}{(4\pi)^2\ R_{RJ}{}^2\ L_J} \tag{11-5}$$

$$10 \log (J) = 10 \log (P_J) + 10 \log (G_{JR}) + 10 \log (G_{RJ})$$
$$+ 20 \log (\lambda) - 20 \log (4\pi) - 20 \log (R_{RJ}) \qquad (11\text{-}6)$$
$$- 10 \log (L_J)$$

$$10 \log (J) = 10 \log (P_J) + 10 \log (G_{JR}) + 10 \log (G_{RJ})$$
$$- 20 \log (f_c) - 20 \log (R_{RJ}) - 10 \log (L_J) \qquad (11\text{-}7)$$
$$+ 147.5582$$

where:

ERP_J = Jammer effective radiated power, watts

P_J = Jammer transmitter peak power, watts

G_{JR} = Jammer antenna gain in the direction of the radar, no units

L_{Jt} = Jammer transmit loss, no units

R_{RJ} = Radar-to-jammer slant range, meters

A_e = Radar receive antenna effective area in the direction of the jammer, square meters (m^2)

G_{RJ} = Radar antenna gain in the direction of the jammer (where ever the jammer is in the radar antenna pattern: mainbeam, sidelobes, backlobe), no units

λ = Wavelength, meters

L_J = Total jammer-related losses, no units

L_{JRa} = Jammer-to-radar atmospheric attenuation loss, no units

L_{Jpol} = Jammer-radar polarization mismatch loss, no units

L_{Rr} = Radar receive loss, no units

L_{Rsp} = Radar signal processing loss, no units

J = Received support jammer peak power, watts

f_c = Radar carrier frequency, hertz

Since we now have an equation for the received support jammer power, we can compare it with the target signal power and radar receiver thermal noise power (both from Chapter 2). The single-pulse/sample received target signal power (S), radar receiver thermal noise (N), and received support jammer power (J) as a function of radar-to-target range and radar-to-jammer range are shown in Figure 11-3. As seen in this figure and Equation (11-5), the received support jammer power varies as a function of $1/R^2$. Thus, the received jammer power decreases by 6 dB (1/4) when the radar-to-jammer range doubles and increases by 6 dB (4 times) when the radar-to-jammer range halves. In contrast, the received target signal power varies as a function of $1/R^4$ (see Chapter 2). Thus, the received target signal power decreases by 12 dB (1/16) when the radar-to-target range doubles and increases by 12 dB (16 times) when the radar-to-target range halves. Note that the y-axis grid for this figure is in 10 dB steps, one order of magnitude (power of 10), so the reader can easily interpret the change in power with range. Other figures in this chapter will use the same approach.

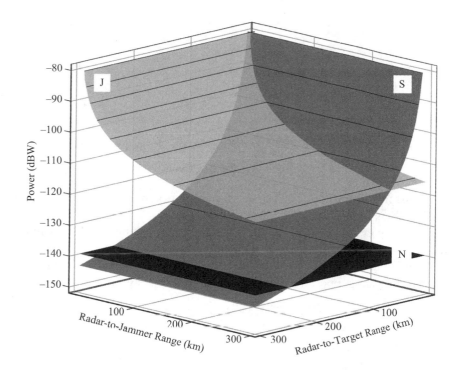

FIGURE 11-3 ▪
Target Signal (S),
Receiver Thermal
Noise (N), and
Jammer (J) Power
versus Radar-to-
Jammer and Radar-
to-Target Ranges

11.2.1 Constant Gain Jammer

Most jammers have a constant power output, as we have described so far. Constant power jammers are simple to implement, and they are always transmitting the maximum available (saturated) power. Some jammers operate by amplifying the received radar signal power by a constant gain and transmitting it back to the radar. This type of jammer is called a constant gain or repeater jammer [Chrzanowski, 1990, Sections 2.10, 2.11]. A constant gain jammer is analogous to the combination of a microphone, fixed gain amplifier, and speaker. Constant power and constant gain jammers are shown in Figure 11-4.

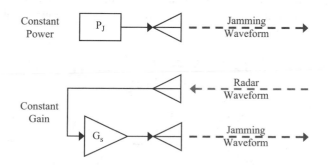

FIGURE 11-4 ▪
Constant Power and
Constant Gain
Jammers

FIGURE 11-5 ■
Incremental Build-
Up of Constant Gain
Support Jammer
Effective Radiated
Power

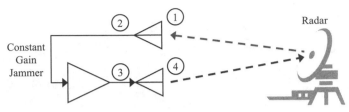

(1) Incident Radar Power Density   (2) Received Radar Power
(3) Amplified Radar Power   (4) Jammer Effective Radiated Power

The transmitted output power for a constant gain jammer is directly proportional to the power of the intercepted radar signal.

The support jamming equation for a constant gain jammer relates jammer, radar, and geometry/environment characteristics together to allow us to determine the jammer power received by the radar system. As shown in Figure 11-5, the constant gain support jammer ERP is computed using incremental steps:

Step 1: incident radar power density

Step 2: received radar power

Step 3: amplified radar power

Step 4: jammer ERP

The constant gain support jammer ERP is given in Equation (11-8). For clarity, in this and subsequent equations the losses are initially excluded. Total jammer-related losses are included at the conclusion of the support jamming equation development.

A constant gain jammer will linearly amplify the received radar signal until the jammer transmitter reaches its maximum (saturated) power. Thus, the power from a constant gain jammer is limited to less than or equal to the saturation power of its transmitter (Equation 11-9). When the amplifier is driven into saturation, the constant gain jammer becomes a constant power jammer. The constant gain support jammer ERP propagates to the radar resulting in an incident power density (Equation 11-10). The incident constant jammer power density is collected by the radar receive antenna (Equation 11-11). The received constant gain support jammer power including jammer-related losses is given in Equation (11-12).

$$\text{ERP}_{\text{J}} = \frac{P_{\text{R}}\, G_{\text{RJ}}\, G_{\text{JRr}}\, \lambda^2\, G_{\text{s}}}{(4\pi)^2\, R_{\text{RJ}}{}^2}\,(G_{\text{JRt}}) \qquad (11\text{-}8)$$

$$\underbrace{\frac{P_{\text{R}}\, G_{\text{RJ}}\, G_{\text{JRr}}\, \lambda^2\, G_{\text{s}}}{(4\pi)^2\, R_{\text{RJ}}{}^2}\,(G_{\text{JRt}})}_{\leq\, P_{\text{Js}}} \qquad (11\text{-}9)$$

$$\underbrace{\frac{P_R \, G_{RJ} \, G_{JRr} \, \lambda^2 \, G_s}{(4\pi)^2 \, R_{RJ}{}^2}}_{\le \, P_{Js}} (G_{JRt}) \left[\frac{1}{(4\pi) \, R_{RJ}{}^2} \right] \qquad (11\text{-}10)$$

$$\underbrace{\frac{P_R \, G_{RJ} \, G_{JRr} \, \lambda^2 \, G_s}{(4\pi)^2 \, R_{RJ}{}^2}}_{\le \, P_{Js}} \left[\frac{G_{JRt}}{(4\pi) \, R_{RJ}{}^2} \right] \left[\frac{G_{RJ} \, \lambda^2}{(4\pi)} \right] \qquad (11\text{-}11)$$

$$J_{cg} = \underbrace{\frac{P_R \, G_{RJ} \, G_{JRr} \, \lambda^2 \, G_s}{(4\pi)^2 \, R_{RJ}{}^2}}_{\le \, P_{Js}} \left[\frac{G_{JRt} \, G_{RJ} \, \lambda^2}{(4\pi) \, R_{RJ}{}^2 \, L_J} \right] \qquad (11\text{-}12)$$

where:

ERP_J = Jammer effective radiated power, watts

P_R = Radar transmitter peak power, watts

G_{RJ} = Radar antenna gain in the direction of the jammer (where ever the jammer is in the radar antenna pattern: mainbeam, sidelobes, backlobe), no units

G_{JRr} = Jammer receive antenna gain in the direction of the radar, no units

λ = Wavelength, meters

R_{RJ} = Radar-to-jammer slant range, meters

G_s = Jammer system gain, no units

G_{JRt} = Jammer transmit antenna gain in the direction of the radar, no units

P_{Js} = Jammer saturation peak power, watts

J_{cg} = Received constant gain jammer power, watts

L_J = Total jammer-related losses, no units

Normally the constant gain (linear) region of a constant gain jammer occurs at large distances from the radar. This is because at large distances the incident radar power density is small and thus less likely to drive the jammer amplifier into saturation. A constant gain jammer transitions to a constant power (saturated) jammer as the jammer approaches the radar, as shown in Figure 11-6.

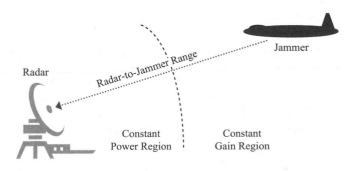

FIGURE 11-6 ■
Constant Power and Constant Gain Regions

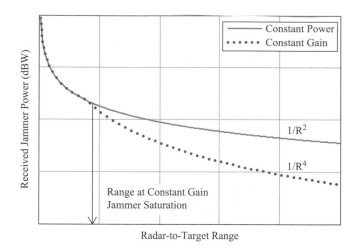

The received support jammer power for constant power and constant gain jammers as a function of radar-to-jammer range is shown in Figure 11-7. As seen in this figure, the received support jammer power for a constant gain jammer varies with range just like a target $(1/R^4)$. When the amplifier is driven into saturation, the constant gain jammer becomes a constant power jammer.

11.3 | SUPPORT NOISE JAMMING

For a noise support jammer, the jammer transmits a noise (random amplitude and phase with a uniform power spectral density) waveform (see Chapter 8) at the radar. The received jammer noise combines with the radar receiver thermal noise (see Chapter 2) to create an interference signal. Noise jamming is a denial/degrade form of electronic attack (EA). Noise jamming can deny detection of the target or can degrade (reduce) the detection range of the radar. Noise jamming can also deny target measurements or can degrade the radar's ability to measure target states. Noise jamming can also introduce noise power of sufficient magnitude into the radar receiver so the noise jamming alone exceeds the detection threshold, resulting in numerous false alarms that degrade the efficient processing of target detections.

11.3.1 Support Noise Jamming in the Radar Receiver

The jammer noise waveform is usually not perfectly matched to the radar waveform (and associated matched filter receiver; see Chapters 2 and 3), as shown in Figure 11-8. The mismatch can be in carrier frequency and/or bandwidth. Since the radar receiver accepts signals only within its bandwidth, only the

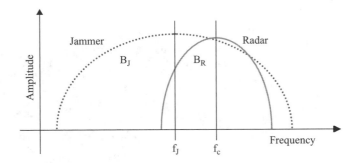

FIGURE 11-8 ■
Jammer Waveform
Mismatch to the
Radar Waveform
and Receiver
Bandwidth

portion of the received jammer noise power within the radar receiver bandwidth passes through the radar receiver.

To account for the mismatch between the radar receiver and jammer noise bandwidths, the received support jammer power is reduced by the ratio of the radar receiver bandwidth to the jammer noise bandwidth (Equation 11-13). This equation assumes the jammer noise bandwidth overlaps the radar receiver bandwidth; it is greater than the radar receiver bandwidth and the carrier frequency of the jammer noise is approximately equal to the radar carrier frequency. This assumption also ensures the jammer noise is wideband with respect to the receiver thermal noise, the receiver bandwidth, and target signal. Inside the radar receiver there is a complex (amplitude and phase) combination of the support jammer noise and radar receiver thermal noise. The result of this combination is called the interference signal (Equation 11-14). The interference signal equation uses the concept of additive means from probability theory, here its mean radar receiver thermal noise plus mean jammer noise.

$$J_N = \frac{P_J \, G_{JR} \, G_{RJ} \, \lambda^2}{(4\pi)^2 \, R_{RJ}^2 \, L_J} \left(\frac{B_R}{B_J}\right) \quad \text{Note: assumes } B_J \geq B_R \text{ and } f_J \approx f_c \quad (11\text{-}13)$$

$$I = N + J_N \qquad (11\text{-}14)$$

where:

J_N = Jammer noise power out of the radar receiver, watts
B_J = Jammer noise bandwidth, hertz
B_R = Radar receiver processing bandwidth, hertz
f_J = Jammer transmitted frequency, hertz
f_c = Radar carrier or transmitted frequency, hertz
I = Interference signal power, watts
N = Radar receiver thermal noise power, watts

FIGURE 11-9 ■
Target Signal (S),
Receiver Thermal
Noise (N), Support
Jammer Noise (J_N),
and Interference (I)
versus Radar-to-
Jammer and Radar-
to-Target Ranges

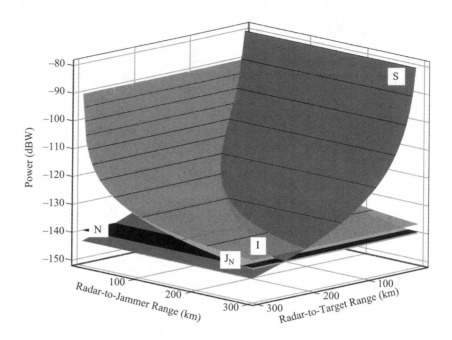

The single-pulse/sample received target signal power, radar receiver thermal noise, support jammer noise, and interference signal power as a function of radar-to-target range and radar-to-jammer range are shown in Figure 11-9. For the support jammer to be effective, its noise must be significantly greater than the radar receiver thermal noise ($J_N \gg N$), at the vast majority of ranges. Thus, the interference signal power equals, or is slight more than, the support jammer noise for the vast majority of ranges. This is clearly the case for the example shown in this figure.

The single-pulse signal-to-interference ratio (S/I) is a measure of the strength of the received target signal power relative to the interference signal. The S/I is similar to the target signal-to-noise ratio (S/N) discussed in Chapters 2 and 3, except the radar receiver thermal noise is combined with the support jammer noise in the denominator. The S/I is the ratio of the power from a single received target pulse to the power in one sample of the interference noise signal (Equation 11-15). Including the equations for the target signal power (see Chapter 2), radar receiver thermal noise (see Chapter 2), and support jammer noise results in Equation (11-16). As discussed in Chapter 3, radar systems improve their detection performance by integrating multiple pulses. The improvement in S/I provided by integration is quantified by the integration gain (see Chapter 3). The S/I after integration of multiple pulses is given in Equation (11-17).

$$\frac{S}{I} = \frac{S}{N + J_N} \tag{11-15}$$

$$\frac{S}{I} = \frac{P_R \, G_{RT}^2 \, \lambda^2 \, \sigma \, G_{sp}}{(4\pi)^3 \, R_{RT}^4 \, L_R \left[F_R \, k \, T_0 \, B_R + \dfrac{P_J \, G_{JR} \, G_{RJ} \, \lambda^2}{(4\pi)^2 \, R_{RJ}^2 \, L_J} \left(\dfrac{B_R}{B_J} \right) \right]} \tag{11-16}$$

$$\left(\frac{S}{I} \right)_n = \frac{P_R \, G_{RT}^2 \, \lambda^2 \, \sigma \, G_{sp} \, G_I}{(4\pi)^3 \, R_{RT}^4 \, L_R \left[F_R \, k \, T_0 \, B_R + \dfrac{P_J \, G_{JR} \, G_{RJ} \, \lambda^2}{(4\pi)^2 \, R_{RJ}^2 \, L_J} \left(\dfrac{B_R}{B_J} \right) \right]} \tag{11-17}$$

where:

S/I = Target signal-to-interference ratio, no units

P_R = Radar peak transmit power, watts

G_{RT} = Radar antenna gain in the direction of the target, no units

σ = Target radar cross section, square meters (m^2)

G_{sp} = Radar signal processing gain, no units

R_{RT} = Radar-to-target slant range, meters

L_R = Radar-related losses, no units

F_R = Radar receiver noise figure (≥ 1), no units

k = Boltzmann's constant, 1.38×10^{-23} watt-seconds/Kelvin

T_0 = IEEE standard reference temperature for radio frequency receivers, 290 Kelvin

$(S/I)_n$ = Target signal-to-interference ratio after integration of multiple pulses, no units

G_I = Integration gain, no units

Figure 11-10 shows the multiple-pulse target signal-to-noise ratio $(S/N)_n$ (see Chapter 3), detection threshold (see Chapter 3), and $(S/I)_n$ as a function of radar-to-target range and radar-to-jammer range. The EA designer often strives for $(S/I)_n < 0$ dB, while the radar engineer would note that a radar is essentially ineffective when the $(S/I)_n$ is less than the detection threshold.

11.3.2 Support Noise Jamming Metrics

We have developed the $(S/I)_n$ and compared it with the $(S/N)_n$ and detection threshold, but the effectiveness of the support noise jammer is now addressed in terms of two support jamming metrics: burnthrough range and J/S. Target detection achieved in the presence of a noise jammer is called "burnthrough," and the radar-to-target range at which it occurs is called the "burnthrough range." Thus, burnthrough range is the detection range of the radar in the presence of the support noise jammer and is the primary support noise jamming

FIGURE 11-10 ■
Multiple Pulse
Signal-to-Noise
Ratio (S/N)n,
Detection Threshold
(SNRdt), and Signal-
to-Interference Ratio
(S/I)n versus Radar-
to-Jammer and
Radar-to-Target
Ranges

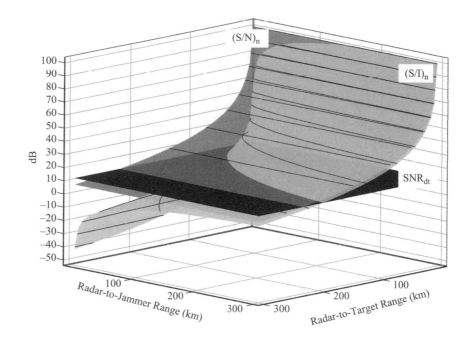

FIGURE 11-10 ■ Multiple Pulse Signal-to-Noise Ratio $(S/N)_n$, Detection Threshold (SNR_{dt}), and Signal-to-Interference Ratio $(S/I)_n$ versus Radar-to-Jammer and Radar-to-Target Ranges

metric. A secondary support noise jamming metric is J/S, the ratio of the jamming noise power to the target signal power. J/S tells us how strong or weak the jamming is relative to the received target signal.

Burnthrough Range

The $(S/I)_n$ equation can be solved for the range at which the radar will detect the target signal in the presence of the noise jammer—the burnthrough range. Solving the $(S/I)_n$ for the burnthrough range, the range at which the target signal exceeds the interference noise by the detection threshold, gives us Equation (11-18). Figure 11-11 shows the burnthrough range and detection range overlaid on plots of the $(S/N)_n$ and $(S/I)_n$.

$$R_{bt} = \sqrt[4]{\frac{P_R \, G_{RT}^2 \, \lambda^2 \, \sigma \, G_{sp} \, G_I}{(4\pi)^3 \, SNR_{dt} \, L_R \left[F_R \, k \, T_0 \, B_R + \frac{P_J \, G_{JR} \, G_{RJ} \, \lambda^2}{(4\pi)^2 \, R_{RJ}^2 \, L_J} \left(\frac{B_R}{B_J} \right) \right]}} \qquad (11\text{-}18)$$

where:

R_{bt} = Radar burnthrough range, meters
SNR_{dt} = Radar detection threshold, no units

The burnthrough range is a function of many target, jammer, and radar parameters. As the reader who does the problems at the end of the chapter will

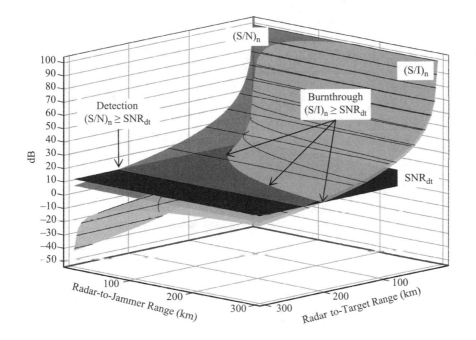

FIGURE 11-11 ▪
Burnthrough Range
and Detection Range

find, only a small jammer ERP (relative to the radar ERP) is sufficient to reduce the detection range of a long-range radar system, particularly if the radar cross section of the target is low and/or the radar antenna gain in the direction of the jammer is high. This is because the jammer signal has a one-way propagation, $1/R^2$, while the target signal has a two-way propagation, $1/R^4$, even though the two R's can be significantly different.

We will look at the sensitivity of the burnthrough range to several main target, jammer, and radar parameters or relationships. The sensitivity of the burnthrough range to target RCS is shown in Figure 11-12. In this figure the burnthrough range is normalized to the value at an RCS of 0 dBsm (1 m^2). The sensitivity of burnthrough range to the jammer-radar bandwidth mismatch (B_J/B_R) is shown in Figure 11-13. In this figure the burnthrough range is normalized to the value at $B_J/B_R = 1$. The sensitivity of burnthrough range to jammer transmit antenna gain in the direction of the radar is shown in Figure 11-14. In this figure the burnthrough range is normalized to the value at the jammer antenna mainbeam gain; the increase in burnthrough range is depicted as a function of reduced jammer antenna gain. The sensitivity of burnthrough range to the difference between the radar antenna gain in the direction of the target (G_{RT}) and the direction of the jammer (G_{RJ}) is shown in Figure 11-15. In this figure the burnthrough range is normalized to the value when the radar antenna gain in the direction of the target and the jammer being equal $(G_{RT} = G_{RJ})$; the increase in burnthrough range is depicted as a function of reduced radar antenna

FIGURE 11-12 ■
Sensitivity of
Burnthrough Range
to Target Radar
Cross Section

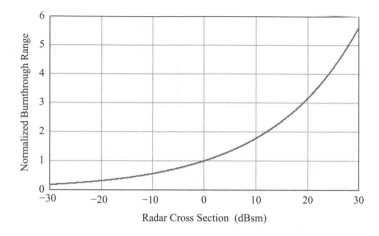

FIGURE 11-13 ■
Sensitivity of
Burnthrough Range
to Jammer-Radar
Bandwidth
Mismatch (assumes
$J_N \gg N$)

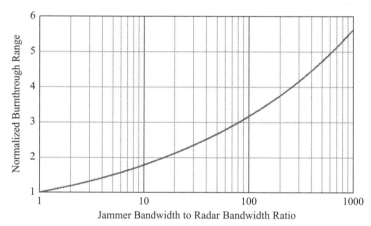

FIGURE 11-14 ■
Sensitivity of
Burnthrough Range
to Jammer Antenna
Gain in the Direction
of the Radar
(assumes $J_N \gg N$)

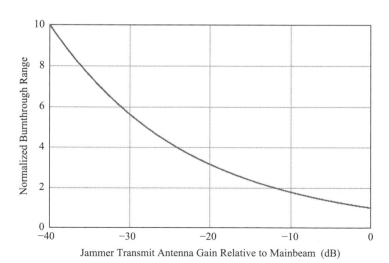

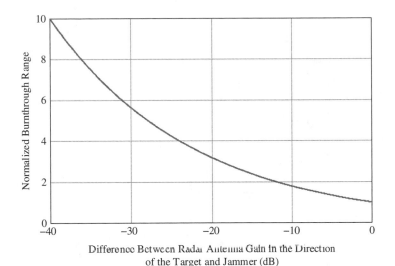

FIGURE 11-15 ■ Sensitivity of Burnthrough Range to the Difference between the Radar Antenna Gain in the Direction of the Target and the Direction of the Jammer (assumes $J_N \gg N$)

Difference Between Radar Antenna Gain in the Direction of the Target and Jammer (dB)

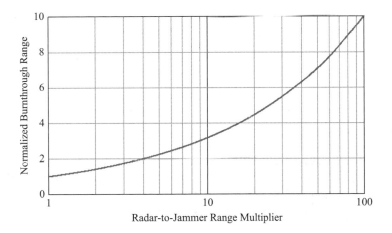

FIGURE 11-16 ■ Sensitivity of Burnthrough Range to Radar-to-Jammer Range Multiplier (assumes $J_N \gg N$)

Radar-to-Jammer Range Multiplier

gain in the direction of the jammer compared with the target. The sensitivity of burnthrough range to the radar-to-jammer range (R_{RJ}) is shown in Figure 11-16. In this figure the burnthrough range is normalized to the value at a radar-to-jammer range multiplier of one. If the radar-to-jammer range is doubled, the range multiplier would have a value of two. The sensitivities in Figure 11-13 through Figure 11-16 assume that the support noise jamming is much greater than the radar receiver thermal noise ($J_N \gg N$).

Jamming-to-Signal Ratio

J/S, the ratio of the jamming noise to the target signal power, is a long-standing EW metric. It quantifies the power relationship between the jamming noise and

FIGURE 11-17 ■
Jamming-to-Signal
Ratio as a Function
of Radar-to-Jammer
and Radar-to-Target
Ranges

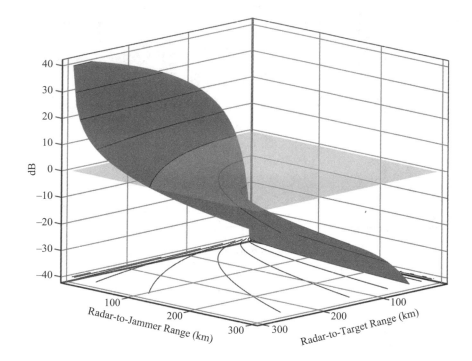

the target signal (see Chapter 2) after the multiple pulses are integrated (see Chapter 3). The multiple-pulse J/S, $(J_N/S)_n$, is given in Equations (11-19), algebraic, and (11-20), logarithmic. A J/S greater than one, or positive in dB, tells us the noise jammer power is higher than the target signal power. A J/S less than one, or negative in dB, tells us the noise jammer power is lower than the target signal power. Figure 11-17 shows the J/S (dB) as a function of radar-to-target range and radar-to-jammer range. In this figure the J/S is often negative, a very common situation with noise jammers.

$$\left(\frac{J_N}{S}\right)_n = \frac{P_J\,G_{JR}\,G_{RJ}}{R_{RJ}^2\,L_J}\left(\frac{B_R}{B_J}\right)\frac{(4\pi)\,R_{RT}^4\,L_R}{P_R\,G_{RT}^2\,\sigma\,G_{sp}\,G_I} \tag{11-19}$$

$$\begin{aligned}
10\log\left[\left(\frac{J_N}{S}\right)_n\right] = {} & 10\log(P_J) + 10\log(G_{JR}) + 10\log(G_{RJ}) \\
& + 10\log(B_R) + 10\log(4\pi) + 40\log(R_{RT}) \\
& + 10\log(L_R) - 20\log(R_{RJ}) - 10\log(L_J) \\
& - 10\log(B_J) - 10\log(P_R) - 20\log(G_{RT}) \\
& - 10\log(\sigma) - 10\log(G_{sp}) - 10\log(G_I)
\end{aligned} \tag{11-20}$$

where:

$(J_N/S)_n$ = Noise jamming-to-target signal ratio after integration of multiple pulses, no units

Summary

Burnthrough range is the principal noise jamming metric. It shows the reduced detection range as a result of the noise jammer. J/S is a secondary noise jammer metric. Many people assume (and you know what is said to happen when we assume) the jamming must be greater than the target signal for the jammer to be effective. As shown in Figure 11-18, we do not necessarily need a positive (in dB) J/S (i.e., the jamming is greater than the target signal) for a noise jammer to be effective in reducing the detection range! This is because the effect of noise jamming is relative to the receiver thermal noise, not the target signal power. Thus, J/S can be a misleading metric for noise jammers. To avoid making an "assumer" out of you or others use burnthrough range as the metric for noise jammers.

Burnthrough range and J/S are both dynamic quantities, dependent on the instantaneous conditions of the radar–target–jammer engagement. Specifically, there is a dynamic radar–target and radar–jammer geometry due to the target and/ or jammer movement and/or maneuvers. Radar mode, target RCS, radar antenna gain, jammer antenna gain, propagation paths, and polarization mismatch can all change over the course of an engagement. Moreover, there are dynamic radar–jammer interactions, such as manual, semiautomatic, and automatic jammer and radar operations taking place over the course of the engagement. Hence, the performance metrics discussed provide us with a "snapshot" of these dynamics.

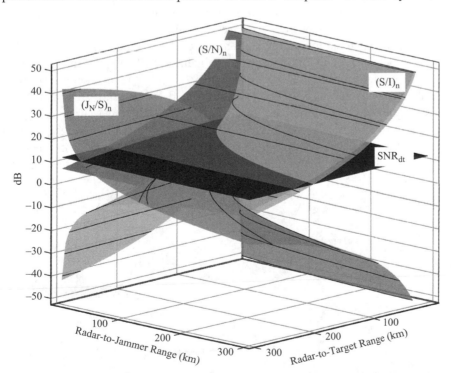

FIGURE 11-18 ■ Burnthrough and Detection Ranges with the Jamming-to-Signal Ratio

Thorough radar or jammer performance predictions account for all of these various dynamics in the form of sophisticated computer models.

11.4 | SUPPORT FALSE TARGET JAMMING

For a support false target jammer, the jammer transmits a false target waveform (see Chapter 8) similar to the actual radar waveform at the radar. The false target jamming waveform is received and processed just like a received target signal. False target jamming is a degrade/deceive type of EA. It degrades the radar's ability to effectively process true targets by introducing multiple false targets with the true targets and deceives the radar to make measurements and tracks of false targets. The degrade/deceive effect of false target jamming is shown in Figure 11-19.

11.4.1 Support False Target Jamming in the Radar Receiver

The support false target jamming signal arrives at the radar's receiver. Sometimes it is more "attractive" (e.g., stronger, nonfluctuating) to the radar than are the true target signals. However, it is very difficult, if not impossible, for the false target jamming signal to perfectly match a true target signal. Most often, this false target jamming mismatch occurs for radar systems employing pulse compression modulation (frequency or phase) waveforms (see Chapter 5). The false target jamming signal mismatch is often reflected in the radar signal processing gain (see Chapter 2) and/or the integration gain (see Chapter 3). In most cases, detailed hardware testing and evaluation (T&E) and/or high-fidelity modeling, simulation, and analysis (MS&A) is required to determine the radar signal processing gain and integration gain for a mismatched false target jamming signal.

In the radar receiver, a perfectly matched false target jamming signal appears the same as a true target signal. Thus, we will describe it in the same way as we do with a true target signal (see Chapter 2). However, instead of a

FIGURE 11-19 ■
Support False Target
Jamming

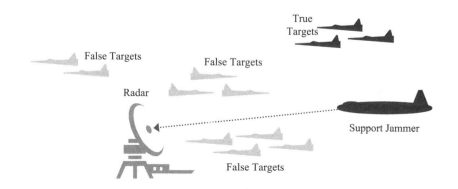

target signal-to-noise ratio we have a jamming-to-noise ratio (J/N), which is the ratio of the power from a single received false target jamming pulse to the power of one sample of radar receiver thermal noise (see Chapter 2) (Equation 11-21). This equation includes the effect of radar signal processing (see Chapter 2) on the J/N. Also as discussed in Chapter 3, radar systems improve their detection performance by integrating multiple pulses. The improvement in J/N provided by integration is quantified by the integration gain (see Chapter 3). The J/N after integration of multiple pulses is given in Equation (11-22).

$$\frac{J}{N} = \frac{P_J \, G_{JR} \, G_{RJ} \, \lambda^2 \, G_{sp}}{(4\pi)^2 \, R_{RJ}{}^2 \, F_R \, k \, T_0 \, B_R \, L_J} \qquad (11\text{-}21)$$

$$\left(\frac{J}{N}\right)_n = \frac{P_J \, G_{JR} \, G_{RJ} \, \lambda^2 \, G_{sp} \, G_I}{(4\pi)^2 \, R_{RT}{}^2 \, F_R \, k \, T_0 \, B_R \, L_J} \qquad (11\text{-}22)$$

where:

J/N = Single-pulse false target jamming-to-noise ratio, no units
G_{sp} = Radar signal processing gain, no units
$(J/N)_n$ — False target jamming-to-noise ratio after integration of multiple pulses, no units
G_I = Integration gain, no units

Figure 11-20 shows the multiple-pulse true target signal-to-noise ratio $(S/N)_n$ (see Chapter 3), the radar detection threshold (see Chapter 3), and the $(J/N)_n$ as a function of radar-to-jammer range and radar-to-target range. Note the change in $(J/N)_n$ is $1/R^2$, whereas the change in $(S/N)_n$ is $1/R^4$. The $(J/N)_n$ decreases by 6 dB (1/4) when the radar-to-jammer range doubles and increases by 6 dB (4 times) when the radar-to-jammer range halves. The $(S/N)_n$ decreases by 12 dB (1/16) when the radar-to-target range doubles and increases by 12 dB (16 times) when the radar-to-target range halves.

11.4.2 Support False Target Jamming Metrics

We have developed the $(J/N)_n$ and compared it with the $(S/N)_n$ and detection threshold, but the effectiveness of the support false target jammer is now addressed in terms of support jamming metrics: detection range and J/S. Detection range is the radar-to-jammer range at which the radar will detect the false target jamming signal in the presence of the radar receiver thermal noise; how far away is the support jammer when the false targets are detected. Detection range is the primary support false target jamming metric. J/S is the ratio of the support false target jamming signal power to the target signal power. It quantifies the power relationship between the false target jamming and target signals, and is the secondary support false target jamming metric.

FIGURE 11-20 ■
Multiple Pulse
Signal-to-Noise
Ratio (S/N)ₙ,
Detection Threshold,
and Jamming-to-
Noise Ratio (J/N)ₙ
versus Radar-to-
Jammer and Radar-
to-Target Ranges

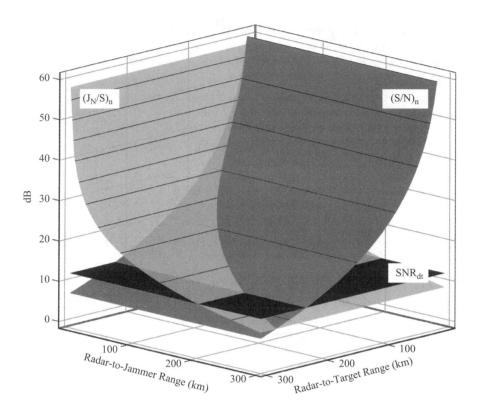

Detection Range

The radar-to-jammer range at which the radar will detect the false target jamming signal is the primary support false target jamming metric. The $(J/N)_n$ equation is solved for the radar-to-jammer range at which the false target jamming signal exceeds the radar receiver thermal noise by the detection threshold. The false target detection range is given in Equations (11-23), algebraic, and (11-24), logarithmic. The logarithmic form of this equation can be simplified by using the carrier frequency instead of wavelength and combining the constants (Equation 11-25).

$$R_{dt} = \sqrt{\frac{P_J \, G_{JR} \, G_{RJ} \, \lambda^2 \, G_{sp} \, G_I}{(4\pi)^2 \, SNR_{dt} \, F_R \, k \, T_0 \, B_R \, L_J}} \qquad (11\text{-}23)$$

$$
\begin{aligned}
20 \log (R_{dt}) = {} & 10 \log (P_J) + 10 \log (G_{JR}) + 10 \log (G_{RJ}) \\
& + 20 \log (\lambda) + 10 \log (G_{sp}) + 10 \log (G_I) \\
& - 20 \log (4\pi) - 10 \log (SNR_{dt}) \\
& - 10 \log (F_R) - 10 \log (k) - 10 \log (T_0) \\
& - 10 \log (B_R) - 10 \log (L_J)
\end{aligned}
\qquad (11\text{-}24)
$$

$$20 \log (R_{dt}) = 10 \log (P_J) + 10 \log (G_{JR}) + 10 \log (G_{RJ})$$
$$- 20 \log (f_c) + 10 \log (G_{sp}) + 10 \log (G_I)$$
$$- 10 \log (SNR_{dt}) - 10 \log (F_R) - 10 \log (B_R) \quad (11\text{-}25)$$
$$- 10 \log (L_J) + 351.5355$$

where:

R_{dt} = Radar detection range, meters
SNR_{dt} = Radar detection threshold, no units

Figure 11-21 shows an example comparison of the false target detection range to the true target detection range. As can be seen in this example, there is generally ample opportunity for the false target jamming signal to be detected by the radar system.

Jamming-to-Signal Ratio

J/S, the ratio of the false target jamming signal power and the target signal power (see Chapter 2) after integration of multiple pulses (see Chapter 3), is a long-standing EW metric. Note that the fundamental concept of J/S assumes that a target signal is in the same resolution cell (see Chapter 5) as the false target jamming signal, which may not be the case for a support jammer! Thus, it

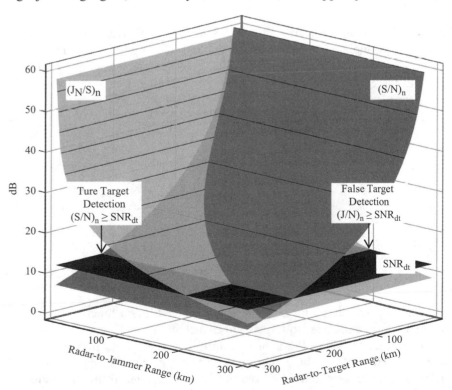

FIGURE 11-21 ■ True Target and False Target Jamming Detection Ranges

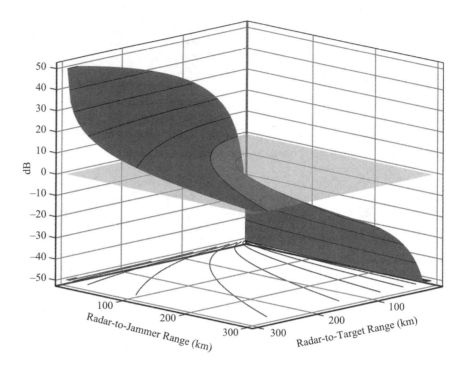

can be challenging to accurately use J/S for a support jammer. The $(J/S)_n$ is given in Equations (11-26), algebraic, and (11-27), logarithmic. Figure 11-22 shows an example of the $(J/S)_n$ as a function of the radar-to-jammer range and radar-to-target range.

$$\left(\frac{J}{S}\right)_n = \frac{P_J \, G_{JR} \, G_{RJ}}{R_{RJ}^2 \, L_J} \frac{(4\pi) \, R_{RT}^4 \, L_R}{P_R \, G_{RT}^2 \, \sigma} \tag{11-26}$$

$$\begin{aligned}
10 \log\left[\left(\frac{J}{S}\right)_n\right] = \; & 10 \log\left(P_J\right) + 10 \log\left(G_{JR}\right) + 10 \log\left(G_{RJ}\right) \\
& + 10 \log\left(4\pi\right) + 40 \log\left(R_{RT}\right) + 10 \log\left(L_R\right) \\
& - 20 \log\left(R_{RJ}\right) - 10 \log\left(L_J\right) - 10 \log\left(P_R\right) \\
& - 20 \log\left(G_{RT}\right) - 10 \log\left(\sigma\right)
\end{aligned} \tag{11-27}$$

where:

$(J/S)_n$ = False target jamming-to-target signal ratio after integration of multiple pulses, no units

Figure 11-23 shows the constant gain jammer J/S as a function of radar-to-jammer range and radar-to-target range, including where a radar-to-jammer range is reached where the jammer transmitter reaches its maximum output power (saturation). At this range, the constant gain jammer can put out only its maximum power and is a constant power jammer from this range on.

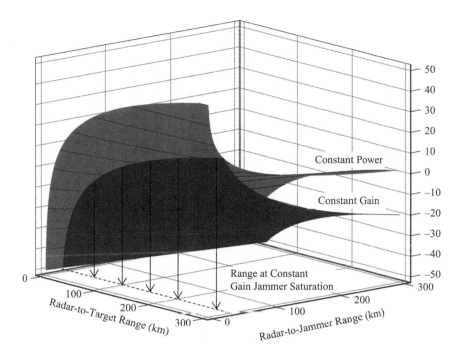

FIGURE 11-23 ■
Comparison of the
Jamming-to-Signal
Ratio from Constant
Power and Constant
Gain Jammers

The constant gain J/S then "varies" the same as the J/S of a constant power jammer, as shown in the figure.

J/S is often used as an indicator of false target jamming effectiveness. The false target jamming is "assumed effective" if the J/S exceeds a minimum value. The minimum J/S required for effectiveness is a complicated function of the particular EA technique, the part of the radar being attacked, and, if present, the operator skill level. The minimum J/S required for effectiveness is determined based on detailed T&E and/or MS&A. If there is sufficient J/S to provide the necessary jamming "effectiveness," increasing the J/S will not necessarily increase the effectiveness and could in fact be harmful if it triggers the radar's EP capabilities or alerts the operator to the presence of jamming (see Chapters 8 and 12).

J/S is often misunderstood and misused. It is easy to calculate but difficult to measure! Because the individual equations are relatively straightforward, it is easy to show charts with the individual target, jammer, clutter, and thermal noise signals. However, the radar receiver contains a complex (amplitude and phase) combination of target, jammer, clutter, and thermal noise signals. It is very hard, if not impossible, to accurately and consistently separate the individual signals from the receiver output in the real world, such as we can with equations and plots. J/S is very difficult to measure in the controlled environment of a laboratory and almost impossible to measure in the dynamic environment of an open-air range. Care must be taken, therefore, in identifying and understanding assumptions made in expressions of the J/S metric.

Summary

Detection range is the principal support false target jamming metric. Detection range shows how far away the support jammer is when the false target jammer signal is detected by the radar system. J/S is a secondary support false target jamming metric. Often, a specific J/S is needed for the false target jammer to be effective. Detection range and J/S are both dynamic quantities dependent on the instantaneous conditions of the radar–target–jammer engagement. There is a dynamic radar–target and radar–jammer geometry due to the radar or targets or jammer movement or maneuvers. Radar mode, target RCS, radar antenna gain, jammer antenna gain, propagation paths, and polarization mismatch can all change over the course of an engagement. Moreover, there are dynamic radar–jammer interactions, such as manual, semiautomatic, and automatic jammer and radar operations taking place over the course of the engagement. Hence, the performance metrics discussed here provide us with a "snapshot" of these dynamics during the course of the overall engagement. Thorough radar and/or jammer performance predictions take into account all of these various dynamics using sophisticated computer models.

■ 11.5 | SUPPORT EXPENDABLES

A growing trend in support jamming is the use of expendable platforms. Expendable support jammer platforms can be used in missions considered too dangerous for manned platforms, for example as stand-in (see Chapter 8) jamming within the lethal range of threat weapon systems. Expendable support jammer concepts include active decoys, noise jammers, and false target jammers. Numerous expendable active decoys would appear to be actual targets to the radar system, degrading the radar system's ability to efficiently process true targets. Expendable noise and/or false target stand-in jammers take advantage of the short radar-to-jammer range to provide high jammer power density to the radar system. As previously discussed, with a short radar-to-jammer range it is easier to provide support jamming though much of the radar antenna sidelobes, providing noise or false targets over wide angular regions about the radar system.

■ 11.6 | SUMMARY

This chapter discussed support jamming and developed the fundamental support jamming equation. We examined noise (wideband random amplitude and phase) support jamming and false target (radar-like waveform) support jamming, their effects on the radar, and the metrics used to describe their performance. We concluded by looking at expendable support jamming platforms.

Of course, the radar community does not sit idly by while the EW community renders its radar system useless. The radar community has developed numerous electronic protection (EP) techniques that attempt to preserve the radar system's target detection, measurement, and tracking capability and performance in the presence of EA. EP techniques are discussed in detail in Chapter 12.

11.7 | EXERCISES

11-1. A support jammer has the following characteristics: peak transmit power $P_J = 25$ watts; transmit antenna gain in the direction of the radar $G_{JR} = 3$ dBi; total jammer-related losses $L_J = 15$ dB; and jammer transmit loss $L_{Jt} = 2$ dB. The jammer is 100 km, R_{RJ}, from the radar system from Exercises 2-5 and 3-15. The radar antenna gain in the direction of the jammer G_{RJ} is 20 dB less than the radar antenna gain in the direction of the target G_{RT}. Compute the following: (a) the transmitted jammer effective radiated power ERP_J (watts and dBW), (b) the received jammer power density at the radar receive antenna, and (c) the received jammer power, J (watts and dBW). Compare the received jammer power to the jammer transmitted effective radiated power. (Hint: use the results from Exercises 2-5 and 3-15.)

11-2. The support jammer from Exercise 11-1 uses a noise waveform with a bandwidth $B_J = 200$ MHz centered about the radar carrier frequency. Compute the following: (a) the jammer noise power in the radar receiver J_N (watts and dBW), (b) the interference signal power I (watts and dBW), and (c) the target signal-to-interference ratio after integration of multiple pulses $(S/I)_n$ (no units and dB). Compare the target signal-to-interference ratio with the target signal-to-noise ratio after integration of multiple pulses from Exercise 3-15. (Hint: use the results of Exercises 2-5 and 3-15.)

11-3. Compute the burnthrough range R_{bt} (meters) and jamming-to-signal ratio after integration of multiple pulses $(J_N/S)_n$ (no units and dB) for the support noise jammer from Exercise 11-2. Compare the burnthrough range to the detection range from Exercise 3-15. (Hint: use the results from Exercises 2-5 and 3-15.)

11-4. The support jammer from Exercise 11-1 uses a false target waveform that perfectly matches the radar waveform. Compute the false target jamming-to-noise ratio after integration of multiple pulses $(J/N)_n$ (no units and dB). Compare the false target signal-to-noise ratio with the target signal-to-noise ratio after integration of multiple pulses from Exercise 3-15. (Hint: use the results from Exercises 2-5 and 3-15.)

11-5. Compute the false target jamming-to-signal ratio after integration of multiple pulses $(J/S)_n$ (no units and dB) and the false target radar detection range R_{dt} (meters) for the support false target jammer from Exercise 11-4. Compare the false target radar detection range to the radar detection range for the true target in Exercise 3-15. (Hint: use the results from Exercises 2-5 and 3-15.)

11-6. In a tactical operation, a support noise jammer is to be used to screen aircraft $\sigma = 1$ m^2 radar cross section attacking a radar system. The attacking aircraft can launch their missiles at $R_{RT} = 10$ km range. The support jammer aircraft must stand off at range $R_{RJ} = 100$ km. The radar system to be attacked is known to have the following characteristics: antenna mainbeam gain in the direction of the target $G_{RT} = 30$ dB gain; peak transmit power $P_R = 1$ MW; integrates 10 1 μsec coherent pulses; signal processing gain $G_{sp} = 1$; and a matched filter receiver. To be conservative, a 0 dB S/J is required for the jammer to be effective. What effective radiated power ERP_J (watts) does the support jammer need if the jammer noise bandwidth $B_J = 20$ MHz? Assume that the radar-related and jammer-related losses are the same and the radar antenna gain in the direction of the support jammer is one-half the mainbeam gain.

▮ 11.8 ▮ | REFERENCES

Chrzanowski, Edward, 1990, *Active Radar Electronic Countermeasures*, Norwood MA: Artech House.
   Contains a very detailed description of support jamming concepts, technology, and supporting math. Jamming against search radar systems, the emphasis of most support jamming concepts, is discussed in Chapters 3 and 4.

Neri, Filippo, 2006, *Introduction to Electronic Defense Systems*, 2nd Edition, Raleigh, NC: SciTech Publishing.
   Contains a very detailed description of support jamming concepts, technology, and supporting math in Chapter 5.

Schleher, D. C., 1999, *Electronic Warfare in the Information Age*, Norwood, MA: Artech House.
   Contains a very detailed description of support jamming concepts, technology, and supporting math, including numerous MATLAB files.

Van Brunt, L. B., 1978, *Applied ECM*, Volume I and II, Dunn Luring, VA: EW Engineering, Inc.
   Contains clear descriptions and details on myriad EA-ECM and EP-ECCM tactics and techniques.

Electronic Protection Concepts

HIGHLIGHTS

- Discuss the use of waveform diversity and low probability of intercept concepts
- Discuss antenna based signal processing: sidelobe canceler and sidelobe blanker
- Discuss sophisticated target trackers: correlated measurements and state trackers

Of course, the radar community does not sit by idly while the electronic warfare (EW) community renders its radar system useless. It has developed numerous electronic protection (EP) techniques that attempt to preserve the radar system's target detection, measurement, and tracking capability and performance in the presence of electronic attack (EA) (see the numerous references at the end of this chapter). In discussing what follows, remember that for every EA there is an EP and for every EP an EA, ad infinitum. Consequently, there are continuing efforts in peace and war to keep one step ahead of everyone else.

In previous chapters we described a few EP techniques including constant false alarm rate (CFAR) detection, leading-edge trackers to counter repeater jammers, using a variety of conical scan frequencies, scan on receive only for lobing (conical scan and sequential) angle trackers, monopulse angle tracking, and well-trained experienced operators. Many steps taken to improve radar system capability and/or performance also have EP benefits.

In this chapter we will describe three EP concepts in more detail: radar waveform diversity and low probability of intercept; antenna-based signal processing; and sophisticated target trackers.

■ 12.1 | WAVEFORM DIVERSITY AND LOW PROBABILITY OF INTERCEPT

Waveform diversity is the most effective EP concept because it makes it very difficult for an ES system to detect, identify, and/or keep current with the radar waveform [Chrzanowski, 1999, Section 3.5, Chapter 9; Neri, 2006, Section 6.5; Schleher, 1999, Section 2.5; Stimson, 1998, Chapter 42]. Radar waveform polarization (see Chapter 2) diversity can result in a polarization mismatch loss (see Table 9-1) that reduces the detection range of the ES system. The overwhelming majority of EA techniques require measuring at least some of the radar waveform characteristics to be effective. Thus, without good, current radar waveform characteristics the full effect of the EA technique on the radar system may not be obtained. It is important to note that many radar systems change waveform characteristics (e.g., ranging waveforms, Doppler waveforms, moving target indication [MTI] waveforms) as part of implementing different modes, not just as an EP concept.

The carrier frequencies and bandwidths of the jammer and the radar are of vital importance to EA (see Chapters 10 and 11). To extend its bandwidth, thereby forcing the jammer to do the same, the radar can be designed to hop around in frequency, a feature known as "frequency diversity" or "frequency agility." The hopping may be from pulse burst-to-pulse burst, pulse-to-pulse, periodically (fixed or random), or combinations of these. The radar always knows what frequency it is transmitting so it performs as usual. Unless the jammer is combined with an EW receiver (see Chapter 9) to follow the radar's frequency changes, it is forced to transmit a very wideband noise signal (see Chapters 8, 10, and 11). Also, as discussed in Chapter 5, a radar system using pulse compression will have a wide signal bandwidth. The jammer must cover the wider spectrum of the pulse compression radar signal, resulting in a lower jammer power spectral density (watts/Hz).

Because radar systems can be located in angle by their emissions and attacked as a result, there is a continuing need for radar systems that are not easily detectable by EW systems. By employing low probability of intercept (LPI) techniques, a radar system can greatly reduce the range at which it can be detected by an EW receiver. The main idea is to keep the radar off the air as much as possible, emitting only short pulse bursts, each with minimum power, at a different frequency, with a wideband modulation. The radar designer seeks four attributes: narrow beamwidths, ultra-low sidelobes, tailored power output, and a wideband noise-like modulated signal. The last of these has become practical only recently because of the computational capacity digital processing has brought us. Wideband spread spectrum signals with pseudorandom noise modulation are a reality. Heretofore, EW receivers (see Chapter 9) have kept up with the detection and cataloging of radar signals. LPI radar systems may change this situation, but not without difficulty. Consider the requirement of an LPI radar system: to detect targets at substantial ranges without being detected by an EW receiver. The EW receiver has only a one-way propagation path, while the radar has two-way (round-trip). Thus, LPI radar systems have a $1/R^2$ disadvantage to overcome.

12.2 | ANTENNA-BASED SIGNAL PROCESSING

Antenna-based signal processing techniques attempt to reject support jamming signals entering the radar system through its antenna sidelobes (see Chapter 11) [Farina, 1992]. A sidelobe canceler (SLC) attempts to cancel continuous wave (CW) noise jamming. The three main types are closed loop (analog), open loop (digital), and adaptive nulling (electronically steered arrays). A sidelobe blanker (SLB) attempts to blank false target jamming signals. It turns off the radar receiver output (range gate) when an unwanted signal appears in the radar antenna sidelobes.

12.2.1 Sidelobe Canceler

The purpose of an SLC is to suppress CW noise jamming signals entering the radar receiver through the sidelobes of the radar receive antenna pattern, as shown in Figure 12-1 [Farina, 1992, Chapter 4]. A closed-loop (analog processing) or open-loop (digital processing) SLC uses auxiliary antennas to "cover" the sidelobes of the main radar antenna. The auxiliary antenna patterns have the same or slightly higher gain than the sidelobes of the radar receive antenna.

The flow diagram of a radar system with such an SLC is shown in Figure 12-2. The noise jamming signals are received by both the main radar antenna channel and the auxiliary channel. The noise jamming signals in the auxiliary channel are amplified and shifted to be 180° out of phase from the

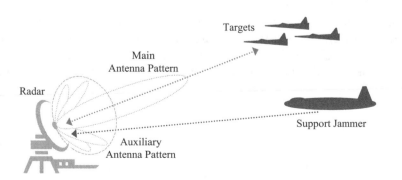

FIGURE 12-1 ■
Radar System with a
Sidelobe Canceler

FIGURE 12-2 ■
Flow Diagram of a
Radar System with a
Sidelobe Canceler

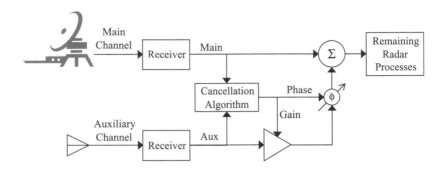

noise jamming signals in the main channel. The auxiliary channel and main channel noise jamming signals are then summed (complex sum: amplitude and phase). The effect of this complex sum is the cancellation of the noise jamming signals in the main channel. Because the target signal present in the auxiliary receiver is much smaller than the target signal in the radar receiver, the radar antenna gain in the direction of the target is much greater than the auxiliary antenna gain in the direction of the target, it can be ignored. Concepts of coherent sidelobe cancelers and adaptive EP were greatly enriched by the late Paul Howells [1976]. Farina [1992, Chapter 4] discusses many SLC concepts and their strengths and weaknesses.

The following is an example of a best-case SLC with one auxiliary antenna and receiver, in which the main and auxiliary receivers are identical and phase is preserved in all stages [Farina, 1992, page 103]. This will allow us to bound SLC performance. Farina [1992, Chapter 4] discusses numerous limitations on SLC performance in detail. The residual jammer and receiver thermal noise power (P_Z) after SLC is given in Equation (12-1). The SLC attempts to minimize P_Z. The ratio of the residual noise power to the receiver thermal noise power (P_Z/N) quantifies residual noise power relative to the receiver thermal noise power naturally in the radar receiver (Equation 12-2). When the $J_N/N \gg 1$ and the radar and auxiliary antenna gains are the same, we have P_Z constrained to twice the receiver thermal noise (Equation 12-3). Since the jammer noise is the same in both channels, it can be canceled. However, the receiver thermal noise in the auxiliary channel is independent of the thermal noise in the main channel. Thus, the receiver thermal noise in the auxiliary channel adds to the thermal noise in the main channel, doubling the thermal noise power of the radar system. The increase in thermal noise serves to provide an upper bound on the effectiveness of an SLC.

$$P_Z = (J_N + N) - \frac{\left(\dfrac{G_{AJ}}{G_{RJ}}\right) J_N{}^2}{\left(\dfrac{G_{AJ}}{G_{RJ}}\right) J_N + N} \qquad (12\text{-}1)$$

$$\frac{P_Z}{N} = \frac{\dfrac{J_N}{N}\left(\dfrac{G_{AJ}}{G_{RJ}} + 1\right) + 1}{\dfrac{J_N}{N}\left(\dfrac{G_{AJ}}{G_{RJ}}\right) + 1} \tag{12-2}$$

$$P_Z = 2N \qquad \text{when } J_N/N \gg 1 \text{ and } G_{AJ} = G_{RJ} \tag{12-3}$$

where:

P_Z = Residual noise, jamming and thermal, power after cancellation, watts
J_N = Jammer noise power in the radar receiver (see Chapter 11), watts
N = Radar main and auxiliary receiver thermal noise power (see Chapter 2), assumed to be the same, watts
G_{AJ} = Auxiliary antenna gain in the direction of the jammer, no units
G_{RJ} = Radar antenna gain in the direction of the jammer, no units

The P_Z/N as a function of J_N/N for various ratios of the auxiliary (G_{AJ}) and radar (G_{RJ}) antenna gains in the direction of the jammer is shown in Figure 12-3. If G_{AJ} is $10 \times G_{RJ}$, the ratio of the two antenna gains is 10 dB. As seen in this figure, the more antenna gain in the direction of the jammer the auxiliary antenna has relative to the radar antenna, the smaller the P_Z/N. However, auxiliary antenna gain cannot become too large in the direction of the jammer or the target signal present in the auxiliary receiver also becomes large relative to the receiver thermal noise. The SLC algorithm can potentially cancel a portion of the target signal. The fact a target signal present in the auxiliary

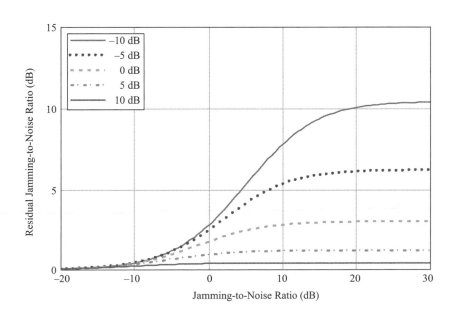

FIGURE 12-3 ■ Residual Noise Power-to-Receiver Thermal Noise Ratio as a Function of Noise Jamming-to-Receiver Thermal Noise Ratio for Various Ratios of the Auxiliary and Radar Antenna Gains in the Direction of the Jammer

receiver is also canceled from the main channel serves to provide an upper bound on the ratio of G_{AJ} to G_{RJ}.

A common metric for SLC performance is the cancellation ratio (CR), the ratio of the output noise power without the SLC to the output noise power with the SLC (Equation 12-4). Including the residual noise power from Equation (12-1) results in Equation (12-5). When $J_N/N \gg 1$ and the radar and auxiliary antenna gains are the same we have the CR constrained to $J_N/2N$ (Equation 12-6).

$$CR = \frac{J_N + N}{P_Z} \tag{12-4}$$

$$CR = \frac{\left(\dfrac{J_N}{N} + 1\right)\left(\dfrac{J_N}{N}\dfrac{G_{AJ}}{G_{RJ}} + 1\right)}{\dfrac{J_N}{N}\left(\dfrac{G_{AJ}}{G_{RJ}} + 1\right) + 1} \tag{12-5}$$

$$CR = \frac{J_N}{2N} \qquad \text{when } J_N/N \gg 1 \text{ and } G_{AJ} = G_{RJ} \tag{12-6}$$

where:

CR = Cancellation ratio, no units

The CR as a function of J_N/N for various ratios of the auxiliary (G_{AJ}) and radar (G_{RJ}) antenna gains in the direction of the jammer is shown in Figure 12-4. From these equations we can see how receiver thermal noise power and jammer noise power provide a bound on the cancellation ratio.

FIGURE 12-4 ■ Cancellation Ratio as a Function of Noise Jamming-to-Receiver Thermal Noise Ratio for Various Ratios of the Auxiliary and Radar Antenna Gains in the Direction of the Jammer

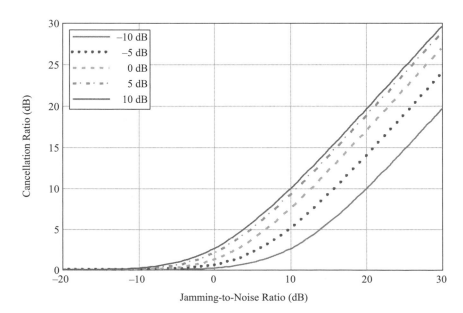

All this discussion on SLC is well and good, but the real bottom line is its effect on the performance of the radar system. The effect of a noise jammer is captured in the target signal-to-interference (sum of the jammer noise and receiver thermal noise) ratio after integration of multiple pulses $(S/I)_n$ (see Chapter 11). With an SLC, the interference term is the residual noise power (Equation 12-1). The resultant $(S/I)_n$ after SLC is given in Equation (12-7). When the $J_N/N \gg 1$ and the radar and auxiliary antenna gains are the same, the $(S/I)_n$ is bound by the increased thermal noise (Equation 12-8). Even with ideal jammer cancellation we don't get something (reduced J_N) for nothing (increased receiver thermal noise).

$$\left(\frac{S}{I}\right)_n = \frac{S\, G_I}{P_Z} = \frac{S\, G_T}{\left(\dfrac{J_N + N}{CR}\right)} \tag{12-7}$$

$$\left(\frac{S}{I}\right)_n = \frac{S\, G_I}{2N} \qquad \text{when } J_N/N \gg 1 \text{ and } G_{AJ} = G_{RJ} \tag{12-8}$$

where:

$(S/I)_n =$ Post-cancellation target signal-to-interference ratio after integration of multiple pulses, no units

$S =$ Target signal power (see Chapter 2), watts

$G_I =$ Integration gain (see Chapter 3), no units

The previous paragraphs discussed an SLC with one auxiliary antenna and receiver. To be effective, an SLC needs one auxiliary antenna and receiver, also referred to as a degree-of-freedom or canceler loop, for each noise jammer in the environment. As the number of canceler loops is increased, the receiver thermal noise powers continue to add, further reducing the SLC effectiveness. When the number of jammers is less than or equal to the number of SLC loops, maximum cancellation will occur. However, when there are more jammers than SLC loops the SLC is "overloaded" and minimal or no cancellation will occur. The maximum CR, when the number of jammers is less than or equal to the number of SLC loops, is approximately the same for all jammers. The minimum CR, when the number of jammers exceeds the number of SLC loops, is dependent on the number of SLC loops. The higher the number of SLC loops, the higher the minimum CR. A discussion and figures showing the relationship between the CR, number of jammers, and number of SLC loops can be found in Farina [1992, Chapter 4]. In addition, there are many other small residuals in the canceling process, such as radar instabilities, filter mismatches, and the SLC response time.

Another category of an SLC is an adaptive array, or adaptive nulling, system for an electronically steered array radar antenna (see Chapter 4). Implicit in the existence of nulls in the radar antenna pattern sidelobes is the

FIGURE 12-5 ■
Canceling A
Sidelobe

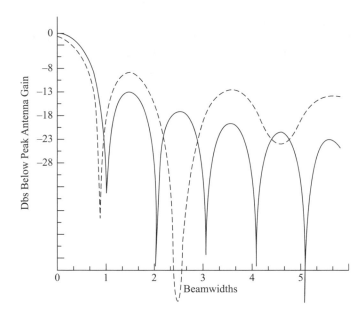

FIGURE 12-5 ■
Canceling A
Sidelobe

notion of adaptively moving those nulls to the angles of arrival of sidelobe jamming, often referred to as "null steering." Recalling the discussion of antenna sidelobes in Chapter 4, the locations of sidelobe minima and maxima depend on the antenna dimensions and the weighting of the signal across the antenna (illumination function). Changes of phase of particular elements or assigning of phase and amplitude characteristics to auxiliary elements can create nulls in arbitrary locations. Figure 12-5 is a computer simulation of a 21-element array in which a null has been placed in the center of the second sidelobe of the normal array pattern by introducing a phasor of equal amplitude and 180° phase difference. Notice, however, how the overall radar antenna pattern is modified, mainbeam gain is reduced, and the farther-out sidelobes become higher (you never get something for nothing). Farina [1992, Chapter 5] discusses adaptive arrays in detail.

Phased arrays, which are designed to respond optimally to a variety of environments, are called "adaptive arrays." Conceptually they have great prowess. In practice, their cost, complexity, intensive processing requirements, and settling time have worked against them. Modern signal processing advances are reducing the processing requirements, though cost and complexity remain.

In addition to sidelobe cancellation, the cancellation concept can also be applied to the mainlobe. The difference beam in a monopulse tracker (see Chapter 5) can perform mainbeam cancellation. Pointing a monopulse tracker at a jammer cancels the noise in the difference beam. Targets slightly separated in angle from the jammer can be picked up in the split beams of the monopulse system.

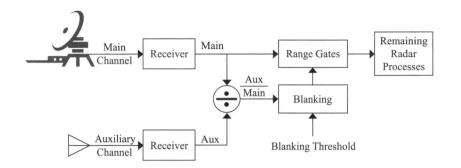

FIGURE 12-6 ■
Flow Diagram of a
Radar System with a
Sidelobe Blanker

12.2.2 Sidelobe Blanker

The purpose of an SLB is to prevent false target (pulsed) jamming signals entering the radar receiver through the sidelobes of the radar receive antenna from appearing in the radar output [Farina, 1992, Chapter 3]. The flow diagram of a radar system with an SLB is shown in Figure 12-6. The SLB uses auxiliary antennas to "cover" the sidelobes of the main radar antenna. The auxiliary antenna patterns have the same or slightly higher gain than the sidelobes of the radar receive antenna. The false target jamming signals are received by both the main radar antenna channel and the auxiliary channel. The SLB creates blanking signals at the corresponding range gates (see Chapter 5) where the signals in the auxiliary channel exceed the signals in the main channel by the blanking threshold. A blanking threshold is used to avoid the accidental blanking by receiver thermal noise of desired target signals. However, it may allow some false target jamming signals to get through. The time-based blanking signal essentially zeros out the range gate where the unwanted signal exists.

An example of the effect of an SLB can be seen on a radar A-scope, radar output and detection threshold as a function of time (see Chapter 5), is shown in Figure 12-7. Numerous false targets exceed the detection threshold with the SLB Off, and the single true target is difficult to notice. With the SLB On, the range gates containing the false targets are now "blanked," and the single true target is clearly visible. Farina [1992, Chapter 3] discusses SLB principles, performance, and design in great detail.

■ 12.3 | SOPHISTICATED TARGET TRACKERS

With advancements in computer performance, sophisticated target trackers have become common in radar systems (see Section 5.6) [Stimson, 1998, Chapter 29; Richards et al., 2010, Chapter 19]. A sophisticated target tracker, such as correlation and state trackers, can minimize the effectiveness of some

FIGURE 12-7 ■
Example Effect of a
Sidelobe Blanker:
Radar A-Scope
(Amplitude vs. Time)

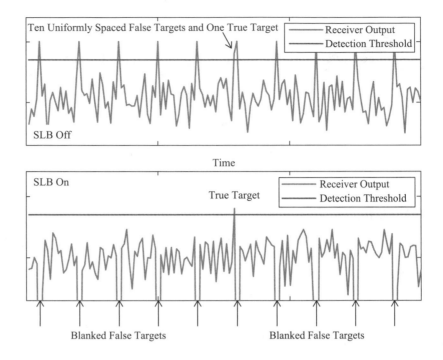

EA techniques. Correlated measurement trackers consider both the range and range rate measurements together when computing the overall range and range rate track, as shown in Figure 12-8. In a correlation tracker the range change must correlate with the range rate, and vice versa. Uncorrelated measurements, often due to false target jamming (see Chapter 10), are "ignored" by the tracker. Some correlated measurement trackers also include the angle measurement. Thus, the EA technique must coordinate the range, range rate, and angle EA together. Often this required coordination constrains the amount of the tracking error produced by the EA technique.

Target state trackers provide estimates of target state vectors (e.g., position, velocity, acceleration) from individual range, range rate, and angle measurements, as shown in Figure 12-9. Target state trackers use current

FIGURE 12-8 ■
Correlated
Measurement
Tracker

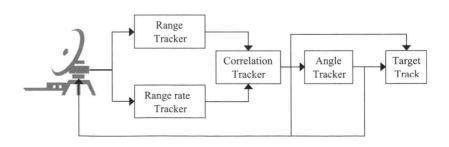

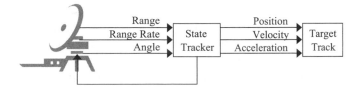

FIGURE 12-9 ■
Target State Tracker

target measurements combined with previous target information and a target state model to compute a target state estimate. Target state trackers operate much like our minds do when we are driving (or at least when we are driving well): observing the current states of the cars around us, remembering their previous states, predicting where they will be in the future. Target state models define the reasonable bounds of target dynamics. This helps the tracker "ignore" unrealistic and/or unexpected measurements from both actual targets and EA signals.

Examples of target state trackers are α-β and Kalman Filters. A classical α-β tracker (see Chapter 5) has been used since the 1960s and is relatively easy to implement. However, it does not handle maneuvering targets well. An adaptive α-β tracker has additional algorithms to handle maneuvering targets.

Kalman filters have been used in radar trackers since the 1970s and can inherently handle maneuvering targets. They also contain a model for measurement errors and target trajectory. Much more computationally demanding than an α-β Tracker, though well within the capabilities of modern computers, Kalman filters have become indispensable when dealing with missing measurements, variable measurement errors, and maneuvering targets. They have traditionally been used for airborne fire control radar systems but are becoming common in modern ground-based radar systems. As computing capacity has become better, faster, and cheaper, a wide range of tracking algorithms such as interacting multiple model (IMM) and multiple hypothesis tracker (MHT) has been applied to radar applications.

12.4 | SUMMARY

There are a wide range of EP concepts and techniques, and we have only touched the surface. The references at the end of this chapter all discuss EP techniques in detail. As long as there is ES and EA, there will be EP, and vice versa. Consequently, a more general approach has long been advocated by many [Skolnik, 1980, pp. 547–553; Howells, 1976]. Paul Howells, particularly, enunciated a philosophy for coping with all EA (in principle, even techniques that had not yet been invented), based on netted high-resolution, multiple-mode radar systems using adaptive processing. Electro-optical (EO), laser, and infrared (IR) sensors have long been used to augment tracking radar systems in the presence of jamming. Networked radar systems, usually across a wide spectrum of transmit frequencies and detection and measurement performance, along with sensor fusion algorithms have been used for many years to form an integrated "system of systems." Net-enabled spectrum warfare is a growing area of EW research and development to attack networked radar systems.

▮▮▮ 12.5 ▮ EXERCISES

12-1. A radar system has a frequency modulated (FM) chirp waveform with a bandwidth of $B_{pc} = 100$ MHz over a transmitted pulse width $\tau = 200$ microseconds. What is the consequence of the pulse compression waveform on (a) a narrowband jammer approximately matched to the transmitted pulse width and (b) a wideband jammer approximately matched to the pulse compression modulation bandwidth?

12-2. A radar system uses a variable power transmitter to provide a low probability of intercept electronic protection (EP) technique; to reduce the detection range of electronic warfare (EW) receivers. The desire is to reduce the EW receiver detection range by half by reducing the transmit power. What is the necessary radar transmitter power level relative to full power?

12-3. A radar system has a sidelobe canceler (SLC) with an auxiliary antenna gain in the direction of the jammer $G_{AJ} = 25$ dB. Without the SLC the noise jamming-to-receiver thermal noise ratio $J_N/N = 20$ dB when the radar antenna gain in the direction of the jammer $G_{RJ} = 25$ dB. (a) What is the residual noise-to-receiver thermal noise ratio P_Z/N (dB)? (b) The radar–jammer geometry changes, and the radar antenna gain in the direction of the jammer is 10 dB less. What is the residual noise-to-receiver thermal noise ratio P_Z/N (dB)? Compare your results with Figure 12-3.

12-4. What are the cancellation ratios (CR) for the two radar–jammer conditions in Exercise 12-3? Compare your results with Figure 12-4.

12-5. The radar system from Exercises 12-3 and 12-4 has a received target signal power $S = -124$ dBW, integration gain $G_I = 20$ dB, and receiver thermal noise power $N = -120$ dBW. (a) What is the target signal-to-noise ratio after integration $(S/N)_n$ (dB)? (b) For the initial radar–jammer geometry from Exercise 12-3, what is the target signal-to-interference ratio after integration $(S/I)_n$ (dB)? (c) What is the target signal-to-interference ratio after integration with the SLC $(S/I)_n$ (dB)? (d) For the radar–jammer geometry from Exercise 12-3 part (b), what is the target signal-to-interference ratio after integration $(S/I)_n$ (dB)? (e) What is the target signal-to-interference ratio after integration with the SLC $(S/I)_n$ (dB)?

12.6 | REFERENCES

Chrzanowski, Edward, 1990, *Active Radar Electronic Countermeasures*, Norwood MA: Artech House.
> Contains a detailed description of EP-ECCM concepts, technology, and supporting math.

Farina, Alfonso, 1992, *Antenna-Based Signal Processing Techniques for Radar Systems*, Norwood MA: Artech House.
> The first place to go for the details of antenna-based EP-ECCM concepts and supporting math.

Howells, Paul H., 1976, "Explorations in Fixed and Adaptive Resolution at GE and SURC," *IEEE Transactions*, Volume AP-24, Sep., pp. 575–584.

Neri, Filippo, 2006, *Introduction to Electronic Defense Systems*, 2nd Edition, Raleigh, NC: SciTech Publishing.
> Contains a detailed description of EP-ECCM concepts, technology, and supporting math.

Richards, M. A., Sheer, J. A., and Holm, W. A. (editors), 2010, *Principles of Modern Radar*, Volume 1: Basic Principles, Raleigh, NC: SciTech Publishing.
> Radar tracking algorithms including Kalman filters are discussed in Chapter 19.

Schleher, D. C., 1999, *Electronic Warfare in the Information Age*, Norwood, MA: Artech House.
> Contains clear discussions as well as mathematical analysis of EA and EP, including numerous MATLAB files.

Stimson, George W., 1998, *Introduction to Airborne Radar*, 2nd Edition, Raleigh, NC: SciTech Publishing.
> Automatic tracking including Kalman filters is discussed in Chapter 29. Low probability of intercept (LPI) concepts are discussed in Chapter 42.

Van Brunt, L. B., 1978, *Applied ECM*, Volume I and II, Dunn Luring, VA: EW Engineering, Inc.
> Contains clear descriptions and details on myriad EA-ECM and EP-ECCM tactics and techniques.

Loose Ends of Radar and/or Electronic Warfare Lore

HIGHLIGHTS

- Picking up on loose ends of radar and/or electronic warfare lore
- The radar line of sight
- The troposphere, complex but definable
- The ionosphere as absorber, refractor, reflector, and polarizer
- Deriving the far field of an antenna and a target
- Some radar and electronic warfare rules-of-thumb

Several other brief quantitative notions and a list of rules-of-thumb should be included in any set of radar and electronic warfare (EW) fundamentals. Some are good for general day-to-day use; others are important to detailed radar and/or EW design calculations. They are pulled together in this chapter without concern for their relevance to each other but only because they are valid radar and EW lore.

13.1 | RADAR LINE OF SIGHT

As advanced as radar systems have become, they cannot see through the earth. The target needs to be above the horizon to be illuminated by the radar system, as shown in Figure 13-1 for a round smooth earth (RSE) geometry. Solving for the optical horizon is a simple problem in trigonometry, as shown in Figure 13-2 and given in Equation (13-1).

$$R_h{}^2 + R_E{}^2 = (R_E + h)^2$$

$$R_h = \sqrt{2\,R_E\,h + h^2}$$

$$(13\text{-}1)$$

where:

R_h = Range to the horizon, meters
R_E = Radius of the earth, 6371 kilometers
  h = Height of the target or height of the radar, meters

Often an approximation to the range to the horizon equation is used because $2R_E h \gg h^2$. This approximation is acceptable for computing the range to the horizon for the vast majority of ground-based and airborne radar geometries: essentially any geometry in the atmosphere. This approximation is *unacceptable* for computing any of the associated angles in the triangle. Thus, it is best to just use the actual range to the horizon equation.

Contrary to all the straight-line propagation paths we have used in this book, radio frequency (RF) waves refract (bend) as they propagate through the

FIGURE 13-1 ■
Radar Line of
Sight—Round,
Smooth Earth

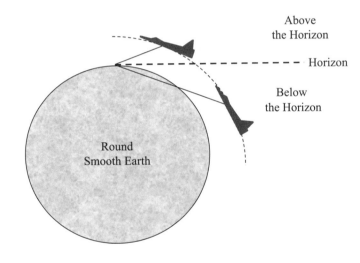

FIGURE 13-2 ■
Range to the Horizon

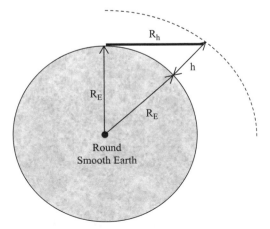

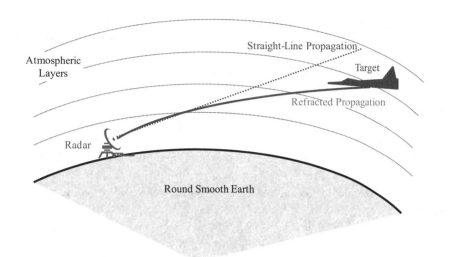

FIGURE 13-3 ■
Refracted
Propagation

many layers of the atmosphere, as shown in Figure 13-3 [Meeks, 1982]. As seen in this figure, refraction results in a different radar-to-target propagation path length and elevation angle than associated with straight-line propagation. To account for refraction, we multiple the earth's radius by a refraction factor (Equation 13-2). Refraction factors that are constant as a function of altitude are often used because of their computational simplicity. Common constant refraction factors are given in Equation (13-3). Since refraction varies with altitude, a refraction factor that varies exponentially altitude is also used (Equation 13-4). Computing the radar horizon with a variable refraction factor requires incrementing in very fine altitude steps through all the atmospheric layers from the radar to the target. Thus, it is a much more computationally demanding approach than for a constant refraction factor.

$$R_h = \sqrt{2\,k_r\,R_E\,h + h^2} \tag{13-2}$$

$$k_r = {}^4\!/_3,\, {}^6\!/_5, \ldots \tag{13-3}$$

$$k_r = \frac{1}{1 + 10^{-6}\,R_E\,\dfrac{dN(h)}{dh}} \qquad N(h) = N_s\,e^{-a\,h} \tag{13-4}$$

where:

k_r = Refraction factor, no units
N_s = Surface refractivity
a = Refractive index gradient

Temperature and/or humidity changes can produce highly variable refraction profiles with altitude. This is most pronounced over shallow water in hot climates but can also occur over land. Highly variable refraction profiles can produce a "duct" that traps the radar wave in a layer of the atmosphere—a condition often called super refraction. Generally, the duct is close to the surface of the earth. Super refraction or ducting results in a refraction factor greater than approximately two. Specific refraction factors are obtained from detailed atmospheric measurements and/or radio wave propagation models.

Two horizon components, the clutter horizon and the target horizon, make up the radar line of sight (LOS), as shown in Figure 13-4. The clutter horizon is how far the radar can illuminate the Earth's surface, thus producing a clutter return (Equation 13-5). The target horizon is how far a target is when it comes over the horizon (Equation 13-6). The radar LOS, or radar horizon, is the sum of the clutter horizon and the target horizon (Equation 13-7).

$$R_{hC} = \sqrt{2\, k_r\, R_E\, h_R + h_R{}^2} \tag{13-5}$$

$$R_{hT} = \sqrt{2\, k_r\, R_E\, h_T + h_T{}^2} \tag{13-6}$$

$$R_{LOS} = R_{hC} + R_{hT} \tag{13-7}$$

where:

R_{hC} = Clutter horizon, meters
h_R = Radar height, meters
R_{hT} = Target horizon, meters
h_T = Target height, meters
R_{LOS} = Radar line of sight, meters

Terrain greatly complicates the radar LOS. With terrain the radar LOS can be significantly smaller when the radar is next to a mountain or greater when

FIGURE 13-4 ■
Radar Line of Sight

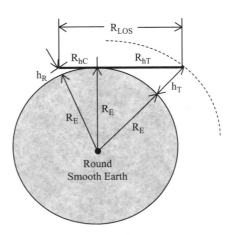

the radar is on a plateau, than for the RSE case. Computer models are used to predict radar LOS using digitized terrain data.

In summary, radar LOS is a function of radar height and target height. Refraction, the bending of the radar waves as they propagate through the layers of the atmosphere, needs to be accounted for when computing the radar LOS. Computing the radar LOS for the RSE case and a constant refraction factor is straightforward, whereas computing the radar LOS for a varying refraction factor and/or terrain is complicated. Radar LOS can be a limiting factor in radar detection performance for low-altitude targets—the radar cannot detect what it cannot illuminate. Everything we have discussed here from a radar–target perspective also applies to EW systems (see Chapter 8).

13.2 | PROPERTIES OF THE PROPAGATION MEDIUM

In all discussions in this book, I have considered the propagation medium lossless and noninterfering. Actually, it is not. In fact, both the ionosphere (for low microwave frequencies) and the troposphere (for high microwave frequencies) cause attenuation, refraction, and dispersion of radar signals. Because the degradation caused by dispersion is usually small compared with the first two and is relatively complex, I shall not discuss it. Tropospheric refraction was discussed in the previous radar LOS section.

13.2.1 Troposphere

Tropospheric or atmospheric attenuation is the loss in power in the radar wave as it travels through the molecules in the atmosphere [Morchin, 1990]. At radar frequencies, oxygen and water vapor are the offending molecules. Atmospheric attenuation is a result of an electromagnetic resonance between the radar wave and oxygen and water vapor molecules. Thus, atmospheric attenuation is a function of wavelength. As shown in Figure 13-5, the lower the radar frequency, the lower the attenuation. Because the distribution of oxygen and water vapor varies with altitude, atmospheric attenuation varies with altitude. As shown in Figure 13-6, the higher the altitude, the lower the attenuation.

Rain and clouds also affect atmospheric attenuation. It is best presented in Figure 13-7, which is based on data and calculations reported by Barton [1979, pp. 468–472] and Nathanson [1969, pp. 193–199]. Included in Figure 13-7 are absorption effects of various atmospheric molecules, especially oxygen and water vapor, as well as rain and clouds.

FIGURE 13-5 ■
Atmospheric
Attenuation Overall
with Oxygen and
Water Vapor
Components

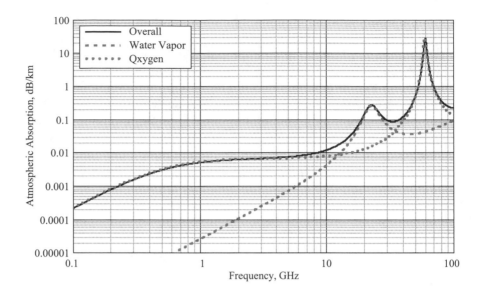

FIGURE 13-6 ■
Atmospheric
Attenuation as a
Function of Altitude

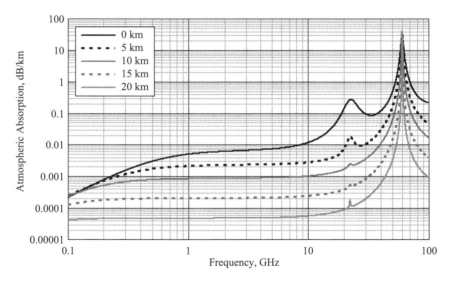

13.2.2 Ionosphere

Electromagnetic waves are attenuated, refracted, reflected, and their polarization
is rotated by the ionosphere [Glasstone and Dolan, 1977]. Ampere's law and
Faraday's law allow us to understand the ionosphere in principle, but its condition of constant fluctuations due to energy deposited continuously by the sun, the
cosmos, and man makes it impossible to predict accurately.

If we think of the ionosphere as a large, imperfectly conducting sheet with
conductivity a variable, depending on the density of free electrons in the
medium, we have a model that will serve our purposes. When there are

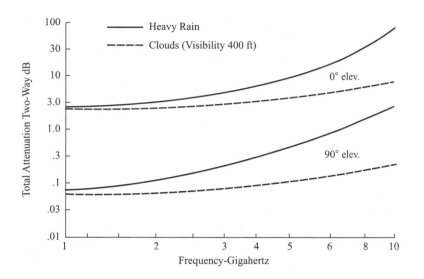

FIGURE 13-7 ■
Atmospheric
Attenuation (dB/
mile): Rain and
Clouds

moderate numbers of free electrons, there is refraction. When these electrons collide with other atoms, there is attenuation; when the electron density is very high, there is reflection. All these interactions are more pronounced at longer wavelengths; in fact, there is a λ^2 dependency.

Attenuation

Attenuation of electromagnetic waves by interaction with an ionized medium is given in Equation (13-8). Because the altitude of the ionosphere exceeds 45 Nmi where $v \leq 6 \times 10^6$, $v^2 \ll \omega^2$, for radar systems of interest and thus "v^2" can be dropped out, yielding Equation (13-9). When the path is oblique through some path length, appropriate trigonometric functions must be applied. For the two equations on attenuation to be useful, we must know ionospheric electron densities at various levels. Unfortunately, these vary enormously from day to night with solar activity and as a function of latitude. In practice, radar systems operating through the ionosphere take frequent soundings as well as subscribe to National Bureau of Standards bulletins on the matter.

$$\alpha = \frac{2.6 \times 10^4 \, N_e \, v}{\omega^2 + v^2} \qquad (13\text{-}8)$$

$$\alpha \cong 4 \times 10^{-3} \frac{N_e}{f^2} \qquad (13\text{-}9)$$

where:

$\alpha =$ Attenuation, dB/Nmi
$N_e =$ Electron density, electrons/cm^3
$v =$ Collision frequency of the electrons with particles
$\omega =$ Radar frequency, radians/second
$f =$ Radar frequency, megahertz

FIGURE 13-8 ■
Refraction Geometry

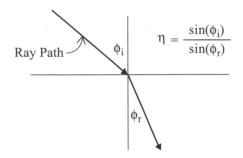

Refraction

Useful refraction relationships for the ionosphere can also be written down. Loss in an ionized medium has an imaginary component producing changes in phase that in turn result in an effective change in direction of the radar beam. An index of refraction based on electron densities is given in Equation (13-10). This equation presumes radar frequencies above 10 MHz and N_e below 10^9 per cm^3. Outside these bounds, the plasma frequency ($f_p \approx 9\sqrt{N_e}$) and the collision frequency modify the relationship substantially. From the geometry of refraction shown in Figure 13-8, calculations for refraction are easy when the electron densities of ionospheric layers are known.

$$\eta \cong \left(1 - \frac{N_e}{10^4\ f^2}\right)^2 \qquad (13\text{-}10)$$

where:

η = Index of refraction

Polarization Rotation

Faraday rotation is another phenomenon affecting linearly polarized electromagnetic waves that propagate through an ionized medium in a magnetic field. Virtually all the arriving energy interacts with the ionized particles. Conceptually, electrons excited by the incident electric field of the radar wave are diverted in their paths by the presence of the earth's magnetic field in accordance with Faraday's law. Thus, they oscillate and reradiate at an angle from the direction of their excitation. With each stage of excitation, oscillation, and reradiation, the angle increases. The result is that the polarization (E-field) of the radiated energy revolves. As with the other ionospheric interactions discussed, Faraday rotation has a λ^2 dependence. The actual rotation depends on the path length, the earth's magnetic field strength, and the angle between the magnetic field lines and the direction of propagation. A worst-case (100 Nmi path, zero elevation angle, maximum interaction angle, high field strength, and electron density) relationship has been worked out, as given in

Equation (13-11). It is apparent that there is a substantial rotation effect (>10 degrees) at frequencies as high as 2.5 GHz.

$$\Omega = \frac{6.3 \times 10^{20}}{f^2} \tag{13-11}$$

where:

Ω = Faraday rotation, degrees

13.3 | FAR FIELD OF AN ANTENNA OR A TARGET

The far field of an antenna can be defined as the distance from the antenna beyond which there is a "negligible" phase difference between electromagnetic waves arriving from the center of the aperture and waves arriving from its edge. The aperture is assumed to be "focused" on infinity, and the far field begins when the radiating signal is essentially a plane wave. The far-field region is easy to derive by referring to Figure 13-9 and Equation (13-12). The key question is, "What value should we assign to arbitrary phase difference?" Conventional wisdom has it when at least half the power arriving from the antenna edge is contributing to the pattern the far field has been reached. This occurs at a phase difference of $45°$ ($\lambda/8$). The far-field distance for a $\lambda/8$ phase difference is given in Equation (13-13). A more conservative engineer might define the phase difference as $\lambda/16$ and a less conservative one as $\lambda/4$. Their "far fields" would vary accordingly.

$$(R+\varepsilon)^2 = R^2 + \left(\frac{D}{2}\right)^2$$

$$R^2 + 2R\varepsilon + \varepsilon^2 = R^2 + \left(\frac{D}{2}\right)^2 \Rightarrow 2R\varepsilon \cong \left(\frac{D}{2}\right)^2 \quad \text{since } 2R\varepsilon >> \varepsilon^2 \tag{13-12}$$

$$R \cong \frac{D^2}{8\varepsilon}$$

$$R_{ff} \cong \frac{D^2}{\lambda} \tag{13-13}$$

where:

R = Distance to the far-field point, meters
ε = Arbitrary phase difference from center to edge of the antenna, meters
D = Antenna dimension, meters
R_{ff} = Far-field distance, meters
λ = Wavelength, meters

FIGURE 13-9 ■
Far-Field Calculation

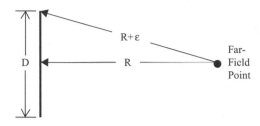

There is also a far field associated with target radar cross section (RCS). This is because most complex radar targets are made of multiple scatterers (see Chapter 6). Each individual scatterer produces a complex (amplitude and phase) reflection; they combine in space as the reflected waves propagate back to the radar. Defining the far-field distance for target RCS is more challenging than for an antenna because it is difficult to know which scatterers are combining to produce the overall RCS, and thus the distance between them. This information can be gained from detailed RCS measurements and/or computer modeling.

■ 13.4 | CONVENIENT RADAR AND ELECTRONIC WARFARE RULES-OF-THUMB

Some results of radar and/or EW design problems occur repeatedly. They seem to represent enduring verities, are recognized as having broad application, and crop up in meetings or are used as springboards for quick extrapolations. A radar and/or EW neophyte needs to memorize a handful of such rules-of-thumb as presented here:

- There are approximately 7 μsec to a kilometer (two-way).
- There are approximately 11 μsec to a statue mile (two-way).
- There are 12 μsec to a nautical mile (two-way).
- The received target signal power varies as a function of $1/R^4$ and thus decreases by 12 dB (1/16) when the radar-to-target range doubles and increases by 12 dB (16 times) when the range halves.
- A 290 Kelvin equivalent noise temperature receiver has a 3 dB noise figure.
- Probability of detection is meaningless without probability of false alarm.
- The gain all the way around any antenna is 1 (0 dBi).
- An array of N half-wavelength spaced isotropic elements has a gain of N.
- 1 μsec is 150 meters, or approximately 500 feet (two-way).
- A 1 MHz bandwidth gives 150 meter, or approximately 500 feet, range resolution.
- A 500 MHz bandwidth gives 0.3 meter, or approximately 1 foot, range resolution.

- The natural limit on range resolution is one radio frequency cycle, or one wavelength.

- 1 hertz of Doppler shift is 1 meter/second of range rate at very high frequency (VHF) ($f_c = 165$ MHz or $\lambda = 1.8$ meter); 4 hertz of Doppler shift is 1 meter/second of range rate at ultra high frequency (UHF) ($f_c = 650$ MHz or $\lambda = 0.5$ meter); 40 hertz of Doppler is 1 meter/second of range rate at C band ($f_c = 6$ GHz or $\lambda = 0.05$ meter); and 67 hertz of Doppler shift is 1 meter/second of range rate at X band ($f_c = 10$ GHz or $\lambda = 0.03$ meter).

- A fixed antenna, focused on infinity, has a finest cross-range resolution equal to its diameter.

- The natural limit on angular resolution is λ^2.

- There are 4π steradians or approximately 41,000 square degrees in a sphere.

- 1/10 milliradian angular accuracies are all the atmosphere will support; this is the same for 0.03 meter or approximately 0.1 foot range accuracies.

- A passive target cannot reradiate more energy than it intercepts.

- Radar maps as good as optical photographs are impossible.

- A strip-map synthetic aperture radar system can have the finest cross-range resolution equal to half its physical aperture, no matter how small.

- For an EW receiver, the received radar signal power varies as a function of $1/R^2$. Thus, the received radar signal power decreases by 6 dB (1/4) when the EW receiver-to-radar range doubles and increases by 6 dB (4 times) when the range halves.

- The received jammer signal power varies as a function of $1/R^2$ and thus decreases by 6 dB (1/4) when the radar-to-jammer range doubles and increases by 6 dB (4 times) when the range halves.

▌ 13.5 │ SUMMARY

This chapter addressed a few brief concepts important to radar and EW fundamentals: radar LOS and refraction; atmospheric attenuation; and the far field of an antenna or target. A list of several radar and EW rules-of-thumb was included, which should be helpful in many radar and/or EW applications.

▌ 13.6 │ EXERCISES

13-1. What is the radar line of sight for a radar height $h_R = 10$ meters and a target height $h_T = 2000$ meters? Use a refraction factor $k_r = 4/3$.

13-2. Fifteen ground-based radar systems stretched across the northern border of the United States will give coverage at 1400 meter altitude and above. How many ground-based radar systems will be required to extend coverage down to 60 meters? Use a refraction factor $k_r = 4/3$.

13-3. How many airborne pulse-Doppler radar systems flying at 6000 meters are required to cover the United States northern border against targets flying down to 30 meters? Use a refraction factor $k_r = 4/3$.

13-4. A radar system has the following characteristics: carrier frequency $f_c = 9.5$ GHz, and radar system losses without atmospheric attenuation $L_R = 6$ dB. A very low-altitude target comes above the radar horizon at a range $R_{RT} = 30$ km. (a) What is the atmospheric attenuation loss L_a (dB)? (b) What is the radar system losses including atmospheric attenuation L_R (dB)?

13-5. Using electron densities of 10^4 and $10^5/cm^3$, calculate some ionospheric effects (attenuation, refraction, and polarization rotation) for a microwave frequency $f_c = 3$ GHz.

13-6. What is the far-field range of the Arecibo antenna from Section 7.7 on its boresight?

13-7. The space track radar in Florida has an antenna aperture diameter of 55 meters at ultra high frequency ($\lambda = 0.6$ meters). What is the range to its far field?

13-8. A homemade telescope has an aperture of 0.2 meters. What is the range to its far field? (Take optical wavelengths to be 0.5 micrometers.)

◾ 13.7 | REFERENCES

Barton, D. K., 1979, *Radar Systems Analysis*, Dedham, MA: Artech House.

Glasstone, S., and Dolan, P. J. (editors), 1977, *The Effects of Nuclear Weapons*, Washington, DC: U.S. Department of Defense and Energy R&D Administration. The presentation on the ionosphere is based on the excellent tutorial on the subject contained in pages 462–466 and 489–494.

Meeks, M. L., 1982, *Radar Propagation at Low Altitudes*, Dedham MA: Artech House. Classic and concise reference on radar wave propagation.

Morchin, William C., 1990, *Airborne Early Warning Radar*, Dedham, MA: Artech House. Chapter 3, pp. 89–98, is the source of the algorithm used to generate the atmospheric attenuation figures.

Nathanson, F. E., 1998, *Radar Design Principles*, Raleigh, NC: SciTech Publishing, IET.

Appendix 1: Decibels

Bels were named after Alexander Graham Bell. They were simply the logarithm of the ratio of power out and power in. The definition of a decibel (dB) is contained in any radio engineer's handbook, as given as follows. Since radar and electronic warfare engineers use power, the $10\log_{10}(x)$ from is by far the most common.

$$dB = 20 \ \log_{10}\left(\frac{V_o}{V_i}\right) = 10 \ \log_{10}\left(\frac{P_o}{P_i}\right)$$

where:

dB = decibel, no units
$\log_{10}$ = Base 10 log, no units
V_o = Voltage out, volts
V_i = Voltage in, volts
P_o = Power out, watts
P_i = Power in, watts

In colloquial usage, decibels are used to express any ratio or often just an absolute value. Decibels are just 10 times the $\log_{10}$ of the ratio or value. A factor of two results in about 3 dB, while a factor of one half results in about −3 dB.

$$dB = 10 \ \log_{10}(\text{absolute}) \qquad \text{absolute} = 10^{\left(\frac{dB}{10}\right)}$$
$$3.0103 = 10 \ \log_{10}(2) \qquad -3.0103 = 10 \ \log_{10}(0.5)$$

Decibels provide a numerical value compression, allowing the very large range of values to be easily represented and plotted. As shown in Table A.1, 12 orders

TABLE A.1 ■ Example of Absolute and Decibel Numerical Values

Absolute	dB	Absolute	dB
$0.000001 = 10^{-6}$	−60	$1,000,000 = 10^6$	60
$0.00001 = 10^{-5}$	−50	$100,000 = 10^5$	50
$0.0001 = 10^{-4}$	−40	$10,000 = 10^4$	40
$0.001 = 10^{-3}$	−30	$1,000 = 10^3$	30
$0.01 = 10^{-2}$	−20	$100 = 10^2$	20
$0.1 = 10^{-1}$	−10	$10 = 10^1$	10
$1 = 10^0$	0	$1 = 10^0$	0

of magnitude (powers of 10) are represented by "only" 120 dB. After working with dB values a while, they are usually memorized by radar and EW people.

Since radar engineers often deal in power, decibels relative to a watt or decibels relative to a milliwatt are used. Another common use of decibels is for radar cross section (RCS), decibels relative to a square meter. Decibels used for antenna are often relative to isotropic, gain of one (1) everywhere.

$$\text{dBW} = 10 \ \log_{10}\left(\frac{P(W)}{1 \ \text{Watt}}\right) \qquad \text{dBm} = 10 \ \log_{10}\left(\frac{P(mW)}{1 \ \text{milliWatt}}\right)$$

$$\text{dBW} = \text{dBm} - 30 \qquad \text{dBm} = \text{dBW} + 30$$

$$\text{dBsm} = 10 \ \log_{10}\left(\frac{\sigma(m^2)}{1\,m^2}\right)$$

$$\text{dBi} = 10 \ \log_{10}\left(\frac{G}{1}\right)$$

where:

dBW = Decibels relative to one watt, carries the units of watts
P(W) = Power, watts
dBm = Decibels relative to one milliwatt, carries the units of milliwatts
P(mW) = Power, milliwatts
dBsm = Decibels relative to one square meter, carries the units of square meters
$\sigma(m^2)$ = Radar cross section, square meters (m^2)
dBi = Decibels relative to an isotropic antenna, no units
G = Antenna gain, no units

The use of decibels can simplify some mathematical equations. As shown below, multiplication becomes addition, division becomes subtraction, and raising to an exponent becomes multiplication. *However, one should never mix decibel and absolute numbers in the same equation, except for some extremely rare exceptions.*

$$10 \ \log(A \times B) = 10 \ \log(A) + 10 \ \log(B)$$

$$10 \ \log\left(\frac{A}{B}\right) = 10 \ \log(A) - 10 \ \log(B)$$

$$10 \ \log(A^B) = 10B \ \log(A)$$

Appendix 2: The Radar Spectrum

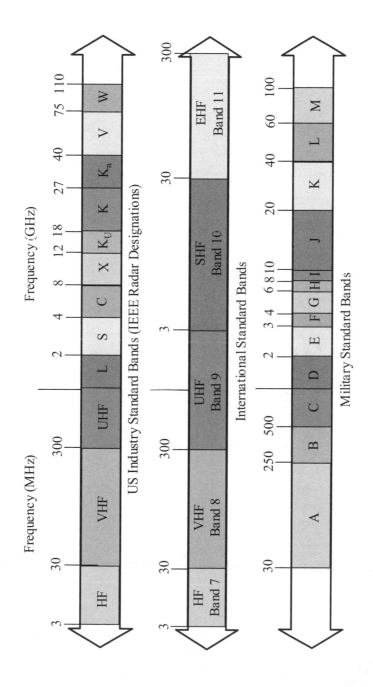

Appendix 3: Fourier Series and Transforms

Working in the field of heat transfer in the early nineteenth century, Jean B. S. Fourier found that virtually all functions of time, particularly repetitive ones, could be described in a series of sine and cosine waves of various frequencies and amplitudes. His work has been described as one of the most elegant developments in modern mathematics. Whatever its stature for the world, the benefits for the radar engineer are epic. The following presentation uses the approach taken by H. H. Skilling in *Electrical Engineering Circuits*, Chapters 14–15 (New York: Wiley, 1957). George Stimson also has an excellent discussion on Fourier series and transforms in *Introduction to Airborne Radar*, 2nd edition, Chapters 17 and 20 (Raleigh, NC: SciTech, 1998).

FOURIER SERIES

The statement for the Fourier series is that any wave may be broken down into the sum of sines and cosines of various amplitudes and frequencies. In mathematical notation:

$$f(t) = \frac{1}{2}a_0 + a_1\cos(\omega t) + a_2\cos(2\omega t) + \ldots$$
$$+ b_1\sin(\omega t) + b_2\sin(2\omega t) + \ldots \tag{A3-1}$$

$f(t)$ is a function of time here (it does not have to be), and the a_i and b_i are constants to make the expression on the right equal to $f(t)$; $1/2\, a_0$ is the dc (direct current, zero frequency) component of the function, if any.

The next step is to find a way to evaluate a_i and b_i. To do this, we use the orthogonality of the sine and cosine function, which is the value of their cross-products integrated from zero to 2π resulting in a value of zero. Furthermore, the sine and cosine each integrated from zero to 2π is zero. These integrations are easy to check by referring to a table of integrals:

$$\int_0^{2\pi} \sin(mx)\,dx = 0 \quad \int_0^{2\pi} \cos(nx)\,dx = 0 \quad \int_0^{2\pi} \sin(mx)\cos(nx)\,dx = 0$$

$$\tag{A3-2}$$

Also

$$\int_0^{2\pi} \sin(mx)\sin(nx)\,dx = 0 \qquad \int_0^{2\pi} \cos(mx)\cos(nx)\,dx = 0 \quad \text{(A3-3)}$$

But

$$\int_0^{2\pi} \left(\sin(mx)\right)^2 dx = \pi \qquad \int_0^{2\pi} \left(\cos(nx)\right)^2 dx = \pi \qquad \text{(A3-4)}$$

Let us pick a coefficient to solve for, say, a_2. Multiply through Equation (A3-1) by $\cos(2\omega t)$, and then multiply both sides by $d(\omega t)$ and integrate:

$$\int_0^{2\pi} f(t)\cos(2\omega t)\,d(\omega t) = \int_0^{2\pi} \frac{1}{2}a_0\cos(2\omega t)\,d(\omega t) +$$
$$\int_0^{2\pi} a_1\cos(\omega t)\cos(2\omega t)\,d(\omega t) + \dots \int_0^{2\pi} b_1\sin(\omega t)\cos(2\omega t)\,d(\omega t) + \dots$$

We can see by inspection that most of these integrals are zero. On the right side of the equation, the first and second terms are zero. However, the third term equals $a_2\,\pi$. In fact, all other terms written down and all implicit terms are zero, giving finally

$$\int_0^{2\pi} f(t)\cos(2\omega t)\,d(\omega t) = a_2\pi \qquad \text{(A3-5)}$$

and

$$a_2 = \frac{1}{\pi}\int_0^{2\pi} f(t)\cos(2\omega t)\,d(\omega t) \qquad \text{(A3-6)}$$

Because we do know $f(t)$, we can find a_2, even if we have to do it by numerical integration on a computer or calculator. In addition, because we did this in general, we can now generalize to

$$a_n = \frac{1}{\pi}\int_0^{2\pi} f(t)\cos(n\omega t)\,d(\omega t) \qquad \text{(A3-7)}$$

and, after multiplying through by a sine function

$$b_n = \frac{1}{\pi} \int_0^{2\pi} f(t)\sin(n\omega t)\,d(\omega t) \tag{A3-8}$$

Equation (A3-1) is the Fourier series, but it can be rewritten for brevity as

$$f(t) = \frac{1}{2}a_0 + \sum_{m=1}^{\infty} \Big(a_m\cos(m\omega t) + b_m\sin(m\omega t)\Big) \tag{A3-9}$$

Writing down a function and evaluating it is the essence of tedium, but by astute observation the process can be shortened. Some of the radar waves we are interested in are not too difficult. For example, a periodic square wave (Figure A3-1) has no even harmonics. It is just an infinite sum of odd harmonics with ever increasing frequencies and ever decreasing coefficients. The answer for a square wave is

$$f(\omega) = \frac{4}{\pi}\Big(\sin(\omega t) + \frac{1}{3}\sin(3\omega t) + \frac{1}{5}\sin(5\omega t) + \ldots\Big) \tag{A3-10}$$

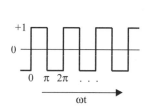 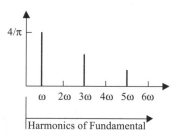

FIGURE A3-1 ■ A Square Wave and Its Fourier Spectrum

FOURIER TRANSFORMS

To arrive at the Fourier transforms, we must take a few preliminary steps. The first is to restate Equation (A3-1) in exponential form. We can do this by substituting for cosines and sines their identities, which are

$$\cos(x) = \frac{e^{jx} + e^{-jx}}{2} \qquad \sin(x) = \frac{e^{jx} - e^{-jx}}{2j}$$

We can now write the Fourier series as

$$f(t) = \ldots + A_{-2}e^{-j2\omega t} + A_{-1}e^{-j\omega t} + A_0 + A_1 e^{j\omega t} + A_2 e^{j2\omega t} + \ldots \tag{A3-11}$$

or

$$f(t) = \sum_{m} A_m e^{jm\omega t}$$

The coefficients are related in this way

$$A_n = \frac{1}{2}(a_n - jb_n) \quad A_{-n} = \frac{1}{2}(a_n + jb_n) \qquad \text{(A3-12)}$$

and the coefficients can be found from the integral

$$A_n = \frac{1}{2\pi} \int_0^{2\pi} f(t) e^{-jn\omega t} \, d(\omega t) \qquad \text{(A3-13)}$$

f(t) is the function of time to be expressed as a Fourier series, and the integer n can now be positive, negative, or zero. The integral (A3-13) is used just like Equations (A3-7) and (A3-8) to obtain coefficients. It is more efficient because there is only a single integral, and it is much more readily integrated. Equation (A-13) can be derived from equations (A3-7) and (A3-8) using Euler's theorem. By matching their infinite series, Euler showed the following:

$$e^{j\alpha} = \cos(\alpha) + j\sin(\alpha)$$

Consider now a square wave, as in Figure A3-2, where we have made the width of the pulse equal the interpulse period divided by a constant. We solve for the coefficients with

$$A_n = \frac{1}{2\pi} \int_{-\pi/k}^{\pi/k} e^{-jnk} \, dx$$

for $n = 0$, $A_0 = 1/k$ and for $n \neq 0$, $A_n = \frac{1}{k}\frac{\sin(n\pi/k)}{n\pi/k}$, giving a spectrum like Figure A3-3. The spectrum of Figure A3-3 has an envelope of the $\sin(x)/x$

FIGURE A3-2 ∎
Another Square
Wave

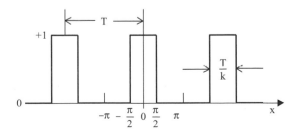

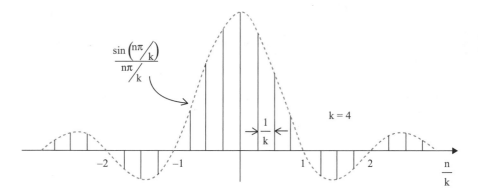

FIGURE A3-3 ▪
Another Spectrum of
a Square Wave

shape, which we know to be the Fourier transform of a single pulse and the larger the value of k, the closer the spectral lines are together and the more closely a continuous function is replicated. No matter how large we make k, however, this envelope is still caused by lines of discrete frequencies. Those discrete frequencies can exist because the time function they are describing goes on forever. If it starts and stops, there will be a necessity of continuum of frequencies; that is, the A_n will blend. There will be an A_n for every incremental distance along the frequency axis, not just at discrete points. The A_n will become a function of frequency, say $g(\omega)$, and the integer n will become an ω on the other side of the equation. The limits of integration must be extended to find the stopping point of the time function. These changes give us

$$g(\omega) = \frac{1}{2\pi} \int_{-\infty}^{\infty} f(t)e^{-j\omega t}\,dt \qquad (A3\text{-}14)$$

for the frequency function and, by symmetry,

$$f(t) = \int_{-\infty}^{\infty} g(\omega)e^{j\omega t}\,d\omega \qquad (A3\text{-}15)$$

These are called the Fourier integral equations or the Fourier transform pair. In this book, waveform time functions are transformed into their spectra. For this transform, Equations (A3-14) and (A3-15) are applicable. Let us do an example with a single square pulse of height one (1) and duration τ

$$g(\omega) = \frac{1}{2\pi} \int_{-\tau/2}^{\tau/2} 1\,e^{-j\omega t}\,dt \qquad (A3\text{-}16)$$

$$g(\omega) = \frac{1}{j\omega 2\pi} \left[-e^{-j\omega\frac{\tau}{2}} + e^{j\omega\frac{\tau}{2}} \right] \qquad (A3\text{-}17)$$

$$g(\omega) = \frac{1}{\pi\omega} \sin\left(\omega\frac{\tau}{2}\right) = \frac{\tau}{2\pi} \frac{\sin\left(\omega\frac{\tau}{2}\right)}{\omega\frac{\tau}{2}} \qquad \text{(A3-18)}$$

FIGURE A3-4 ■ A Single Pulse and Its Fourier Transform

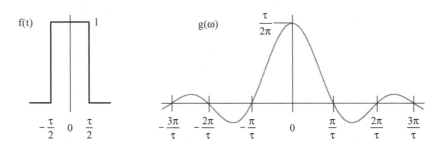

The results are shown in Figure A3-4, and g(ω) can be converted back into f(t) rather easily by using the form g(ω) takes in Equation (A3-17) to make up the integrand and remembering to integrate with dω from −∞ to ∞. Fourier transforms can also be used to transform antenna illumination functions into far-field antenna gain patterns. Note the similarity of Equation (A3-18) with an antenna gain pattern equation.

Appendix 4: Answers to Exercises

The answers listed here are rounded to two decimal places, so do not be concerned if you did not get exactly the same value. A solution set (Mathcad 14 and PDF files) for the vast majority of the exercises is available from the publisher's website.

Chapter 2   Radar Systems

1. R = 100.5 km, Δt = 2.56 seconds
2. Power density = 9.03×10^{-16} W/m^2
3. N $- 7.97 \times 10^{-15}$ watts $= -140.99$ dBW $\equiv -110.99$ dBm, T_e = 864.51 K, $\Delta N = 5.97 \times 10^{-15}$ watts $\equiv -142.24$ dBW $\equiv -112.24$ dBm; the additional noise is the dominant component of the receiver thermal noise power
4. L_R = 9.11 $\equiv$ 9.59 dB
5. (a) ERP$_R$ = 3.18×10^9 W $\equiv$ 95.03 dBW, (b) 0.01 W/m^2, (c) 0.06 W, (d) 1.99×10^{-13} W/m^2, (e) S = 7.94×01^{-14} W $= -130.999$ dBW, (f) the radar received approximately 23 orders of magnitude (power of 10) less power than it transmitted, (g) N = 1.19×10^{-14} W $\equiv -139.22$ dBW, (h) SNR = 6.65 $\equiv$ 8.23 dB
6. R_{dt} = 393.49 km
7. $C_A = 2C_P$

Chapter 3   Target Detection

1. $N_d = 2.7 \times 10^7$, detections per second = 9.6×10^8, $P_{fa} = 1.74 \times 10^{-11}$, FAR = 0.02 false alarms per second, $N_{fa} = 4.17 \times 10^{-4}$, SNR = 24.78 $\equiv$ 13.94 dB
2. D_0 = 13.2 dB using Equations (3-20) or (3-22), same for Figure 3-9 and Table 3-1, D_0 = 13.14 dB using Equations (3-25) and (3-26)
3. D_0 = 11.24 dB, $D_1 = D_2$ = 12.77 dB, $D_3 = D_4$ = 12.01 dB, $R_{dt\_SW1} = R_{dt\_SW2}$ = 91.58 km, $R_{dt\_SW3} = R_{dt\_SW4}$ = 95.7 km
4. D_0 = 14.24 dB, $D_1 = D_2$ = 22.4 dB, $D_3 = D_4$ = 18.32 dB, $R_{dt\_SW1} = R_{dt\_SW2}$ = 62.5 km, $R_{dt\_SW3} = R_{dt\_SW4}$ = 79.06 km

5. $P_d < 5\%$, $n_p = 32$, coherent; $G_I = 32 \equiv 15.05$ dB, $SNR_n = 20.16 \equiv$
   13.05 dB, $85\% < P_d < 90\%$, noncoherent; $G_I = 13.93 \equiv 11.44$ dB,
   $SNR_n = 8.79 \equiv 9.44$ dB, $15\% < P_d < 20\%$

6. $n_p = 5$ coherent pulses (rounded up to an integer from 4.68), $n_p = 8$ non-coherent pulses (rounded up to an integer from 7.61)

7. $G_{I0} = 12.16 \equiv 10.85$ dB, $G_{I1} = 11.97 \equiv 10.78$ dB, $G_{I2} = 17.1 \equiv 12.33$ dB,
   $G_{I3} = 12.07 \equiv 10.82$ dB, $G_{I4} = 14.42 \equiv 11.56$ dB, range factors;
   SW1 = 0.91, SW2 = 0.997, SW3 = 0.96, SW4 = 0.999

8. $G_{I0} = 15.33 \equiv 11.86$ dB, $G_{I1} = 14.08 \equiv 11.49$ dB, $G_{I2} = 94.51 \equiv 19.75$ dB,
   $G_{I3} = 14.69 \equiv 11.67$ dB, $G_{I4} = 38.07 \equiv 15.81$ dB, range factors;
   SW1 = 0.61, SW2 = 0.98, SW3 = 0.78, SW4 = 0.99

9. $R_{dt\_SW1} = 97.9$ km, $R_{dt\_SW2} = 157.6$ km, $R_{dt\_SW3} = 98.94$ km, $R_{dt\_SW4} =$
   125.53 km

10. $P_{dc} = 0.994$, $P_{fac} = 10 \times 10^{-6}$, range factor = 1.20

11. $P_{fa} = 10^{-12}$, $N_{fa} = 24$, 24 additional or 3.75% more antenna beam positions, 0.6 additional seconds or 3.75% more time

12. $P_d = 0.3456$, 0.3456, 0.1536, and 0.0256 for exactly 1-out-of-4, 2-out-of-4, 3-out-of-4, and 4-out-of-4, respectively; $P_{fa} = 3.988 \times 10^{-3}$,
    5.988×10^{-6}, 3.996×10^{-9}, and 1×10^{-12} for exactly 1-out-of-4, 2-out-of-4, 3-out-of-4, and 4-out-of-4, respectively; $P_d = 0.8704$, 0.5248,
    0.1792, and 0.0256 for at least 1-out-of-4, 2-out-of-4, 3-out-of-4, and 4-out-of-4, respectively; and $P_{fa} = 3.994 \times 10^{-3}$, 5.992×10^{-6},
    3.997×10^{-9}, and 1×10^{-12} for at least 1-out-of-4, 2-out-of-4, 3-out-of-4, and 4-out-of-4, respectively

13. $P_d = 0.5931$, $P_{fa} = 9.9985 \times 10^{-12}$, and $SNR_{dt} = 14.3$ dB

14. $R_{dt} = 76.74$ km

15. $G_I = 9.75 \equiv 9.89$ dB, $(S/N)_n = 64.79 \equiv 18.12$ dB, $R_{dt} = 213.29$ km

Chapter 4   Radar Antennas

1. IR, a = 5.64 millimeters; RF, a = 28.21 meters

2. N = 400 elements, A = 100 m^2

3. No answer; exercise is a proof

4. $G = 6.28 \times 10^3 \equiv 37.98$ dBi, $\theta_{3dB\_az} = 1.02°$, $\theta_{3dB\_el} = 2.55°$

5. Normalized gain at $5° = 4.34 \times 10^{-3} \equiv -23.63$ dBi; gain at $5° = 11.37$ dBi

6. $\theta_0 = 4.78°$

7. $\theta_1 = 14.48°$, wrap-up factor L/d = 34.64

8. $\Delta\theta_0 = 0.196°$

9. $\theta_{3dB} = 2.76°$ for 25°, 2.89° for 30°, and 3.54° for 45°

10. G = 31.07 dBi for 25°, 30.88 dBi for 30°, and 29.99 dBi for 45°

Chapter 5   Radar Measurements and Target Tracking

1. $\Delta R = 75$ m, $R_u = 500$ km
2. Dead zone $= 75$ km
3. $B_{pc} = 5$ MHz, $T_I = 1 \times 10^{-3}$ sec
4. PRI $= 2.67 \times 10^{-4}$ sec, PRF $= 3.75$ kHz, $\tau = 2 \times 10^{-7}$ sec, $T_I = 1.67 \times 10^{-3}$ sec
5. (a) PRF $= 2.4 \times 10^{4}$ Hz, (b) $R_u = 6.25$ km
6. (a) D $= 17.72$ meters; (b) $\tau = 0.1$ μsec, $T_I = 0.0167$ sec
7. (a) $\tau B_{pc} = 20{,}000$; (b) $\Delta R_{pc} = 1.5$ meters
8. No answer, exercise is a proof
9. $\Delta R = 300$ m, which does not meet the requirement, $N\phi = 60$
10. $n_{rg} = 1000$, $n_{df} = 40$, $n_b = 360$
11. (a) $\tau = 2 \times 10^{-7}$ sec, $T_I = 3.33 \times 10^{-3}$ sec, and PRF $= 3.6 \times 10^{4}$ Hz, (b) $R_u = 4.17$ km, (c) At UHF $\Delta R = 30$ m (meets the requirement), $\Delta R_{dot} = 90$ meters/sec (does not meet the requirement), and $R_{dotu} = \pm 5400$ meters/second (exceeds the requirement)

Chapter 6   Target Signature

1. (a) Use Equation (6-5) or Table 6-1 for the RCS of a flat plate, (b) use the effective area in place of the physical area
2. $f_c = 22.77$ GHz (average of the two frequencies)
3. $\theta = 5.56 \times 10^{-3}$ radians $\equiv 0.32$ degrees
4. The RCS of the trihedral corner reflector (339.29 m^2) is 50% more than the dihedral corner reflector (226.19 m^2). Thus, the RCS statement is true. The detection range of the radar for the trihedral corner reflector (11.07 km) is 11% more than for the dihedral corner reflector. Thus, the detection range statement is false.
5. $\bar{\sigma} = 1.35 \times 10^{3}$ m^2
6. $A_c = 5.89 \times 10^{4}$ m^2, $\sigma_c = 589.41$ m$^2 \equiv 27.7$ dBsm

Chapter 7   Advanced Radar Concepts

1. S/C $= 0.02 \equiv -17.19$ dB
2. No answer, exercise is a proof
3. For airborne pulse-Doppler radar systems, the clutter spectrum is complex (see Figure 7-8). The rotating helicopter blades and tail rotor and the vibration of the airframe result in complex target signal at varying Doppler shifts (see Section 6.3), some lost in the clutter and some separated from the clutter.

4. (a) $f_{dmax} = \dfrac{2\,V_{ac}}{D}$ for both ranges; (b) $T_t = \dfrac{R\,\lambda}{V_{ac}\,D}$ and $\dfrac{2\,R\,\lambda}{V_{ac}\,D}$;

(c) $N_f = \dfrac{2\,R\,\lambda}{D^2}$ and $\dfrac{4\,R\,\lambda}{D^2}$; (d) $d_a = D/2$ for both ranges

5. (a) $\Delta f_d = 1.75$ Hz, (b) $T_I = 0.57$ sec, (c) L = 171.89 m, (d) $d_a = 4.36$ m; with squint and elevation angles (c) L = 119.696 m, (d) $d_a = 6.27$ m

6. 9.93×10^5 cells/second

7. (a) $PRF_{max} = 1.67$ kHz, (b) $V_R = 416.67$ m/sec; with squint and elevation angles (b) $V_R = 482.96$ m/sec

8. (a) $d_a = 1.28$ m, (b) $PRF_{min} = 376.99$ Hz, (c) $R_{sw} = 397.89$ km

9. (a) For $\theta_p = 60°$ DBS_a limited to 23.31 m by having to be unfocused. For $\theta_p = 30°$ DBS_a limited to 40.37 m by having to be unfocused. (b) For $\theta_p = 60°$ $DBS_a = 29.62$ m. For $\theta_p = 30°$ $DBS_a = 51.31$ m. All these results are the same as in Figure 7-14.

10. $R_1 R_2 = 730.58$ km^2

11. (a) $T_I = 0.032$ seconds, (b) PRF = 2.381 kHz and $T_I = 0.0134$ seconds, (c) 0.0186 seconds additional time and 2.381 additional search waveforms (beam positions)

12. (a) Isolation = $1.58 \times 10^{-9} \equiv -88.03$ dB, (b) Isolation = $4.46 \times 10^{-8} \equiv -73.51$ dB, (c) direct signal power = 0.16 W $\equiv -8.03$ dBW, (d) direct signal out of FM/CW processor = -81.53 dBW, (e) needed, the target signal is -51.47 dB under the direct signal

13. D = 0.0134 meters

Chapter 9   Electronic Warfare Receivers

1. $ERP_R = 9.98 \times 10^8$ W $\equiv 89.99$ dBW, 3.53×10^{-3} W/m^2, $S_{RWR} = 4.02 \times 10^{-7}$ W $\equiv -63.95$ dBW, the received radar power is approximately 15 orders of magnitude less than the transmitted ERP_R – mainly due to the $1/R^2$ propagation

2. $F_{RWR} = 416.46 \equiv 26.196$ dB, $(S/N)_{RWR} = 40.24 \equiv 16.05$ dB

3. $R_{dtRWR} = 169.21$ km

4. $R_{dtRWR} = 37.02$ km

5. $R_{RT} = 26.75$ km

Chapter 10   Self-Protection Jamming Electronic Attack

1. (a) $ERP_J = 31.47$ W $\equiv 14.98$ dBW, (b) jammer power density = 1.11×10^{-10} W/m^2, (c) J = 2.8×10^{-11} W $\equiv -105.53$ dBW, the received jammer power is approximately 12 orders of magnitude less than the transmitted ERP_J – mainly due to the $1/R^2$ propagation

2. (a) $J_N = 1.05 \times 10^{-13}$ W $\equiv -129.79$ dBW, (b) I $= 1.17 \times 10^{-13}$ W $\equiv$ -129.32 dBW, (c) $(S/I)_n = 6.62 \equiv 8.21$ dB, because of the noise jammer the $(S/I)_n$ is about an order of magnitude less than the $(S/N)_n$ from Exercise 3-15

3. $R_{bt} = 102.29$ km which is about half the detection range from Exercise 3-15, $(J_N/S)_n = 0.14 = -8.67$ dB

4. $(J/N)_n = 2.28 \times 10^4 \equiv 43.59$ dB, which is almost 3 orders of magnitude higher than $(S/N)_n$ from Exercise 3-15

5. $(J/S)_n = 352.596 \equiv 25.47$ dB, $R_{dt} = 5.69 \times 10^3$ km, which is remarkably greater than the true target detection range from Exercise 3-15

6. $J/N = 24.9 \equiv 13.96$ dB, and thus the radar receiver thermal noise does not need to be included

7. $S/J = 1.25 \equiv 0.98$ dB for 1 m^2 RCS, $S/J = 12.52 \equiv 10.98$ dB for 10 m^2 RCS

8. $R_{bt} = 112.17$ km for 1 m^2 RCS, $R_{bt} = 354.72$ km for 10 m^2 RCS

9. $(J_{PE}/S)_n = 1.06 \times 10^3 \equiv 30.26$ dB

Chapter 11   Support Jamming Electronic Attack

1. (a) $ERP_J = 1.26$ kW $= 31$ dBW, (b) jammer power density $= 1.002 \times 10^{-8}$ W/m^2, (c) $J = 2.52 \times 10^{-11}$ W $\equiv -105.98$ dBW, the received jammer power is approximately 14 orders of magnitude less than the transmitted ERP_J – mainly due to the $1/R^2$ propagation

2. (a) $J_N = 9.45 \times 10^{-14}$ W $\equiv 130.24$ dBW, (b) I $= 1.06 \times 10^{-13}$ W $\equiv 129.73$ dBW, (c) $(S/I)_n = 7.27 \equiv 8.62$ dB, because of the noise jammer the $(S/I)_n$ is about an order of magnitude less than the $(S/N)_n$ from Exercise 3-15

3. $R_{bt} = 123.4$ km, which is about half the detection range from Exercise 3-15, $(J_N/S)_n = 0.12 = -9.13$ dB

4. $(J/N)_n = 2.06 \times 10^4 \equiv 43.13$ dB, which is almost 3 orders of magnitude higher than $(S/N)_n$ from Exercise 3-15

5. $(J/S)_n = 317.33 \equiv 25.02$ dB, $R_{dt} = 3.6 \times 10^3$ km, which is remarkably greater than the true target detection range from Exercise 3-15

6. $ERP_J = 31.83$ kW

Chapter 12   Electronic Protection Concepts

1. (a) The jammer bandwidth must increase by a factor of 20,000 thus reducing the power spectral density; (b) No consequence; see the solution sheet on the website for the details

2. 25% of full power

3. (a) $P_Z/N = 1.99 \equiv 2.99$ dB; (b) $P_Z/N = 1.01 \equiv 0.41$ dB

4. $CR = 50.75 \equiv 17.05$ dB; $CR = 10.01 \equiv 10.004$ dB

5. (a) $(S/N)_n = 39.81 \equiv 16$ dB; (b) $(S/I)_n = 0.39 \equiv -4.04$ dB; (c) $(S/I)_n = 20.005 \equiv 13.01$ dB; (d) $(S/I)_n = 3.62 \equiv 5.59$ dB; (e) $(S/I)_n = 36.22 \equiv 15.59$ dB

Chapter 13   Loose Ends of Radar and/or Electronic Warfare Lore

1. $R_{LOS} = 197.38$ km
2. 73 ground-based radar systems (rounded up from 72.46)
3. 7 airborne radar systems (rounded up from 6.77)
4. (a) $L_a = 0.6$ dB, (b) $L_R = 6.6$ dB
5. For $N_e = 10^4/cm^3$: $\alpha = 4.44 \times 10^{-6}$, $\eta = 999.9998 \times 10^{-3}$; for $N_e = 10^5/cm^3$: $\alpha = 44.44 \times 10^{-6}$, $\eta = 999.9978 \times 10^{-3}$; $\Omega = 7 \times 10^{13}$; atmospheric attenuation in the troposphere $= 7 \times 10^{-3}$ dB/km
6. $R_{ff} = 133.34$ km
7. $R_{ff} = 5.04$ km
8. $R_{ff} = 80$ km

Appendix 5: Glossary

This glossary provides assistance in understanding terms used in this book. There may be more general, more specific, or entirely different meanings for these terms when they are used elsewhere.

2D: two-dimensional; radar system provides range or range rate and either azimuth or elevation angle measurements, for a total of two measurements.

3D: three dimensional; radar system provides range or range rate and both azimuth and elevation measurements, for a total of three measurements.

α-β Tracker: a target state tracker based on a linear target model, provides estimates of current target states and predicted target states.

Active expendable: a powered (active) jammer ejected from the platform being protected, a form of electronic attack.

Adaptive array: an electronically steerable antenna designed to respond automatically and optimally to a variety of situations.

Adaptive receiver: a receiver with circuitry or perhaps programming enabling it to adjust to the characteristics of the incoming signal.

AESA: active electronically steered (or scanned) array, a type of radar antenna.

AGC: automatic gain control, keeping the excursions of the received signal within bounds with an automatic negative feedback loop.

Ambiguous angle: an angle where the characteristics of an antenna pattern are such that there is uncertainty about the angular location of a target (as with an interferometer).

Ambiguous range: the condition where the round-trip time between pulses is insufficient to accommodate all the targets the radar will see, resulting in uncertainty about target range.

Ambiguous range rate: the situation that occurs when the Doppler processing scheme of the radar is such that target range rates fold over, making those range rates uncertain.

Ampere's law: current passing through a conductor creates a magnetic field around the conductor; named after the man who discovered the phenomenon.

Amplitude: the "height" of a trace on a display or record, usually referring to a measurement of voltage or power.

Analog error signals: indicators derived from a continuous process that permits correction while the process is going on, as with a radar mainbeam tracking a target.

Analog signal processing: performing various operations on signals while they are still in their received form, that is, continuous rather than quantized or digitized.

Analog-to-digital (AD) converter: electronic circuitry that samples an incoming signal and assigns a number (or a computer word) that describes each sample.

Angle resolution: the capability of radar system to separate two objects in angle.

Angle tracking: to follow a target in angle, by whatever means.

Angular error: the amount by which an antenna or other angle measuring system fails to indicate the exact angle of a target.

Angular rate: the rapidity with which a target's position changes in angle.

Antenna pattern: the "intensity" of the field at locations around the antenna at long distances from it; usually established by taking a series of measurements. Also, the distribution on antenna gain as a function of angle, usually with respect to the boresight (broadside) of the antenna.

Anti-jam gain or margin: the ratio by which a particular waveform or other technique is able to mitigate the effects of jamming.

AOA: angle of arrival of a radar pulse, associated with an electronic support receiver.

Aperture: literally an "opening"; its size determines the amount of the electromagnetic wave intercepted; thus, any antenna is called an aperture.

Arecibo: the location in Puerto Rico of a thousand-foot-aperture ionospheric and astronomic radar.

Array factor: the component of an array antenna gain pattern due to the array of individual elements.

Aspect angle: the angle made by some arbitrary characteristic of the target, say, its axis, with the axis of the radar antenna mainbeam.

Auroral ionosphere: the ionosphere near the earth's magnetic poles when excited by magnetic storms.

Azimuth: the angle from a fixed point, say, due north, in the plane of the earth's surface (horizontal plane).

Barrage noise jamming: wideband disrupting and interfering noise jamming continuously covering all the radar frequencies being radiated.

Beam steering: moving an antenna pattern about the sky, usually by electronic means.

Beamwidth: the lateral dimension (in angle) of the principal lobe (mainlobe or main beam) of an antenna pattern.

Binomial (M-out-of-N) detection: a radar detection theory approach using multiple detection attempts based on the binomial theorem.

Bistatic radar: a radar system whose transmit and receive antennas are substantially separated.

Blanking circuit: an electronic scheme by which particular range or angle locations in the radar coverage are wiped out.

Blass array: an electronically steered array consisting of stacked beams that are turned on and off with a waveguide matrix; named after its inventor.

Blind ranges: ranges where a filter that suppresses clutter at one range also suppresses signals at other ranges.

Blind speeds: speeds at which a filter that suppresses clutter at one range rate also suppresses signals at other range rates.

Boltzmann's constant: the number by which temperature (in Kelvin) is related to energy (in joules or watt-seconds) per hertz of bandwidth; it is 1.38×10^{-23} joules/Kelvin, named after the man who developed the phenomenon.

Boresight: the forward direction on the axis of an antenna; loosely, the direction a reflector antenna is pointing or the direction perpendicular to the surface of any antenna.

Burnthrough range: the radar detection range is the presence of a noise jammer.

Butler array or matrix: an electronically steered array with the characteristics that there are as many beams formed as there are elements, that they are orthogonal, and that the output is the Fourier transform of the input; named after its inventor.

Carrier frequency: the rate of oscillation of the radio frequency waves that carry a signal through space.

Cassegrain feed: an antenna feed patterned after the optical telescope feed of the same name; the feed is located at the center of a parabolic reflector, reflecting onto the reflector via a hyperboloid at its focus.

CFAR: constant false alarm rate; keeping noise level constant by normalizing to a noise sample taken near the target.

Chaff: small, light pieces of material that collectively have high radar cross section.

Circular polarization: a characteristic of radio frequency waves whose electric vector rotates 360° during each radio frequency cycle.

Clutter: unwanted and interfering radar returns from objects in the environment other than targets.

Clutter coefficient: the ratio of the return received from clutter to what would have been received from a perfectly conducting isotropic radiator of the same physical area.

Clutter fence: a screen around a radar site that prevents low-angle clutter sources from being illuminated.

Coherence: the preservation of fixed-phase relationships over time in the conduct of radar operations.

Collision frequency: the rate at which the free electrons in an ionized medium collide with ions or atoms.

Comb filters: an array of filters, arranged like the teeth of a comb, whose response frequencies are close together.

Complex: signal description including both its amplitude and phase components.

Cone sphere: the shape created by a sphere capping a truncated cone; similar in shape to an ice cream cone.

Conical scan: a tracking technique in which an antenna feed or the antenna itself makes a small, circular motion, sequentially comparing returns to obtain more accurate angle information about the target.

Convolution: multiplying the overlapping portions of two functions continuously as one is moved across the other.

Corner reflector (dihedral and trihedral): radar targets designed to have high retrodirective radar cross section. The dihedral has two faces and the trihedral has three.

Constant gain jammer: an active jammer that amplifies the received radar pulse by a constant gain and transmits it back to the radar system.

Constant power jammer: an active jammer that transmits a fixed power signal to the radar system.

CPI: coherent processing interval, also called the coherent integration time or pulse burst duration.

Cross-range: a measurement orthogonal (perpendicular) to the radar-to-target range vector (axis of an antenna), differentiated from arc length derived from an angle measurement.

Cumulative detection: a radar detection theory approach based on cumulative probability theory using multiple detection attempts.

CW: continuous wave, a radio frequency wave that is on all the time.

Data link: a communications channel for information, usually digitized and wide bandwidth.

dBi: decibels relative to an isotropic (gain of one everywhere) antenna.

dBm: decibels relative to one milliwatt (mW), often used for receiver sensitivity.

dBsm: decibels relative to one square meter (m^2), used for target radar cross section.

DBS: Doppler beam sharpening, a form of imaging radar system using Doppler signal processing to provide fine cross-range resolution.

dBW: decibels relative to one watt, often used for transmitter power and effective radiated power.

Dead zone: the region near a radar system from which returns are not received because the receiver is either turned off or not connected to the antenna while the transmitter is radiating.

Deceive: electronic attack concepts or techniques intended to confuse or mislead an adversary. From a radar perspective we often look at this in terms of inducing target measurement or track errors, thus deceiving the radar as to the actual target measurements or track.

Decibel (dB): ten bels, a bel being the logarithm to the base 10 of the ratio of output power to input power; named after Alexander Graham Bell, who developed them while inventing the telephone.

Degrade: electronic attack concepts or techniques intended to interfere with the adversary's efficient processing of information to limit attacks on friendly forces. From a radar perspective we often look at this in terms of degrading (e.g., slow down, miss) the effective processing of target detections, measurements, and tracks.

Deny: electronic attack concepts or techniques intended to control the information an adversary receives and to prevent the adversary from gaining accurate information about friendly forces. From a radar perspective we often look at this in terms of denying target detection and/or associated measurements.

Difference (Δ) beam: the remainder or residue that results when the voltage from a signal in one beam is subtracted from that in an adjoining beam, as with monopulse angle tracking.

Diffraction grating: a matrix of narrow slits that breaks light up into fringes caused by the constructive and destructive interference of light waves.

Diffuse: to break up and distribute on reflection, as with an incident electromagnetic wave.

Digital signal processing: performing various operations on signals after they have been converted to digital form, as differentiated from analog signal processing.

Directional coupler: a switch that connects one electric circuit with another in such a way that signals move easily in one direction but not in the other.

Distributions: various probability density functions of mathematical statistics (exponential, gamma, Gaussian (normal), and Rayleigh) that find applications in radar theory.

Doppler ambiguity: a condition in which Doppler data fold over so that there is uncertainty about the true Doppler frequency.

Doppler sidelobes: the residue of a filter's frequency response that appears in adjacent filters.

Doppler spread: the band of frequencies within which Doppler returns might occur.

Downrange: away from the radar along the axis of its antenna.

DRFM: digital radio frequency memory, a form of electronic attack that captures a digital representation of the radar signal.

E-field: electric field, one of the two components of an electromagnetic wave, perpendicular to the magnetic field (H-field).

Effective radiated power (ERP): the product of the peak transmit power and the antenna gain. ERP is the cousin of the power-aperture product.

EHF: Extremely high frequency; radio frequency band.

Electric vector: the direction and magnitude of the voltage measured in an electromagnetic field.

Electron density: the number of free electrons per unit volume of an ionized gas or plasma.

Electronic attack (EA): using electronic techniques to disrupt radar and communications; a component of electronic warfare.

Electronic countermeasures (ECM): using electronic techniques to disrupt radar or communications; a component of electronic warfare.

Electronic counter-countermeasures (ECCM): activities to counter electronic contermeasures; a component of electronic warfare

Electronic protection (EP): are activities to counter electronic attacks; a component of electronic warfare.

Electronic support (ES): electronic techniques used to intercept, identify, and locate radar systems; a component of electronic warfare.

Electronic support measures (ESM): electronic techniques used to intercept, identify, and locate radar systems; a component of electronic warfare.

Electronic warfare (EW): the reaction to an adversary's use of radar systems; divided into three components: electronic support; electronic attack; and electronic protection.

Element factor: the pattern of individual elements of an array antenna.

Energy: power times time (watt-seconds or joules), also called power spectral density (watt/hertz).

Envelope: the shape, amplitude, or modulation of a signal after the radio frequency carrier has been removed; the post-detection content of the signal.

Envelope integration: building up the signal by summing two or more signal envelopes; post-detection integration.

erf(x): a form of the integral of the standard Gaussian distribution in which the standard deviation has been made narrower by the square root of two, also called the "error function."

erfc(x): the complementary error function, used in radar detection theory.

ERP: effective radiated power of a radar or jammer (peak transmitter power times transmit antenna gain).

ESA: electronically steered (or scanned) array, a type of radar antenna.

Ether: an imagined medium by which, it was thought until the nineteenth century, light waves were able to propagate in space.

False-alarm probability: the likelihood that noise alone will cross a threshold and be erroneously accepted as a signal.

False-alarm rate: the frequency with which noise alone crosses a threshold and is erroneously accepted as a signal.

False target jamming: an electronic attack waveform similar to the radar waveform.

Fan beams: antenna patterns whose mainlobes are in the shape of a fan.

Faraday's law: a time-varying magnetic field will induce a voltage in a circuit immersed in that field; also known as the induction law; named after its discoverer.

Faraday rotation: the turning of the electric vector of an electromagnetic wave as it passes through an ionized medium, an action that occurs due to Faraday's law.

Far field: the region sufficiently far from an antenna that the phases of wavelets arriving from the antenna edges will be negligibly different from those arriving from the center; similar relationship for the radar cross section of a target.

Fast Fourier transform (FFT): an algorithm for efficiently calculating the frequency content of a digitized time function.

Fat beams: antenna patterns whose mainlobes occupy a relatively large solid angle.

Feedhorn: the expanding end of a waveguide that acts as a launcher of electromagnetic waves.

FM chirp: a frequency-modulated signal that smoothly changes frequency upward or downward during its transmission; if it were a sound wave, it would be heard as a chirp.

FM/CW: frequency modulated continuous wave modulation.

FM ramp: the ramp like shape of the plot of a waveform whose frequency is changing either upward or downward during transmission.

Fourier transforms: a mathematical operation that changes a function to reveal its characteristics in a different dimension; named after its inventor.

Fraunhoffer diffraction: the far-field patterns that appear when light is propagated through a narrow slit or grating.

Frequency ambiguity: uncertainty about the true Doppler shift of a target because of fold over in the Doppler processor.

Frequency diversify: a characteristic of radar systems that can radiate at any one of a large number of frequencies; a design to reduce jamming vulnerability.

Frequency domain: the dimension in which a function is evaluated for its spectral content.

Frequency scanning: a technique by which the mainbeam of an array antenna surveys a solid angle by changing its carrier frequency.

Gain: the focusing power of an antenna relative to an isotropic (gain of one (1) everywhere) antenna; or the power of a processor to build up the signal-to-noise ratio by applying various techniques.

Galactic noise: unwanted and interfering electromagnetic radiation that enters the radar from the cosmos.

Gaussian: refers to the normal probability distribution; phenomena whose events are normally distributed.

Geometric optics region: the region where the scattering of electromagnetic waves is from objects whose characteristic dimension is much larger than a wavelength.

Geosynchronous: synchronized with the turning of the earth; satellites in orbits whose period is 24 hours and whose inclination is zero degrees are geosynchronous.

gigahertz (GHz): billions (10^9) of cycles per second.

GMTI: ground moving target indication, a form of signal processing that indicates moving targets against a stationary background.

Grating lobes: patterns in which signals appears at almost equal amplitudes in several locations, as with diffraction through multiple slits or interferometry.

Grazing angle: the angle an antenna mainbeam makes with the surface of the earth.

Gun barrel analogy: the inference that the trajectories of radar targets can be compared to those of bullets emerging from a gun barrel.

Half-wave dipole: a radiating or receiving element consisting of a straight conducting wire one-half as long as the wavelength of the associated radio frequency waves.

Hamming weighting: tailoring the amplitude of an antenna illumination function to reduce antenna sidelobes, or a waveform to reduce measurement sidelobes, so that it is the shape of a cosine on a pedestal; named after its designer.

Haystack radar: a very large, high-precision reflector antenna radar atop Haystack Hill in Massachusetts.

HF: High frequency, radio frequency band.

Horizontal polarization: the condition of a radio wave whose electric vector is in the plane of the earth's surface.

Huggins beam steering: a method of moving a radar mainbeam about the sky by adding and then subtracting the correct phases from the carrier frequency; named after its inventor.

Hybrid: a device that employs a mixture of two or more electronic technologies, such as mixing tubes and solid-state devices, analog and digital processing, and waveguide and electronic circuits.

Hypothesis: test a method of decision-making in mathematical statistics in which a threshold is set at a predetermined level, fixing the probability of incorrectly accepting or rejecting the hypothesis.

I/Q: in-phase (I) and quadrature (Q) representation of signals.

IF: intermediate frequency, associated with mixers in receivers, 10's MHz to 100's MHz.

Illumination function: the voltage or power pattern with which an antenna is excited, thereby producing an electric field (E-field).

Incremental sources: imaginary small radiators of electromagnetic waves used as a convenience for developing theory.

Inertial guidance: a completely autonomous system of navigation that measures accelerations and deduces other quantities from those measurements and original position.

Interferometer: a system that uses phase differences (constructive and destructive interference) to determine angular position to high accuracy.

Interpulse period: the time interval between radar pulses, pulse repetition interval, the reciprocal of the pulse repetition frequency.

Inverse synthetic aperture: the use of the deterministic rotation of an object in a radar mainbeam to derive differential Doppler information and thereby resolve the object in angle; the "inverse" of having the radar mainbeam pass across the target.

Ionospheric radar systems: radar systems that irradiate the ionosphere so that scientific information can be derived from the noncoherent backscatter received.

Ionospheric sounders: systems that probe the ionosphere by transmitting rapidly varying frequencies toward it; returns are evaluated to obtain estimates of electron density as a function of height.

Isotropic radiator: an imaginary source of electromagnetic waves that radiates equal amplitudes and phases in all directions.

Jammer: a system designed to provide electronic attack techniques directed against radar systems.

Jammer receiver processor: a form of electronic support used to sense radar systems to support electronic attack.

JEM: jet engine modulation, a frequency domain characteristic of a target signature.

Jitter: small, rapid, perhaps random fluctuations about an intended point or location.

Kalman filter: a well-known tracking algorithm (named after its creator) that weights measurements according to their quality to optimize results.

Keplerian motion: movement dictated only by the forces of gravity, such as the movement of planets and satellites.

kilohertz (kHz): thousands (10^3) of cycles per second.

kilometers (km): thousands (10^3) of meters.

kilowatts (kW): thousands (10^3) of watts.

Klystrons: high-power amplifiers of radio frequency energy, capable of coherent operation.

Laser radar: a radar system at optical, infrared, or ultraviolet frequencies.

LFMOP: linear frequency modulation on pulse, a form of frequency modulation pulse compression.

Lidar: for "light detection and ranging"; a laser radar; sometimes Ladar.

Linear array: an array antenna consisting of radiating elements arranged in a line.

Line feed: a feed whose elements are arranged in a line; they may also be phased, as in a line feed that removes spherical aberration.

LOS: line of sight, accounts for the radar horizon associated with a round earth.

LPI radar: low probability of intercept radar; a radar system with waveform and antenna designed to minimize the power radiated in both spatial and spectral domains, often used as an electronic protection technique.

H-field: magnetic field, one of the two components of an electromagnetic wave, perpendicular to the electric field (E-field).

Mainbeam: that part of an antenna pattern that contains the major portion of the radio frequency signal.

Marcum: radar detection theory for a constant target signal in the presence of receiver noise, named after the person who developed the approach.

Matched filter: a filter in a radar receiver whose spectral response is matched to the spectral content of the transmitted waveform.

MDS: minimum detectable signal, a measure of the sensitivity of a receiver.

MDS: minimum discernible signal, a measure of the sensitivity of a receiver.

megahertz (MHz): millions (10^6) of cycles per second.

megawatt (MW): millions (10^6) of watts.

microseconds (μsec): one one-millionth (10^{-6}) of a second.

microwatt (μW): one one-millionth (10^{-6}) of a watt.

milliradian: an angle measure of one one-thousandth of a radian, i.e., 0.0573 degrees.

millisecond (msec): one one-thousandth (10^{-3}) of a second.

milliwatts (mW): one one-thousandth (10^{-3}) of a watt.

MMIC: monolithic microwave integrated circuit; refers to an all solid-state array element integrated on a single piece of substrate.

Modulation: the impression upon the radio frequency carrier of the signal fluctuations.

Monopulse tracking: deriving out of a single pulse all the information necessary to obtain angle measurements.

Monostatic radar: radar systems with collocated transmit and receive antenna.

Moving target indication (MTI): the use of the Doppler content of the radar returns to achieve cancellation of clutter at zero range rate (or some other specific range rate).

MPM: microwave power module, a hybrid tube solid-state radio frequency transmitter.

MSA: modeling, simulation, and analysis.

MTD: moving target detection, a form of clutter cancellation using MTI processing followed by pulse-Doppler processing.

MTI cancelers: the electronic circuits or signal processing algorithms that accomplish the clutter cancellation for MTI.

Newtonian trajectories: flight paths that have no forces acting upon them except the forces of gravity.

Nyquist theorem: to preserve information samples need to be made at least twice the rate (hertz) as the highest frequency present in the signal; named after the person who developed the theorem.

Noise: unwanted random amplitude and phase signals that mixes with and interferes with the wanted signal.

Noise jamming: a random amplitude and phase electronic attack waveform.

Noncoherent integration: the adding together of the envelopes or modulation of signals without regard for the phase of the carrier frequency; post-detection integration.

Normalize: to refer data to a convenient or common reference point and apply a standard interval from that point.

North filter: a filter that gives an optimum response for signal against Gaussian noise; also called a matched filter; named after the person who first analyzed it.

Nutate: to nod or wobble slightly.

Nutation: a slow or small rotation superimposed on a more rapid or larger one.

Order of magnitude: a power of 10, e.g., 100 (10^2) is two orders of magnitude.

Orthogonal polarization: orientation of the electric field of electromagnetic radiation (including light) at right angles to a reference electric field.

Oscillators: electric circuits that produce sinusoidal waves (waves that rise and fall smoothly and harmonically).

Over-the-horizon radar: radar that uses ionospheric reflection to detect targets beyond the horizon, operating in the radio frequency band that supports the phenomenon: the high frequency band.

Parabolic antenna: a device that radiates and focuses electromagnetic waves by use of the shape of the curve of a parabola.

Paraboloid: a surface made up by revolving a parabola about its axis; the shape of a parabolic antenna.

Parameter: an arbitrary constant that may take on or be assigned various values.

Passive expendable: the use of such things as chaff and decoys to defeat radar or communications passively (without power), a form of electronic attack.

PCL: passive covert radar, a form of bistatic radar systems using non-cooperative transmitters of opportunity.

PCR: passive covert radar, a form of bistatic radar systems using non-cooperative transmitters of opportunity.

PCR: pulse compression ratio, the ratio of the transmitted pulse width to the compressed pulse width.

PDW: pulse descriptor word, parameters (e.g., carrier frequency, pulse width) of a radar pulse detected by an electronic support receiver.

Pedestal: the structure that supports and moves a radar antenna.

Pencil beam: electromagnetic wave focused to a narrow angle in two dimensions as with a searchlight beam.

Phase array: an antenna that forms a mainbeam by assigning phases to a number of separate radiating elements.

Phase locked: phase held at a constant phase relationship by being tied electrically to a reference oscillator, usually a stable local oscillator.

Phase modulation: a form of pulse compression modulation where the phase of the carrier frequency is modulated over the duration of the radar pulse.

Phase shifter: electronic circuits that shift phase in discrete, predetermined steps.

Plan view: the perspective from above.

Plasma frequency: the rate of vibration of the electrons in an ionized medium.

Polarization: the alignment of the electric field with respect to the propagation vector.

Polarize: to align the electric vectors of electromagnetic radiation.

Potted: physically fixed in position by being embedded in a resin or other seal.

Power-aperture: the product of a radar system's transmit power and the physical aperture of its antenna, watts: square meter (m^2); cousin of effective radiated power.

Power density: power per unit area, watts per square meter (m^2).

Power spectral density: the power present in the frequency constituents of a function (usually a plot of these quantities).

Propagation: the outward spreading of electromagnetic waves.

Pseudorandom code: a series of random quantities (numbers or levels) whose randomness is unproven; or, random quantities that are replicated when generated for later autocorrelation.

Pulse-burst waveform: a finite train of radar pulses.

Pulse compression: a technique by which more bandwidth is inserted into a pulse than its duration would imply it could contain; used to provide fine range resolution.

Pulse-Doppler: a radar system or a waveform that uses a series of pulses that are processed for their range rate content.

PRF: pulse repetition frequency, the number of pulses transmitted per unit time (pulses per second or hertz), the reciprocal of pulse repetition interval.

PRI: pulse repetition interval, the time between transmitted pulses (seconds), the reciprocal of pulse repetition frequency.

Quanta: small discrete packages of energy.

Quantum mechanics: a theory of physics that treats the interactions of radiation and matter; its name derives from the observation that these interactions take place only in discrete packages.

Radar: an instrument for radio detection and ranging.

Radar altimeter: a radar system that measures the altitude of the platform carrying the radar.

Radar cross section: a measure of the amount of the electromagnetic wave a radar target intercepts and scatters back toward the radar.

Radar signature: identifying features or patterns in a target's radar cross section.

Radial velocity: the component of velocity vector on the radial toward or away from a point, for example, along the line of sight from a radar system.

Radian: form to express an angle in radians.

Radian frequency: to express frequency in radians per second rather than in cycles or degrees per second.

Radio frequency (RF): the portion of the electromagnetic spectrum, approximately 3 MHz to 300 GHz.

Random variable: in statistics, a function defined over a sample space; a nondeterministic variable.

Range error: the inaccuracy in a range measurement.

Range rate: the magnitude of the projection of the radar-to-target velocity vector (including both radar and target velocity vectors) on the radar-to-target range vector.

Range rate ambiguities: fold over of range rates in the signal processor, requiring additional processing to determine the actual range rates of the targets.

Range rate resolution: the ability to differentiate two targets in range rate.

Range rate spectrum: the frequencies generated by a moving target.

Range rate tracking: following a target in range rate.

Range resolution: the ability to differentiate two targets in range.

Range sidelobes: the residues of a pulse compression waveform that spill over into and contaminate adjacent range cells.

Range tracking: following a target in range.

Rayleigh probability density function (PDF): describes the receiver noise; named after Lord Rayleigh, who developed it; used in radar detection theory.

Rayleigh region: a region where electromagnetic waves are scattering from targets that are smaller than a wavelength; named after Lord Rayleigh, who calculated the magnitude of that scattering.

Rectification: processing of electric waves that swing both positive and negative into waves that swing only positive.

Refraction: the bending of electromagnetic waves that takes place as the medium varies over the propagation path.

Repeaters: equipment that receives and transmits a signal; it may reradiate a version of the received signal enhanced with new information (e.g., jamming modulation).

Resolution bin or cell: the extent of the radar resolution in angle, range, or range rate.

Resonance region: the region where the wavelength of the scattered electromagnetic wave is of the same order as the characteristic dimension of the target.

Rician probability density function (PDF): describes the combination of a constant amplitude target and receiver noise; named after the person who developed it; used in radar detection theory.

Root mean square (RMS): the square root of the average of the sum of the squares of a series of values.

Root sum square (RSS): the square root of the sum of the squares of a series of values.

RWR: radar warning receiver, a form of electronic support used to warn operators of hostile radar systems.

RSE: round smooth earth, a simple model of the earth used to determine the radar horizon and line of sight.

SA: situational awareness, one of the functions provided by electronic support.

Scanning: moving a radar antenna mainbeam around the sky to cover a prescribed region.

Scattering: the reflection of electromagnetic waves from a target.

Scintillation: rapid variations in the level of scattering from a target; also called target radar cross section fluctuation.

Self-protection jamming: electronic attack protecting the platform carrying the electronic attack system.

Semi-isotropic radiation: emission from a point source of equal levels in all directions within a hemisphere.

Sensitivity time control (STC): attenuating the radar return signal exponentially as a function of time to keep near-in returns from saturating the receiver.

Sequential detection: a radar detection theory approach based on multiple event probability theory using multiple radar waveforms and associated detection attempts; sometimes called alert–confirm.

Servo drive: the power provided to equipment by a technique that uses system output to determine partially what the input will be.

Servo loops: the electric circuits that sample system output and refer it back to the input.

Servomechanism: an automatic device that uses feedback to control systems, usually by inserting at the input control signals derived from samples of the output.

SHF: super high frequency; radio frequency band.

Sidelobe: the unwanted, out-of-place residue of an antenna pattern or waveform.

Sidelobe blanker (SLB): an electronic protection technique used to blank received false target jamming electronic attack signals in the radar receiver.

Sidelobe canceler (SLC): an electronic protection technique used to cancel received noise jamming electronic attack signals in the radar receiver.

Sidelobe jamming: the act of sending interfering and disrupting jamming signals into the radar antenna sidelobes.

Side-looking radar: a radar system that points at an angle perpendicular to the velocity vector of its carrier platform, hence, another name for some synthetic aperture radar systems.

Signal-processing gain: the improvement in signal-to-noise ratio that results when various processing techniques are applied.

Signal-to-noise ratio: the ratio of the target signal power to the noise power at the output of a radar receiver.

Sinusoid: sine or cosine plots.

Skip distance: the range at which a radio wave propagated from the earth toward the ionosphere returns to the earth's surface.

Solid angle: an area in angle, a number of square degrees or steradians.

Spark gap: a mechanism by which electromagnetic waves are radiated by building up the field intensity across a gap until the intervening medium breaks down and a spark occurs.

Specular returns: radar reflections of high amplitude and short duration, like flashes from a mirror.

Spherical aberration: distortion in a wave front, resulting when it is reflected off a spherical surface rather than a parabola.

Spherics: bursts of electromagnetic interference caused by disturbances in the atmosphere.

Spot noise jamming: narrowband disrupting and interfering noise jamming continuously covering all the bandwidth of a radar system.

Squint angle: the angle off the velocity vector of its carrier platform that the antenna mainbeam of a synthetic aperture radar system may be pointed.

Standard reference temperature (T_0): IEEE standard reference temperature for receivers, 290 Kelvin (≈ 62 °F).

Steradian: the solid angle subtended by an area on the surface of a sphere equal to its radius squared.

Subarray factor: the component of an array antenna gain pattern due to a group of elements (the subarray).

Subcutter visibility: the capability of a radar signal processor to suppress clutter.

Sum (Σ) beam: the adding together of two slightly displaced antenna beams single beam that is the sum of the two, as with monopulse angle tracking.

Support jamming: electronic attack protecting the platforms other than the one carrying the electronic attack system.

Surveillance: providing coverage of or keeping watch over.

Swerling target fluctuation models: the common approach for addressing the impact of fluctuating target radar cross section on radar detection theory, named after Dr. Peter Swerling who developed the approach.

Synthetic aperture radar (SAR): a system that uses movement of an antenna mainbeam across an area to synthesize a very large aperture and provide very fine angle, and thus cross-range, resolution.

Synthetic display: an uncluttered presentation obtained by distilling essential formation from noisy data and rejecting the latter.

T&E: test and evaluation.

Tapering: varying the density of the elements in an array or the power by the array elements to obtain a tailored aperture illumination control the array far-field pattern.

Target signature: the combination of the radar cross section (amplitude) and spectrum (frequency domain) characteristics of a target.

Target states: target position, velocity, and acceleration (e.g.) vectors.

Thermal noise: unwanted random amplitude and phase signals generated by the heat inherent in the radar receiver, a major factor in the first stage of radio frequency amplification.

Threshold: a level established for decision-making as to whether or not a desired signal present.

Time-bandwidth product: the product of the time duration and the bandwidth of a signal.

Time-delay networks: circuits in array radar systems that point the antenna various directions by progressively delaying the radiation of the signal array elements.

Time domain: viewing a multidimensional function in its time dimension.

Time sidelobes: the spreading residues in range of pulse compression forms, also called range sidelobes.

TOA: time of arrival of a radar pulse to an electronic support receiver.

T/R module: transmit/receive module, the combination of a transmitter and receiver in a single unit, used for array antennas.

Tracking: following selected targets over time, whether in range, angle, or range rate.

Tracking gate: a region of special attention around a target being tracked with appropriate logic to keep the gate moving with the target.

Track-while-scan (TWS): the radar operation in which targets arc followed by the routine scanning function, differentiated from an operation where changes its routine to do tracking.

Transponders: equipment that receives, stores, and transmits a signal; it may reradiate a version of the received signal enhanced with new information (e.g., jamming modulation).

Truncated cone: a conical shape that has been cut off at right angles to the axis of the cone at some arbitrary point.

TWT: traveling wave tube, a tube radio frequency transmitter.

UHF: ultra high frequency; radio frequency band.

Unambiguous range: the range associated with the time between radar pulses, that is, the maximum distance at which the round trip to a target can be completed before the next pulse is sent.

Unambiguous range rate: a waveform design feature that provides that the range rates of the targets of interest will not fold over in the signal processor.

Variance: a statistical quantity indicating the spread of a distribution about its mean.

Vertical polarization: by convention, an electromagnetic wave whose electric vector is perpendicular to the earth's surface.

Vertical return: that part of an airborne radar signal that returns from directly beneath the aircraft where the angle of incidence is 90 degrees.

VHF: very high frequency; radio frequency band.

Video integration: adding up signals after they have been through the envelope detector when only the envelope of the original signal remains.

Waveforms: various shapes of radar pulses or groups of pulses designed to accomplish particular objectives.

Waveform diversity: changing of the radar waveform parameters, e.g., carrier frequency, pulse width, to allow for different radar modes or as an electronic protection technique.

Weighting: changing the shapes of pulses or the envelopes of groups of pulses to tailor their sidelobes, usually by rounding the ends with a slowly varying function, such as a cosine.

Woodward ambiguity function: the surface that results when responses to waveforms are mapped in both range rate and time; named after the person who led in analyzing ambiguity functions.

Index